T0330974

# QUANTUM MECHANICS
## An Accessible Introduction
### Second Edition

# QUANTUM MECHANICS
## An Accessible Introduction
### Second Edition

## Robert Scherrer

Vanderbilt University, USA

 **World Scientific**

EW JERSEY · LONDON · SINGAPORE · BEIJING · SHANGHAI · HONG KONG · TAIPEI · CHENNAI · TOKYO

*Published by*

World Scientific Publishing Co. Pte. Ltd.

5 Toh Tuck Link, Singapore 596224

*USA office:* 27 Warren Street, Suite 401-402, Hackensack, NJ 07601

*UK office:* 57 Shelton Street, Covent Garden, London WC2H 9HE

Library of Congress Control Number: 2023947989

**British Library Cataloguing-in-Publication Data**
A catalogue record for this book is available from the British Library.

This book was previously published by Pearson Education, Inc.

**QUANTUM MECHANICS**
**An Accessible Introduction**
**Second Edition**

ISBN 978-981-12-8665-0 (hardcover)
ISBN 978-981-12-8729-9 (paperback)
ISBN 978-981-12-8673-5 (ebook for institutions)
ISBN 978-981-12-8674-2 (ebook for individuals)

For any available supplementary material, please visit
https://www.worldscientific.com/worldscibooks/10.1142/13687#t=suppl

Desk Editor: Carmen Teo Bin Jie

Printed in Singapore

To Elizabeth, who missed the first edition

# Preface

For the Instructor

The world is full of quantum mechanics textbooks. Does it really need another one? This book is designed to be an "accessible introduction" to quantum mechanics, suitable for students with a wide range of abilities and backgrounds. It grew out of a quantum mechanics course that I taught at Ohio State University. The chief problem I encountered in teaching this course was the enormous diversity of the student body, ranging from students headed toward careers as academic physicists to students who had never encountered the basic properties of complex numbers. Teaching students with such a wide range of abilities is a fundamental challenge, and it is to this challenge that this book is addressed. It assumes very little knowledge or mathematical background on the part of the students as it takes them through the major topics of quantum mechanics.

Much of the trouble that students encounter in their study of quantum mechanics centers around mathematics. A strong facility with complex numbers is required; although most of this material is at the level of high-school algebra, many students have forgotten it by the time they take a quantum mechanics course. Similarly, linear algebra is normally a prerequisite for any course in quantum mechanics. Often a single course in linear algebra is more than adequate preparation, but sometimes such courses spend weeks on matrix manipulation and linear equations, neglecting the more abstract elements of linear algebra that are necessary to fully understand quantum mechanics.

This book, therefore, contains three "math interludes" (Chapters 2, 5, and 7) which develop the mathematics needed in the book. Physicists have a long tradition of teaching mathematics in their courses; these math interludes simply formalize that tradition. In presenting this material, I have avoided formal proofs. Instead, my approach has been to present an intuitive motivation and then simply state the relevant results. This allows a large amount of mathematics to be presented compactly. Instructors of students with a stronger math background can simply skip this material, while others will want to go over it in detail.

This book takes a similar approach to many of the physics topics, assuming almost no initial knowledge on the part of the student and providing detailed motivation and explanation before launching into the calculations. Unlike many other areas of physics, it is difficult to develop intuition regarding the phenomena of quantum mechanics, so I have made free use of classical analogies to help the student get a better understanding of what is happening. I have pushed some of these analogies into a regime in which classical physics clearly does not apply; but quantum mechanics began as an outgrowth of classical mechanics, so it does no service to the student to treat quantum mechanics as though it popped spontaneously out of the vacuum with no connection to classical physics.

The one area in which I have neglected detailed explanations is the first topic of this book: blackbody radiation. A rigorous treatment of this subject requires a knowledge of statistical physics that students often acquire later than their study of quantum mechanics. Rather than attempt a one-chapter course in statistical physics, the necessary results are simply stated without proof. These results are, in any case, used nowhere else in the book.

With regard to the content of the book, I have omitted one topic which is often considered standard in undergraduate quantum mechanics textbooks: degenerate perturbation theory. Experience has taught me that students rarely grasp this topic during their first exposure to quantum mechanics, so their time in a first course is better spent in getting a firm grasp of nondegenerate perturbation theory. I conclude the book with an elementary introduction to relativistic quantum mechanics. This material is beyond the scope of most standard texts, but it is always of interest to the more advanced students, and it represents the next step in their study of quantum physics.

This book is intended for a standard, full-year, first course in quantum mechanics. A one-semester treatment should bring students through the end of Chapter 8, allowing a discussion of spins and concluding with the interesting topic of measurement theory. With the inclusion of the math interlude chapters, the mathematical background assumed is a good knowledge of calculus and a rudimentary exposure to differential equations. All of the more difficult integrals (both indefinite and definite) in the text and in the end-of-chapter problems can be found in a standard table of integrals or derived using Mathematica or an equivalent program. The required physics background is a standard, first-year sequence in mechanics and electromagnetism. The first chapter will be a review for students with a previous course in modern physics, but it will be useful for most students.

A flowchart of the prerequisite structure of the book is given here, where I have indicated the math interlude chapters as circles and the other chapters as squares. This diagram shows that the first part of the book is organized in a linear fashion, while the second part allows more freedom for the instructor to pick and choose chapters of greatest interest.

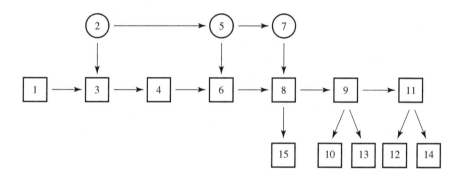

I have used MKS units throughout the book, particularly in the equations for electromagnetism. Although this often makes the derivations clumsier, the use of MKS units has become standard throughout the undergraduate curriculum, and it only confuses students to study electromagnetism with one set of equations, and then to see a different set appear in quantum mechanics. While I have sometimes introduced different units in the book (eV, Å, etc.), the basic equations in the text are all in MKS units.

I would like to thank my colleagues at Ohio State University, where this project was begun, and at Vanderbilt University, where it was completed. Among the latter, I owe particular thanks to John Gore of the Vanderbilt University Institute of Imaging Science for his insights into magnetic resonance imaging and to Donald Pickert for administrative assistance. I am also grateful to my own undergraduate (1979–80) quantum mechanics instructor, P.J.E. Peebles of Princeton University, for first piquing my interest in this subject. And of course, thanks to my wife Sarah for putting up with my many late nights editing at the kitchen table.

Robert Scherrer
Nashville, Tennessee
2005

## Notes on the Second Edition

Like all first editions, this one contained errors, typos, and omissions that colleagues were kind enough to point out to me over the years. I have attempted to correct all of the errors and have added sections that were requested by various instructors, particularly the Heisenberg uncertainty principle and the use of ladder operators to solve the harmonic oscillator. Otherwise, the changes to the first edition are relatively minor.

Robert Scherrer
Nashville, Tennessee
2023

# For the Student

Quantum mechanics is the gateway to modern physics. By the time most undergraduates come to quantum mechanics, they have spent one year, or possibly two, on the intricacies of classical physics. But classical mechanics is nearly 400 years old, and classical electromagnetism is more than 100 years old; most students do not decide to study physics because of a deep-rooted interest in blocks sliding down inclined planes. Instead, they go into this field because of the exciting modern developments: subatomic particles, cosmology, and all of the other exotic discoveries and theories that physics has produced over the past few decades.

Quantum mechanics provides the foundation for all of these subjects. Atomic, nuclear, and particle physics are all based on quantum mechanics, as is condensed matter physics. Indeed, the story of 20th century physics is, to a large extent, the story of quantum mechanics and its various offshoots. While this book will not venture into all of these subjects, it will provide the basis for understanding the ideas underlying all of them.

Quantum mechanics requires a radical shift in the way that we view reality, and this is a shift that some students find troublesome. Furthermore, unlike classical mechanics, the results of quantum mechanics frequently run counter to our own physical intuition. Because of these issues, I have attempted to provide plenty of motivation throughout the book, as well as to make extensive use of analogies with classical physics where possible. Of course, these analogies have their own limitations; much of quantum mechanics runs counter to our experience of how the world behaves at the macroscopic level, and no explanation or analogy can make it conform to classical ideas. On the other hand, it is precisely the more bizarre aspects of quantum mechanics which make it such a fascinating subject.

Quantum mechanics uses several areas of mathematics that are not frequently encountered in classical mechanics or electromagnetism. In particular, complex numbers are central to the mathematics of quantum mechanics, as is the branch of mathematics called linear algebra; the latter is a generalization of our familiar ideas of vectors into the realm of abstract "vector spaces." Because much of this mathematics may be unfamiliar, I have included three chapters (2, 5, and 7) which develop all of the necessary mathematical machinery. In these chapters, I have tried to avoid formal proofs and have presented just the essential results. A previous course in linear algebra is certainly helpful, but it is not required for using this book.

On the other hand, a good knowledge of calculus is essential. Note, however, that many of the more complicated integrals in this book can be found in standard tables of integrals, and I have assumed that they will be solved that way. (Practicing physicists rarely perform more than the most elementary integrals "by hand"; they look them up or use a program like Mathematica. This is akin to using a calculator instead of a pencil to do arithmetical calculations.) While I have also assumed at least a rudimentary exposure to differential equations, the solutions for all of the differential equations encountered in this book are derived in the book itself.

In terms of physics prerequisites, this book assumes at least a standard, first-year, one-semester course in classical mechanics and a one-semester course in electricity and magnetism. Most of the applications of quantum mechanics presented in this book are at the atomic level, where the primary interactions of interest are electromagnetic (e.g., the attraction between the proton and electron in the hydrogen atom and the multiple magnetic interactions in hydrogen). A prior introduction to modern physics is not essential; the first chapter covers much of this material in condensed form.

This book is an "accessible introduction." In other words, it starts at the very beginning, but it progresses through all of the major ideas of quantum mechanics. In making the book accessible, I have tried to provide the maximum amount of explanation, leaving out as few intermediate steps in derivations as possible. Most of all, I have attempted to convey some of the fun and bizarreness of this subject, which revolutionized physics and fundamentally changed our views of physical reality. A student completing this book will develop a good grasp of many of the most important ideas developed by physicists in the first half of the 20th century, laying the groundwork for understanding much of modern physics.

# Contents

*Preface*                                                                              vii

1.  The origins of quantum mechanics                                                     1

    1.1  Introduction . . . . . . . . . . . . . . . . . . . . . . . . . .                1
    1.2  Blackbody Radiation . . . . . . . . . . . . . . . . . . . . .                   3
    1.3  The Nature of Light  . . . . . . . . . . . . . . . . . . . . .                 11
    1.4  The Wave Nature of Matter . . . . . . . . . . . . . . . . .                    17
    1.5  The Bohr Atom . . . . . . . . . . . . . . . . . . . . . . . .                  19
    1.6  Where Do We Stand? . . . . . . . . . . . . . . . . . . . .                     22

2.  Math interlude A: Complex numbers and linear operators                             27

    2.1  Complex Numbers . . . . . . . . . . . . . . . . . . . . . .                    27
    2.2  Operators . . . . . . . . . . . . . . . . . . . . . . . . . . .                35

3.  The Schrödinger equation                                                           41

    3.1  Derivation of the Schrödinger Equation . . . . . . . . . . .                   42
    3.2  The Meaning of the Wave Function . . . . . . . . . . . . .                      51
    3.3  The Time-Independent Schrödinger Equation: Qualitative
         Solutions and the Origin of Quantization . . . . . . . . .                     58

4.  Solutions of the one-dimensional time-independent
    Schrödinger equation                                                               71

    4.1  Unbound States: Scattering and Tunneling . . . . . . . .                       72
    4.2  Bound Systems . . . . . . . . . . . . . . . . . . . . . . . .                  88

5.  Math interlude B: Linear algebra                               107

    5.1  Properties of Linear Operators . . . . . . . . . . . . . .   107
    5.2  Vector Spaces . . . . . . . . . . . . . . . . . . . . . . .   112

6.  Solutions of the three-dimensional time-independent
    Schrödinger equation                                          125

    6.1  Solution in Rectangular Coordinates . . . . . . . . . . .   126
    6.2  Angular Momentum . . . . . . . . . . . . . . . . . . . . .   129
    6.3  The Schrödinger Equation in Spherical Coordinates . . . .   141
    6.4  The Hydrogen Atom . . . . . . . . . . . . . . . . . . . .   149

7.  Math interlude C: Matrices, Dirac notation, and the
    Dirac delta function                                          163

    7.1  The Matrix Formulation of Linear Operators . . . . . . .   163
    7.2  Dirac Notation . . . . . . . . . . . . . . . . . . . . . .   170
    7.3  The Dirac Delta Function . . . . . . . . . . . . . . . . .   172

8.  Spin angular momentum                                         179

    8.1  Spin Operators . . . . . . . . . . . . . . . . . . . . . .   179
    8.2  Evidence for Spin . . . . . . . . . . . . . . . . . . . . .   180
    8.3  Adding Angular Momentum . . . . . . . . . . . . . . . . .   185
    8.4  The Matrix Representation of Spin . . . . . . . . . . . .   187
    8.5  The Stern–Gerlach Experiment . . . . . . . . . . . . . .   194
    8.6  Spin Precession . . . . . . . . . . . . . . . . . . . . . .   197
    8.7  Spin Systems with Two Particles . . . . . . . . . . . . .   202
    8.8  Measurement Theory . . . . . . . . . . . . . . . . . . . .   208

9.  Time-independent perturbation theory                          219

    9.1  Derivation of Time-Independent Perturbation Theory . . .   220
    9.2  Perturbations to the Atomic Energy Levels . . . . . . . .   229
    9.3  The Atom in External Electric or Magnetic Fields  . . . .   239

10. The variational principle                                     251

    10.1  Variational Principle: Theory . . . . . . . . . . . . . .   252
    10.2  Variational Principle: Application to the Helium Atom . .   256

11. Time-dependent perturbation theory 265

    11.1 Derivation of Time-Dependent Perturbation Theory . . . 266

    11.2 Application: Selection Rules for Electromagnetic Radiation 276

12. Scattering theory 287

    12.1 Definition of the Cross Section . . . . . . . . . . . . . . 287

    12.2 The Born Approximation . . . . . . . . . . . . . . . . . 292

    12.3 Partial Waves . . . . . . . . . . . . . . . . . . . . . . . 303

13. Multiparticle Schrödinger equation 313

    13.1 Wave Function for Identical Particles . . . . . . . . . . . 313

    13.2 Multielectron Atoms . . . . . . . . . . . . . . . . . . . . 326

14. Modern applications of quantum mechanics 339

    14.1 Magnetic Resonance Imaging . . . . . . . . . . . . . . . 339

    14.2 Quantum Computing . . . . . . . . . . . . . . . . . . . . 344

15. Relativistic quantum mechanics 351

    15.1 The Klein–Gordon Equation . . . . . . . . . . . . . . . . 351

    15.2 The Dirac Equation . . . . . . . . . . . . . . . . . . . . . 355

*Index* 363

Chapter 1

# The origins of quantum mechanics

## 1.1 Introduction

The story of the development of quantum mechanics has attained mythic stature in the history of physics. By the turn of the $20^{\text{th}}$ century (c. 1900), physics encompassed the fields of classical mechanics, electricity and magnetism, and thermodynamics, a set of subjects now known collectively as "classical physics." Indeed, with a few minor exceptions, the concepts of classical physics seemed capable of explaining all known physical phenomena. In a few decades, this entire framework would be superseded by the development of quantum mechanics. The importance of this development can hardly be exaggerated. The term "modern physics" is practically a synonym for the areas of physics which grew out of quantum mechanics: atomic physics, nuclear physics, particle physics, and condensed matter physics.

The predictions of quantum mechanics are rather bizarre from a classical point of view. Consider the following propositions, which are all postulates of classical physics:

1. The physical universe is deterministic, i.e., given enough information about a physical system, its future evolution can be predicted exactly. Who would dispute this obvious point? The entire function of classical mechanics is to derive such predictions.

2. Light consists of waves, while ordinary matter is composed of particles. The former statement is one of the triumphs of classical electromagnetism, while the latter seems self-evident.

3. Physical quantities, such as energy and angular momentum, can be treated as continuous variables. Again, this assumption is built into the structure of classical mechanics.

4. There exists an objective physical reality independent of any observer. If a tree falls in the woods, of course it makes a sound.

All of these ideas seem obvious. In fact, we know from quantum mechanics that none of them is completely accurate:

1. The physical universe is not deterministic. At the subatomic level, we can assign probabilities to the outcomes of certain experiments but never predict the exact result with certainty. Uncertainty is an intrinsic property of matter at this scale.

2. Both light and matter exhibit behavior that seems characteristic of both particles and waves.

3. Under certain circumstances, some physical quantities are *quantized*, i.e., they can take on only certain discrete values.

4. Finally, it appears that the observer always affects the experiment; it is impossible to disentangle the two.

Why would anyone believe such a preposterous set of ideas? For the only reason that any theory in physics is given credence: because it works. Quantum mechanics allows us to explain physical phenomena, primarily at very small length scales, for which classical physics simply offers no explanation. Physicists in the first half of the 20th century were themselves often hesitant to accept many of the more bizarre consequences of quantum mechanics, but the theory ultimately prevailed because it agreed with experiment.

Because quantum mechanics is so counterintuitive, so strange in many of its predictions, we will begin by examining some of the experiments which led to its birth. In each of the experiments examined below, physicists had developed a classical theory which failed to correctly account for the experimental results. These experiments led to the idea that light could behave as both a particle and a wave and then to the more radical suggestion that matter also had both particle and wave properties. These ideas set the stage for the development of quantum mechanics.

## 1.2    Blackbody Radiation

It is one of the ironies of physics that the greatest discoveries often occur in the areas where one would least expect to find them. This is particularly true of quantum mechanics, whose origins are often traced back to perhaps the least glamorous subject in all of physics: thermodynamics. Specifically, quantum mechanics was launched by a solution to a nagging problem in thermodynamics: the behavior of blackbody radiation. Although the solution to this problem is not the most compelling argument for quantum mechanics, it is of such historical importance that we examine it in some detail.

### *The problem with blackbody radiation*

Blackbody radiation seems like a contradiction: objects are black precisely because they absorb radiation, so how can a blackbody be said to emit radiation? The confusion arises because there are two different ways that we can detect radiation from an object: it can reflect light, or it can emit light from its own internal energy. Almost all objects in our everyday environment give off visible light by reflection, and it is in this sense that a black object absorbs everything and reflects nothing. However, we are interested in the second case, the emission of radiation from an object's internal energy. When objects are heated to a high enough temperature, they emit visible light. Familiar examples include the filament in an electric light bulb or the burner element in an electric stove. The actual spectrum of radiation produced at a given temperature will vary from one object to the next, but a blackbody is unique in this regard: it emits radiation with equal efficiency at all frequencies.

The reason for this lies in a concept from thermodynamics called Kirchhoff's law: a body at the same temperature as its surroundings will emit radiation with the same efficiency at which it absorbs it. Imagine what would happen if this was not the case; if the body absorbed radiation at a different rate than it emitted it, over time the object would gradually heat up or cool down catastrophically! However, this argument can be made stronger: a body must absorb and give off radiation at the same rate *at every frequency*. Again, imagine an object at the same temperature as its environment but now surrounded with a filter which allows only a narrow range of frequencies of radiation to pass through. Emission and absorption must be exactly balanced within this range of frequencies in order for the object to remain at the same temperature as its surroundings.

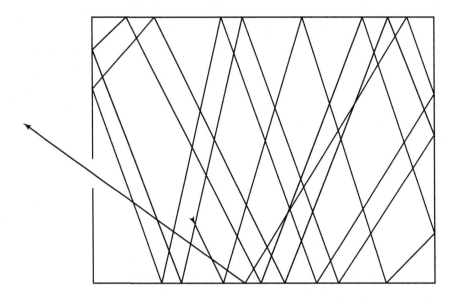

Fig. 1.1   A cavity with a small hole behaves like a blackbody.

So Kirchhoff's law says that a good absorber of radiation will also be a good emitter. Hence, a body which absorbs radiation with perfect efficiency at all frequencies (a "blackbody") must also emit radiation with perfect efficiency at all frequencies. In practice, most things in nature are not perfectly black; they reflect some light and are therefore not perfect blackbodies. To construct a practical blackbody emitter, we can use a cavity with a very small hole (Figure 1.1). Then light inside the cavity will reflect many times before it can leave the cavity. Even if the walls of the cavity are not perfectly absorbing, the probability that the light will escape becomes smaller with each reflection and can be made infinitesimally small. Then the hole of the cavity acts like a blackbody, and the cavity itself is filled with blackbody radiation.

The experimental properties of blackbody radiation were well established in the 19th century. First consider the *total* power given off by a blackbody. This was first measured by J. Stefan in 1879. The power emitted scales as the surface area $A$ of the blackbody, and Stefan discovered that it also scales as the fourth power of the temperature $T$, measured in Kelvin:

$$P = \sigma A T^4 \qquad (1.1)$$

where $\sigma$ is a constant called the Stefan–Boltzmann constant, and Equation (1.1) is called the *Stefan–Boltzmann law*. (Boltzmann showed how the $T^4$ dependence could be derived theoretically.) The Stefan–Boltzmann constant is measured to be

$$\sigma = 5.67 \times 10^{-8} \text{ J sec}^{-1} \text{ m}^{-2} \text{ K}^{-4}$$

A more convenient quantity to work with is the total energy density $\rho$ of radiation inside a blackbody cavity. In terms of the power radiated, this is given by $\rho = (4/c)(P/A)$, where $c$ is the speed of light, so that

$$\rho = aT^4 \tag{1.2}$$

with

$$a = 7.56 \times 10^{-16} \text{ J m}^{-3} \text{ K}^{-4}$$

The spectrum of radiation, i.e., the energy density at a given frequency, can also be measured. This spectrum is expressed as $\rho(\nu)\,d\nu$, the total energy density in blackbody radiation between $\nu$ and $\nu + d\nu$, where $\nu$ is the frequency of the radiation in Hz. (Of course, $\rho(\nu)\,d\nu$ will also be a function of temperature; it is understood that the spectrum is measured at a fixed value of $T$.) A plot of the measured spectrum is shown in Figure 1.2 for three different temperatures.

The spectrum shows two obvious features. First, the amplitude of the spectrum increases with temperature. This is not surprising, since we know that the total energy density must increase as $T^4$, and this total energy density is just the spectrum integrated over all frequencies:

$$\rho = \int_0^\infty \rho(\nu)\,d\nu = aT^4$$

In Figure 1.2, the total energy density is just the area under each curve. A second interesting feature is that the curves shift to the right (higher frequencies) as we go to higher temperatures. More quantitatively, it is observed that the frequency of the peak energy density (or peak emission from the blackbody) scales linearly with the temperature: $\nu_{peak} \propto T$. Since the wavelength scales inversely with frequency, this relation is often expressed as

$$\lambda_{peak} = w/T$$

where $w$ is a constant ($w = 2.90 \times 10^{-3}$ m K). This empirical result is known as *Wien's displacement law*. At room temperature, blackbodies radiate primarily in the infrared (hence the usefulness of night vision goggles).

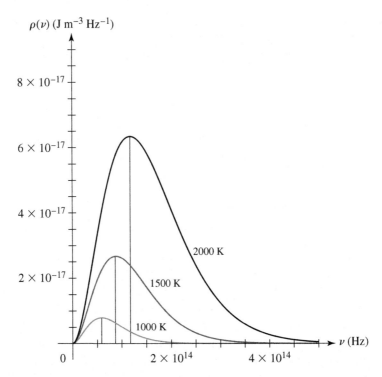

Fig. 1.2    The energy density $\rho(\nu)$ of blackbody radiation as a function of the frequency $\nu$ at temperatures of $T = 1000$ K, 1500 K, and 2000 K. The vertical lines show the peak in the energy density at each temperature.

At temperatures of several thousands of degrees, the radiation shifts into the optical range, giving the familiar phenomenon of a heated object first glowing "red-hot" then "white-hot." Physicists of the 19th century had a theory to explain the observed behavior of the blackbody spectrum; the only problem was that their theory failed to produce the observed spectrum.

To understand this classical theory, we need several ideas from thermo-dynamics, which we will quote without derivation. Consider a collection of electromagnetic waves inside a blackbody cavity at a temperature $T$. The energy density of the radiation is just the average energy of the waves multiplied by their number density. The average energy $\bar{E}$ of a set of classical oscillators is proportional to the temperature $T$:

$$\bar{E} = kT$$

where the constant $k = 1.38 \times 10^{-23}$ J K$^{-1}$ is called Boltzmann's constant.

Classical physics also predicts the number density of waves $n(\nu)\,d\nu$ with frequencies between $\nu$ and $\nu + d\nu$ to be

$$n(\nu)\,d\nu = \frac{8\pi}{c^3}\nu^2\,d\nu \tag{1.3}$$

Then the total energy density is just the number density of waves multiplied by the average energy per wave:

$$\rho(\nu) = n(\nu)\bar{E}$$

giving

$$\rho(\nu)\,d\nu = \frac{8\pi kT}{c^3}\nu^2\,d\nu \tag{1.4}$$

Equation (1.4) is called the *Rayleigh–Jeans formula* for blackbody radiation, and it is based on correct arguments from thermodynamics. The only problem is that it doesn't work. In Figure 1.3, we compare the predictions of this formula to an actual blackbody spectrum at a temperature $T = 2000$ K.

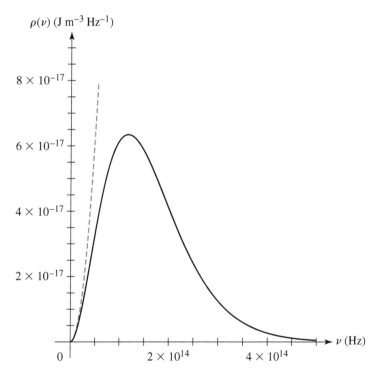

Fig. 1.3  The solid curve gives the observed spectrum of blackbody radiation at $T = 2000$ K, and the dashed curve gives the prediction of the Rayleigh–Jeans formula.

The Rayleigh–Jeans formula is not a complete failure. At the low-frequency end of the spectrum, it gives good agreement with the observations. But it fails at the high-energy end of the spectrum, and here it fails in a spectacular fashion. According to Equation (1.4), the energy density of radiation should increase with increasing frequency all of the way up to infinity! A heated object would give off more ultraviolet light than visible light and more X-rays than either; a person sitting in front of a fireplace would be killed by radiation! This problem came to be known as the *ultraviolet catastrophe*. (In physics, it is common to use the terms "infrared" and "ultraviolet" to refer generically to the low-frequency and high-frequency limits of any spectrum.)

It was Max Planck who found the correct formula for the blackbody radiation spectrum, and in so doing, inadvertently developed the beginnings of quantum mechanics. In order to understand Planck's reasoning, we need to make use of another result from thermodynamics. Consider a collection of interacting particles such as gas molecules in a box or radiation in a blackbody cavity. We have already seen that the *average* energy of these particles $\bar{E}$ is proportional to their temperature: $\bar{E} = kT$. Now we need an expression for the *distribution* of energies for these particles, i.e., the probability that a randomly-chosen particle will have a particular energy. If we pick a random particle in our distribution, then we will define $P(E)\,dE$ to be the probability that it has energy between $E$ and $E+dE$. It is observed that for a wide variety of systems, $P(E)$ has a universal form called the *Boltzmann distribution*, given by

$$P(E) = \frac{e^{-E/kT}}{kT} \tag{1.5}$$

Using this expression, the mean energy of the particles is given by its classical value:

$$\bar{E} = \frac{\int_0^\infty P(E)E\,dE}{\int_0^\infty P(E)\,dE} = kT \tag{1.6}$$

So if we blindly use the Boltzmann distribution given by Equation (1.5), we will still end up with the Rayleigh–Jeans formula, Equation (1.4), which fails at high frequencies.

Planck's idea was to modify the theory to make $\bar{E}$ a function of frequency in such a way that $\bar{E} = kT$ at low frequencies (where the Rayleigh–Jeans formula works well), while $\bar{E} \neq kT$ at high frequencies (where the Rayleigh–Jeans formula fails). In order to do this, Planck assumed that $E$ could no longer take on arbitrary values but only discrete multiples of some fundamental energy. Further, he took this fundamental energy to be proportional to the frequency $\nu$, with a constant of proportionality $h$. (The value of $h$ is not predicted by Planck's theory but must be chosen to fit the observations.) Therefore, according to Planck, the allowed values for $E$ are simply

$$E = 0, \ h\nu, \ 2h\nu, \ 3h\nu, \ldots$$

Clearly, $h$ must have units of energy/frequency, or J·sec. With this assumption, Equation (1.6) can no longer be taken to be an integral over a continuous range of values for $E$, but instead is a sum over the allowed discrete values:

$$\bar{E} = \sum_{E=0,\ h\nu,\ 2h\nu,\ldots} P(E)E \bigg/ \sum_{E=0,\ h\nu,\ 2h\nu,\ldots} P(E)$$

$$= \sum_{n=0}^{\infty} \frac{nh\nu}{kT} e^{-nh\nu/kT} \bigg/ \sum_{n=0}^{\infty} \frac{1}{kT} e^{-nh\nu/kT}$$

$$= \frac{h\nu}{e^{h\nu/kT} - 1}$$

which is obviously different from Equation (1.6). Taking this expression for $\bar{E}$ and multiplying by the number density of waves in Equation (1.3), we obtain Planck's expression for the spectrum of blackbody radiation:

$$\rho(\nu)\, d\nu = \frac{8\pi h}{c^3} \frac{\nu^3}{e^{h\nu/kT} - 1}\, d\nu \tag{1.7}$$

---

### Example 1.1. The Classical Limit of the Planck Blackbody Spectrum

Show that the Planck blackbody spectrum reduces to the Rayleigh–Jeans formula in the limit of low frequencies.

We begin with Equation (1.7) and assume that $h\nu/kT \ll 1$. Recall that for small $x$,

$$e^x \approx 1 + x$$

Thus, Equation (1.7) becomes, for $h\nu/kT \ll 1$,

$$\rho(\nu)\, d\nu = \frac{8\pi h}{c^3} \frac{\nu^3}{1 + h\nu/kT - 1}\, d\nu$$

$$= \frac{8\pi kT}{c^3} \nu^2\, d\nu$$

which is the Rayleigh–Jeans formula.

This result explains why the Rayleigh–Jeans formula works well at low frequencies but fails at high frequencies. It also indicates the frequency range over which the Rayleigh–Jeans formula works well: $\nu \ll kT/h$.

---

The Planck spectrum can be used to derive both the Stefan–Boltzmann law and Wien's law, and these results can be used, in turn, to calculate the value of $h$. The total energy density of blackbody radiation is simply the integral of the Planck spectrum over all frequencies:

$$\rho = \int_0^\infty \rho(\nu)\, d\nu$$

$$= \int_0^\infty \frac{8\pi h}{c^3} \frac{\nu^3}{e^{h\nu/kT} - 1}\, d\nu$$

The integral can be put into simpler form by making the change of variables $x = h\nu/kT$; this gives

$$\rho = \frac{8\pi(kT)^4}{c^3 h^3} \int_{x=0}^\infty \frac{x^3}{e^x - 1}$$

The integral can be evaluated exactly:

$$\int_{x=0}^\infty \frac{x^3\, dx}{e^x - 1} = \frac{\pi^4}{15}$$

giving

$$\rho = \frac{8}{15}\pi^5 \frac{k^4}{c^3 h^3} T^4 \tag{1.8}$$

A comparison of Equation (1.8) with Equation (1.2) shows the correct $T^4$ dependence of $\rho$.

Now consider the frequency at which the maximum emission occurs. We begin with Equation (1.7) for $\rho(\nu)$, and set $d\rho/d\nu = 0$ to find frequency $\nu_{peak}$ at which $\rho(\nu)$ is a maximum:

$$\frac{d\rho(\nu_{peak})}{d\nu} = \left(3 - \frac{h\nu_{peak}}{kT}\right) e^{h\nu_{peak}/kT} - 3 = 0$$

This has a trivial solution at $\nu_{peak} = 0$. The nontrivial solution, which is the one we want, cannot be calculated algebraically, but it can be found numerically: $h\nu_{peak}/kT \approx 2.8$, so

$$\nu_{peak} = 2.8kT/h$$

This result is consistent with Wien's law: the value of $\nu_{peak}$ is proportional to the temperature $T$.

Not only have we shown that the Planck spectrum predicts both the Stefan–Boltzmann law and Wien's law, but we have derived expressions for the corresponding constants of proportionality in both laws. Since the speed of light $c$ is known from other experiments, it is possible to compare these two expressions with the experimentally-measured constants of proportionality to derive values for both $h$ and $k$; Planck did exactly that, obtaining very accurate results (in terms of modern measurements) for both constants. The best current measurements of $k$ and $h$ give

$$k = 1.381 \times 10^{-23} \text{ J K}^{-1}$$

and

$$h = 6.626 \times 10^{-34} \text{ J sec}$$

When these values are inserted into Equation (1.7), the result is a predicted blackbody spectrum in excellent agreement with the observed spectrum.

## 1.3 The Nature of Light

Planck showed that the spectrum of blackbody radiation could be explained only if the energies of light waves in a blackbody cavity were restricted to discrete values proportional to their frequency, $E = nh\nu$. Planck's solution to the problem of blackbody radiation clearly indicated something odd about the nature of the radiation, but the physical interpretation of his proposal was not at all obvious. In the decade after Planck made this proposal, several subsequent experiments clarified what was really going on: light behaves like a gas of particles called photons, and the energy of each photon is given by $h\nu$.

This idea hearkens back to the original theory of light proposed by Newton in the 1600's. In Newton's "corpuscular theory," light consisted of particles, which obeyed the laws of classical mechanics. This theory could explain, for instance, the way in which light reflects from surfaces. However, by the 1800's, it was clear that many observations could be explained only if light consisted of waves rather than particles. These observations included well-known phenomena such as diffraction and interference (Figure 1.4). Maxwell's equations, derived from observations of electromagnetic phenomena, actually predict the wave nature of light in the form of oscillating electric and magnetic fields. By the beginning of the 20th century, the wave nature of light was well established. We will now consider two experiments that reestablished the interpretation of light as a collection of particles.

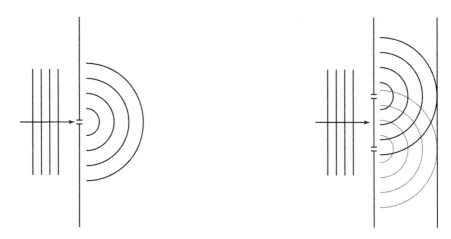

Fig. 1.4   Diffraction (left) and interference (right) indicate that light consists of waves.

### The photoelectric effect

Light shining on a metal plate is observed to produce a current; this effect is called the *photoelectric effect* (see Figure 1.5). The current and maximum energy of the photoelectrons can be measured as the intensity and frequency of the light are varied. [The maximum energy is measured by putting a potential across the circuit until the current stops; if $\Phi_0$ is the potential which stops the current from flowing, then the corresponding maximum electron energy is $E_{max} = e\Phi_0$.]

Here is what is observed:

1. The current is proportional to the *intensity* of the light (i.e., the power per unit area falling on the plate, measured in $W/m^2$).

2. The maximum energy of the photoelectrons $E_{max}$ is proportional to the *frequency* of the light (Figure 1.6). Furthermore, there is a minimum frequency below which no current is observed; this minimum frequency depends on the composition of the metal plate.

It is the second of these observations which is puzzling. In the classical theory of electromagnetic radiation, the energy put into the system should be proportional to the intensity of the light, and the frequency should play no role at all in determining $E_{max}$.

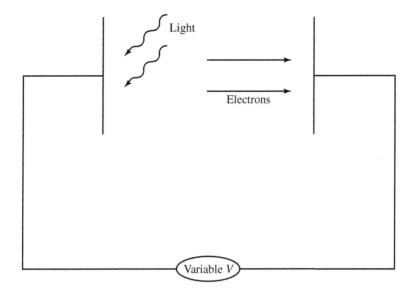

Fig. 1.5   A schematic representation of the photoelectric effect.

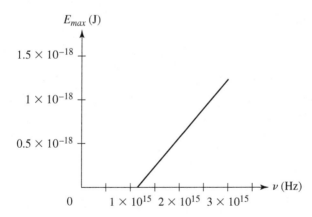

Fig. 1.6   The maximum electron energy $E_{max}$ measured in the photoelectron effect, as a function of the frequency of light $\nu$.

It was Albert Einstein who, in 1905, came up with the explanation for the photoelectric effect. He proposed that light, in this case, behaves as a collection of particles called *photons*. Furthermore, the energy of each photon is given by $h\nu$. When a photon strikes an electron in the metal plate, it can transfer a maximum energy of $h\nu$ to the electron. This would suggest that $E_{max} = h\nu$. However, there is an additional complication. The electrons are bound in the metal with some binding energy $E_B$. The photon must transfer enough energy to remove the electron from the metal; whatever energy is left over then goes into the kinetic energy of the electron. We therefore have

$$E_{max} = h\nu - E_B$$

Although $E_B$ will vary from one metal to the next, Einstein's theory makes one universal prediction: the slope of the graph of $E_{max}$ versus $\nu$ should be given by Planck's constant $h$. This is exactly what is observed. In fact, although Einstein is most famous for the theory of relativity, he won his Nobel Prize for this explanation of the photoelectric effect.

The fact that Planck's constant appears in two very different phenomena (blackbody radiation and the photoelectric effect) suggests that it has a fundamental physical significance. Furthermore, Einstein's theory indicates something very radical about the nature of light: it behaves as a particle with energy $h\nu$. Further confirmation of this behavior was provided by the Compton effect.

### The Compton effect

One of the characteristics of particles is that they can scatter off of each other, conserving both energy and momentum in the scattering process. If light truly does behave like a particle, it should be possible to observe such scattering processes and to predict the change in the energy and momentum of the light when it scatters. One such process that is observed to occur is *Compton scattering*, which refers to the scattering of X-rays or gamma rays off of the electrons in a metal (Figure 1.7). Experimentally, it is observed that the wavelength of the scattered radiation $\lambda_f$ is larger than that of the incident radiation $\lambda_i$, and the change in wavelength is well fit by the relation

$$\lambda_f - \lambda_i = \lambda_C(1 - \cos\theta) \tag{1.9}$$

where $\lambda_C$, called the *Compton wavelength* of the electron, is a constant with units of length

$$\lambda_C = 2.4 \times 10^{-12} \text{ m} \tag{1.10}$$

and $\theta$ is the scattering angle shown in Figure 1.7. Note from Equation (1.9) that the change in wavelength is actually independent of the initial wavelength $\lambda_i$; it depends only on the scattering angle $\theta$.

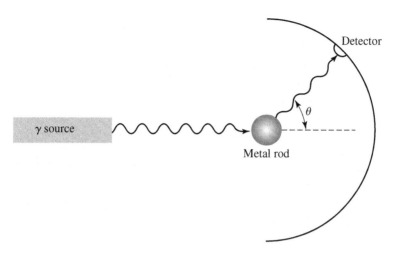

Fig. 1.7 An example of the Compton scattering experiment. Gamma rays from a radioactive source scatter off of electrons in a cylindrical metal target. The wavelength of the scattered radiation is measured as a function of the scattering angle $\theta$.

The Compton effect cannot be explained by the classical wave theory of light. Classically, a light wave scattering off of an electron excites the electron to oscillate at the same frequency as the incident wave, and the oscillating electron produces radiation with the same frequency. Hence, light scattering from an electron undergoes *no* change in frequency. However, if we treat the light as consisting of particles, we will not only be able to derive the correct behavior given in Equation (1.9), but also to obtain the correct value for $\lambda_C$.

To derive Equation (1.9), we will assume that the radiation consists of particles with energy $h\nu$, and we will treat the Compton effect as a scattering problem in classical mechanics (see Figure 1.8). We need to be careful to use the correct relativistic expressions for energy and momentum here. Recall that special relativity gives

$$E^2 = p^2 c^2 + m_0^2 c^4$$

for a particle with mass $m_0$. For the electron we simply take $m_0 = m_e$. For a photon we know that $E = h\nu$, but what do we assume for the rest mass? Any particle moving at the speed of light must have zero rest mass, so that $E^2 = p^2 c^2$, and

$$E = pc$$

for photons.

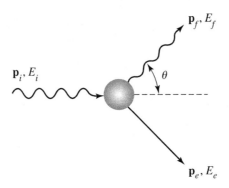

Fig. 1.8 A photon is incident with energy $E_i$ and momentum $\mathbf{p}_i$. It scatters off of an electron, emerging at an angle $\theta$ with final energy $E_f$ and final momentum $\mathbf{p}_f$. The electron ends up with final energy $E_e$ and final momentum $\mathbf{p}_e$.

Applying conservation of momentum to the system in Figure 1.8 gives

$$\mathbf{p}_i = \mathbf{p}_f + \mathbf{p}_e \qquad (1.11)$$

while energy conservation gives

$$E_i + m_e c^2 = E_f + E_e \qquad (1.12)$$

where $i$ and $f$ refer to the incoming and outgoing photon respectively, and $e$ refers to the final state of the electron after scattering (see Figure 1.8). Note that since we are using a fully relativistic treatment, we include the rest energy of the electron on the left-hand side of the equation. We want to eliminate the electron energy and momentum from both equations, so we begin by rewriting Equation (1.11) as $\mathbf{p}_i - \mathbf{p}_f = \mathbf{p}_e$ and squaring both sides (i.e., taking the dot product of each side with itself) to get

$$p_i^2 + p_f^2 - 2p_i p_f (\cos\theta) = p_e^2 \qquad (1.13)$$

Similarly, rearranging terms in Equation (1.12) and squaring, we get

$$(E_i - E_f + m_e c^2)^2 = E_e^2 \qquad (1.14)$$

We now make the appropriate substitutions $E_i = p_i c$, $E_f = p_f c$, and $E_e^2 = p_e^2 c^2 + m_e^2 c^4$ into Equation (1.14), and simplify to obtain

$$(p_i - p_f)^2 + 2(p_i - p_f)m_e c = p_e^2 \qquad (1.15)$$

We can now equate the right-hand sides of Equations (1.13) and (1.15) and reduce the resulting equation to the form

$$\frac{1}{p_f} - \frac{1}{p_i} = \frac{1}{m_e c}(1 - \cos\theta) \qquad (1.16)$$

At this point, we need to express the photon momenta in terms of their wavelengths. Since we have $E = h\nu$, $\nu = c/\lambda$, and $E = pc$, we obtain

$$p = h/\lambda \tag{1.17}$$

Then Equation (1.16) reduces to

$$\lambda_f - \lambda_i = \frac{h}{m_e c}(1 - \cos\theta)$$

This is exactly the same form as Equation (1.9) with

$$\lambda_C = h/m_e c$$

Substituting the values for $h$, $m_e$, and $c$, we indeed obtain the measured value of $\lambda_C$ ($= 2.4 \times 10^{-12}$ m). Thus, the Compton effect can be explained by assuming that the X-rays or gamma rays act as particles with energy $E = h\nu$ and momentum $p = h/\lambda$.

### Is it a Particle or a Wave?

Both the photoelectric effect and the Compton effect provide evidence that light acts like a particle with energy given by $E = h\nu$. But this does not eliminate the results of classical optics in which light behaves like a wave. Instead, we are forced to accept the idea of wave-particle duality: light behaves sometimes as wave and sometimes as a particle. This provides the stepping-stone to a more radical idea: if light can exhibit both wave and particle properties, what about matter?

## 1.4   The Wave Nature of Matter

We saw in Section 1.3 that light can behave as both a particle and a wave. On this basis, Louis de Broglie made a much more startling proposal: that matter, which is composed of particles, might also behave like a wave. (Louis de Broglie was one of the last surviving founders of quantum mechanics, passing away in 1987 at the age of 95.) In particular, de Broglie proposed using the relation between momentum and wavelength appropriate for a photon (Equation (1.17)) and applying it to matter. The *de Broglie wavelength* for a particle is then

$$\lambda = h/p \tag{1.18}$$

This proposal seems absurd. If someone's body behaved like a wave, it would diffract every time that person walked through a door. He or she could even walk through a classroom wall having two doors and form an interference pattern on the front wall! But there is a good reason that these phenomena are not observed.

---

**Example 1.2. The de Broglie Wavelength of a Walking Human**

Consider a 70 kg human being walking at 1 m sec$^{-1}$. The momentum is

$$p = mv = 70 \text{ kg m sec}^{-1}$$

and Equation (1.18) gives a de Broglie wavelength of

$$\lambda = h/p = 9 \times 10^{-36} \text{ m}$$

This is a tiny wavelength compared to ordinary human length scales; in fact, it is tiny compared to atomic or nuclear scales! Therefore, it is not surprising that we do not see any wave-like effects at human length scales.

---

In order to see some effect from the wave nature of matter, it is necessary to conduct an experiment in which $\lambda$ is significant compared to the characteristic size of the physical system. To maximize $\lambda$, it is desirable to use the smallest possible momentum $p$. For a nonrelativistic particle of mass $m$, the relation between (kinetic) energy $E$ and momentum $p$ is $p = \sqrt{2mE}$. Hence, at fixed energy, $p$ will be minimized and $\lambda$ maximized when $m$ is as small as possible. This makes the electron (which has a much smaller mass than the proton or neutron) a convenient particle to use, and in order for the electrons to behave like waves, they must be scattered through an "aperture" with a size comparable to or smaller than their wavelength. This can be achieved by scattering electrons from a crystal lattice, since the separation of the atoms in the lattice is on the order of $10^{-10}$ m.

In an experiment called the *Davisson–Germer experiment* (Figure 1.9), Davisson and Germer scattered electrons off of a nickel sheet and measured the intensity of the electrons as a function of the scattering angle. An interference pattern was produced, which can be explained by taking the electrons to behave like waves with wavelength given by Equation (1.18).

Although we have applied the de Broglie postulate only to electrons, it was proposed for all matter. Thus, we are left with an interesting symmetry between radiation and matter: both can be considered to behave as both particles and waves.

Incident electrons                    Scattered electrons

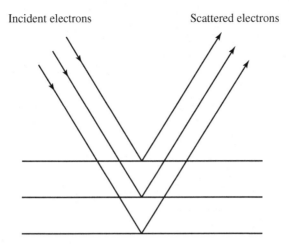

Fig. 1.9  In the Davisson–Germer experiment, electrons are scattered from a crystal lattice. Constructive interference occurs when the path difference between rays scattering from adjacent planes is an integer multiple of the wavelength of the electrons.

## 1.5   The Bohr Atom

We now examine the most important result from the early stages of quantum mechanics: Neils Bohr's model of the atom. Ernest Rutherford's experiments in the scattering of alpha particles from atoms led to our familiar modern picture of the atom: a small, positively-charged, central nucleus surrounded by negatively-charged electrons orbiting the nucleus. There are, however, two problems with this picture. According to classical electromagnetic theory, any accelerating charge will emit electromagnetic radiation. Since the electrons must undergo centripetal acceleration in order to orbit the nucleus, they should give off radiation. Not only is this radiation not observed, but if it did occur, the electrons would lose energy and spiral into the nucleus; every atom would be unstable!

The second problem with this classical picture is that it cannot explain one of the most striking features of hot gasses. If hydrogen gas, for example, is heated, the radiation it emits is not a continuous spectrum of wavelengths. Instead, the radiation is confined to discrete wavelengths, producing a series of bright lines in the spectrum. These lines were measured in the 19th century and found to obey a regular pattern. For hydrogen, for example, the wavelengths at which radiation is emitted are given by the formula

$$\frac{1}{\lambda} = R \left( \frac{1}{m^2} - \frac{1}{n^2} \right) \tag{1.19}$$

where $m = 1, 2, 3, \ldots$ and $n = 2, 3, 4, \ldots$ with $n > m$, and $R$ is a constant (called the Rydberg constant) with a value of $R = 1.097 \times 10^7$ m$^{-1}$. This was a purely empirical relation discovered (for the special case $m = 2$) by Johann Balmer in 1885. (In his honor, the $m = 2$ series of spectral lines is called the *Balmer series*.) Later, spectral lines were found for the other values of $m$; these are called the Lyman series ($m = 1$), the Paschen series, ($m = 3$), and so on. The Rutherford model of the atom provides no explanation for this observed numerological result.

Neils Bohr proposed a model for the atom which solves both of these problems. Bohr assumed that the angular momentum of the electron in a hydrogen atom could not take on arbitrary (continuous) values but instead was *quantized*, i.e., constrained to take on only discrete values, which he took to have units of $h/2\pi$: $L = nh/2\pi$, $n = 1, 2, 3, \ldots$. We now introduce a new version of Planck's constant, $\hbar$ (pronounced "h-bar"), given by

$$\hbar = h/2\pi$$

In terms of $\hbar$, Bohr's quantization condition becomes

$$L = mvr = n\hbar, \quad n = 1, 2, 3, \ldots \qquad (1.20)$$

Although Bohr expressed this condition in terms of angular momentum, it follows immediately from de Broglie's later hypothesis of matter waves. If the electron behaves like a standing wave around the hydrogen nucleus, then the circumference of its orbit must correspond to an integer number of wavelengths (Figure 1.10).

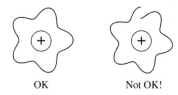

OK                Not OK!

Fig. 1.10    The circumference of the electron orbit must correspond to an integer number of wavelengths of the electron.

Taking

$$2\pi r = n\lambda = nh/p$$

we get

$$L = pr = nh/2\pi = n\hbar$$

which is just the Bohr quantization condition in Equation (1.20).

The rest of Bohr's calculation is purely classical. For a classical orbit, the Coulomb force on the electron must be equal to the centripetal force:

$$F_{Coulomb} = F_{centripetal}$$

$$\frac{e^2}{4\pi\epsilon_0}\frac{1}{r^2} = \frac{mv^2}{r} \tag{1.21}$$

Solving Equations (1.20) and (1.21) for $v$ and $r$, we find that the electron orbiting the hydrogen atom can have only certain discrete values for its orbital radius and velocity, expressed in terms of the integer $n$. Specifically,

$$r = \frac{4\pi\epsilon_0\hbar^2}{me^2}n^2, \quad n = 1, 2, 3 \ldots \tag{1.22}$$

and

$$v = \frac{e^2}{4\pi\epsilon_0\hbar}\frac{1}{n}, \quad n = 1, 2, 3 \ldots \tag{1.23}$$

The total energy of an electron with orbital radius $r$ and velocity $v$ is just the sum of its kinetic and potential energies:

$$E = \frac{1}{2}mv^2 - \frac{e^2}{4\pi\epsilon_0}\frac{1}{r} \tag{1.24}$$

Substituting the allowed discrete values for $r$ and $v$ from Equations (1.22) and (1.23) into Equation (1.24) gives a set of discrete allowed energy levels $E_n$:

$$E_n = -\frac{m}{2\hbar^2}\left(\frac{e^2}{4\pi\epsilon_0}\right)^2\frac{1}{n^2}$$

$$= -13.6 \text{ eV}\frac{1}{n^2}$$

where we have expressed the hydrogen energy levels in units of electron-volts (eV); 1 eV $= 1.6 \times 10^{-19}$ J (Figure 1.11). The energies in Equation (1.25) are negative because the electron is in a bound state.

The origin of the discrete spectral lines in the Bohr model arises from the discrete nature of the allowed energy levels. When an electron makes a transition from $n_1$ to $n_2$, it gives off a photon with energy $E_{n_1} - E_{n_2}$, and frequency $\nu = (E_{n_1} - E_{n_2})/h$. So the integers $m$ and $n$ which appear in Equation (1.19) have a real physical significance: they give the final and initial energy levels, respectively, for the electron, as parametrized in Equation (1.25). It is straightforward to use Equation (1.25) to derive the wavelengths of the spectral lines given in Equation (1.19) (see Problem 1.15).

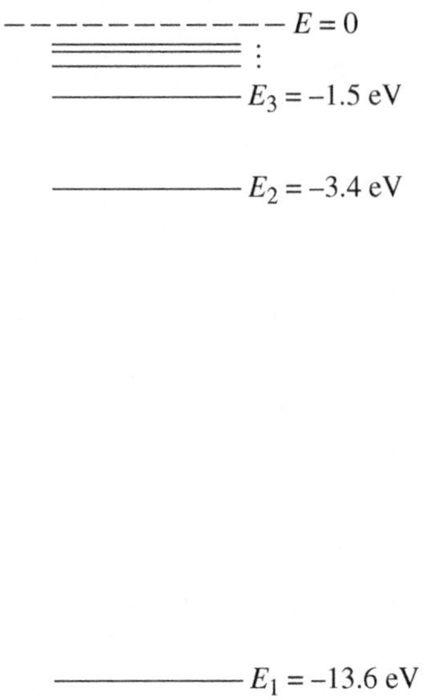

$$------------ E = 0$$
$$\overline{\phantom{xxxxxxx}} \;\; \vdots$$
$$----------------- E_3 = -1.5 \text{ eV}$$

$$----------------- E_2 = -3.4 \text{ eV}$$

$$----------------- E_1 = -13.6 \text{ eV}$$

Fig. 1.11   In the Bohr model of the atom, the energy levels are given by $E_n = -13.6 \text{ eV}(1/n^2)$.

The Bohr model was a triumph of the early stages of quantum mechanics. Since the $n = 1$ state in Equation (1.25) has the lowest possible energy, an electron in this state cannot lose more energy via radiation, so the Bohr model explains the stability of the atom. Further, the Bohr model predicts the existence and correct wavelengths for the discrete lines observed in the spectrum of hydrogen. However, the Bohr model leaves much to be desired. It mixes classical mechanics and quantum mechanics in a regime for which, as we shall see later, classical mechanics does not apply at all. In particular, in the fully quantum mechanical theory of the atom, the electrons do not have a well-defined radius or velocity. Nonetheless, the full quantum mechanical theory of the hydrogen atom (discussed in Chapter 6) predicts the same energy levels as those given by the Bohr theory.

## 1.6   Where Do We Stand?

In this chapter, we have explored the development of quantum theory which took place up to the 1920's. In order to explain various experimental results, it was necessary to postulate that both matter and light could behave sometimes like a particle and sometimes like a wave. An application of the wave nature of matter could then explain the discrete energy levels in the hydrogen spectrum, as well as the stability of the hydrogen atom. This collection of ideas did not really form a coherent theory, but it became the basis for a more complete quantum theory that began to be developed in the late 1920's, based on the Schrödinger equation. It is this theory of quantum mechanics which is the primary topic of the remainder of this book.

## PROBLEMS

**1.1** Assume that a human body emits blackbody radiation at the standard body temperature.
(a) Estimate how much energy is radiated by the body in one hour.
(b) At what wavelength does this radiation have its maximum intensity?

**1.2** A distant star is observed to have a blackbody spectrum with a maximum at a wavelength of 350 nm [1 nm = $10^{-9}$ m]. What is the temperature of the star?

**1.3** The universe is filled with blackbody radiation at a temperature of 2.7 K left over from the Big Bang. [This radiation was discovered in 1965 by Bell Laboratory scientists, who thought at one point that they were seeing interference from pigeon droppings on their microwave receiver.]
(a) What is the total energy density of this radiation?
(b) What is the total energy density with wavelengths between 1 mm and 1.01 mm? Is the Rayleigh–Jeans formula a good approximation at these wavelengths?

**1.4** Over what range in frequencies does the Rayleigh–Jeans formula give a result within 10% of the Planck blackbody spectrum?

**1.5** Let $\rho(< \nu_0)$ be the total energy density of blackbody radiation in all frequencies less than $\nu_0$, where $h\nu_0 \ll kT$. Derive an expression for $\rho(< \nu_0)$.

**1.6** Suppose we want to measure the total energy density in blackbody radiation above some cutoff frequency $\nu_0$. Let $\rho(> \nu_0)$ be the total radiation density in all frequencies greater than $\nu_0$. Using the Planck blackbody spectrum show that $\rho(> \nu_0) =$

$(8\pi/c^3)kT\nu_0^3 e^{-h\nu_0/kT}$ is a good approximation when $h\nu_0$ is much larger than $kT$.

**1.7** (a) Express the Planck spectrum (Equation (1.7)) as a function of the wavelength $\lambda$ of the radiation, rather than the frequency $\nu$.
(b) Use this expression to derive the wavelength $\lambda_{peak}$ at which the spectrum is a maximum.
(c) Does $\lambda_{peak}\nu_{peak} = c$?

**1.8** In a photoelectric experiment, electrons are emitted from a surface illuminated by light of wavelength 4000 Å, and the stopping potential for these electrons is found to be $\Phi_0 = 0.5\ V$. What is the longest wavelength of light that can illuminate this surface and still produce a photoelectric current?

**1.9** A laser emits 40 W of power at a wavelength of $6.0 \times 10^{-7}$ m.
(a) What is the total number of photons emitted per second?
(b) What is the energy of each photon?

**1.10** (a) Using the Planck blackbody spectrum, and the fact that a photon with a frequency $\nu$ has an energy of $h\nu$, derive an expression for $n(\nu)\,d\nu$, the total number density of photons with frequencies between $\nu$ and $\nu + d\nu$ in blackbody radiation.
(b) Using the expression from part (a), show that the total number density of photons in blackbody radiation is given by

$$n = \beta(kT/hc)^3$$

where $\beta$ is a constant given by $\beta \approx 60$. [Note that the integral $\int_0^\infty x^2\,dx/(e^x - 1)$ cannot be done analytically, so use the numerical result that $\int_0^\infty x^2\,dx/(e^x - 1) \approx 2.4$.]

**1.11** A gamma ray with energy 1 MeV is scattered off of an unknown particle which is at rest. The gamma ray is reflected directly backward with a final energy of 0.98 MeV. What is $m_0 c^2$ for the unknown particle? (Express your answer in MeV.)

**1.12** Calculate the de Broglie wavelength of a proton ($mc^2 = 938$ MeV) with (a) a kinetic energy of 0.1 MeV (b) a total energy of 3 GeV.

**1.13** The Balmer series (the $m = 2$ case in Equation (1.19)) was discovered before the other series of spectral lines ($m = 1$, $m = 3$, etc.).

Why? (Hint: Plug in some numbers and calculate wavelengths for $m = 1$, $m = 2$, and $m = 3$.)

**1.14** Verify that $\hbar$ has units of angular momentum.

**1.15** Beginning with the Bohr energy levels (Equation (1.25)), derive the expression for the wavelengths of the spectral lines in hydrogen (Equation (1.19)) and use this result to express $R$ as a function of $m$, $e$, $\hbar$, $c$, and $\epsilon_0$. Plug in values for these constants and verify that the correct result for $R$ is obtained.

**1.16** Suppose that the attractive force between the electron and proton in the hydrogen atom is given by some power law other than the inverse square law, i.e., assume that the magnitude of the force is given by $F = kr^\beta$, where $k$ is a constant, and $\beta$ is an arbitrary number with $\beta \neq 1$. [For example, the ordinary Coulomb law corresponds to the case $\beta = -2$. The harmonic oscillator corresponds to $\beta = 1$.] Use the Bohr quantization rule to show that for $\beta \neq -1$, the energy levels of the atom are give by

$$E = \left(\frac{\hbar^2 n^2}{m}\right)^{(\beta+1)/(\beta+3)} k^{2/(\beta+3)} \left(\frac{1}{2} + \frac{1}{\beta + 1}\right)$$

This formula gives an absurd answer when $\beta = -3$; why?

Chapter 2

# Math interlude A: Complex numbers and linear operators

## 2.1 Complex Numbers

Classical physics, with a few exceptions, relies on real numbers for its mathematical basis. Quantum mechanics marked the entry of complex numbers, in a fundamental way, into physics. Here we review the main properties of complex numbers for use in the remainder of this book.

Consider a set of numbers $0, 1, 2, \ldots$ and some operation such as addition. The set is said to be *closed* under the operation if whenever the operation is applied to the numbers in the set, the result is also in the set. For instance, the set of integers is closed under addition and multiplication. However, to get a set which is closed under subtraction, we need to include the negative numbers. Closure under division requires fractions, and the taking of various roots forces us to include the irrational numbers.

Complex numbers arise when we try to take the square root of negative numbers. The square roots of the negative numbers are said to be *imaginary*, beginning with the square root of $-1$:

$$\sqrt{-1} = i$$

It is convenient to think of the imaginary numbers as occupying a second number line perpendicular to the line occupied by the real numbers. The real and imaginary numbers can then be added just like two-dimensional vectors, resulting in the *complex numbers* which occupy this two-dimensional number plane (Figure 2.1). In general, an arbitrary complex number $z$ can be written as the sum of a real number $a$ and an imaginary number $bi$:

$$z = a + bi \tag{2.1}$$

In Equation (2.1), $a$ and $b$ are both real numbers.

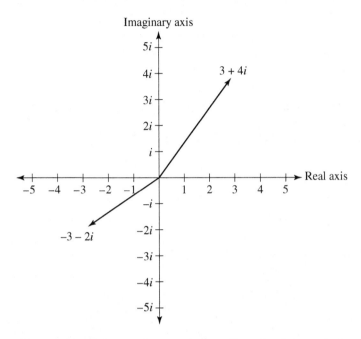

Fig. 2.1   The complex numbers can be treated as two-dimensional vectors in the complex plane; the real part of a complex number gives the horizontal coordinate in this plane, and the imaginary part gives the vertical coordinate.

Addition and subtraction of complex numbers is easy; just as for two-dimensional vectors, the real and imaginary parts are added or subtracted separately:

$$(a + bi) + (c + di) = (a + c) + (b + d)i$$
$$(a + bi) - (c + di) = (a - c) + (b - d)i$$

Multiplication and division are more subtle. For multiplication, it is possible to simply multiply out two complex numbers using the distributive property:

$$(a + bi)(c + di) = ac + bci + adi + bd(i)(i)$$
$$= (ac - bd) + (bc + ad)i$$

However, there is another way to implement complex multiplication based on a different way to represent complex numbers. Recall that the complex numbers form a two-dimensional plane, and there are two types of coordinate systems in the plane: Cartesian (or rectangular) coordinates and polar coordinates. The representation of a complex number as $z = a + bi$ corresponds to Cartesian coordinates; we will now derive a polar representation.

To do this, first consider exponentials of imaginary numbers. For an imaginary number $i\theta$ (where $\theta$ is real), we can use the Taylor expansion of the exponential to give

$$e^{i\theta} = 1 + (i\theta) + \frac{1}{2}(i\theta)^2 + \frac{1}{6}(i\theta)^3 + \frac{1}{24}(i\theta)^4 + \frac{1}{120}(i\theta)^5 + \cdots$$

The terms on the right-hand side of this equation that are even powers of $(i\theta)$ will give real numbers, while the odd powers will give imaginary numbers. Collecting the even and odd powers separately, and factoring out $i$ from the latter, gives

$$e^{i\theta} = \left(1 - \frac{1}{2}\theta^2 + \frac{1}{24}\theta^4 - \cdots\right) + i\left(\theta - \frac{1}{6}\theta^3 + \frac{1}{120}\theta^5 - \cdots\right)$$

But the two sums on the right-hand side of this equation are the Taylor expansions for cosine and sine, respectively. Hence, this equation can be written as

$$e^{i\theta} = \cos\theta + i\sin\theta \tag{2.2}$$

As $\theta$ increases from 0 to $2\pi$, the function $e^{i\theta}$ traces out a unit circle in the complex plane with angle $\theta$ relative to the positive real axis (Figure 2.2). Multiplying by the real number $R$ then gives a complex number at a distance $R$ from the origin and at an angle $\theta$ relative to the positive real axis (Figure 2.3), written as $Re^{i\theta}$. This gives the polar representation of a complex number. Any complex number can be written in either Cartesian or polar form. When a complex number $z$ is written in the form $z = Re^{i\theta}$, then $R$ is called the modulus or the absolute value of $z$, and $\theta$ is called the argument of $z$. Note that $\theta$ must be expressed in radians. Using the standard notation for absolute value, we can write $|z| = R$.

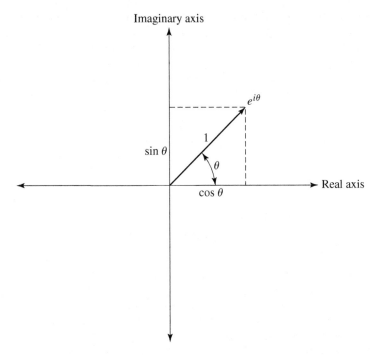

Fig. 2.2   Since $e^{i\theta} = \cos\theta + i\sin\theta$, the complex number $e^{i\theta}$ has unit absolute value and lies at an angle $\theta$ relative to the real axis.

To convert from the Cartesian form of a complex number, $a + bi$, to polar form, $Re^{i\theta}$, and vice versa, we use Equation (2.2) to give

$$Re^{i\theta} = R\cos\theta + iR\sin\theta$$

so that

$$a = R\cos\theta$$
$$b = R\sin\theta$$

and conversely,

$$R = \sqrt{a^2 + b^2}$$
$$\theta = \tan^{-1}(b/a)$$

Note that there is a subtlety in determining the value of $\theta$ for a given $a$ and $b$, because $\tan^{-1}(x)$ actually has two different values for a given choice of $x$, separated from each other by the angle $\pi$. This ambiguity is resolved

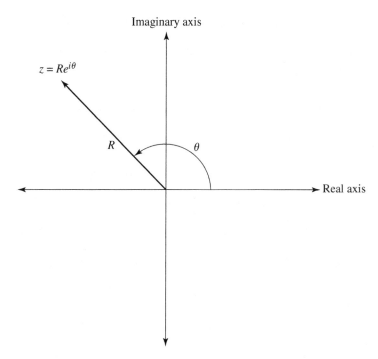

Fig. 2.3   The complex number $z = Re^{i\theta}$ is located at a distance $R$ from the origin and at an angle $\theta$ relative to the positive real axis.

by noting that $\theta$ must be chosen so that the complex number lies in the correct quadrant of the complex plane:

$$0 < \theta < \pi/2 \to a > 0,\ b > 0$$
$$\pi/2 < \theta < \pi \to a < 0,\ b > 0$$
$$\pi < \theta < 3\pi/2 \to a < 0,\ b < 0$$
$$3\pi/2 < \theta < 2\pi \to a > 0,\ b < 0$$

---

**Example 2.1. Converting a Complex Number to Polar Form**

Express $2 + 2i$ and $1 - i$ in polar form.

For $2 + 2i$, we have

$$R = \sqrt{2^2 + 2^2} = 2\sqrt{2}$$

and

$$\theta = \tan^{-1}(2/2) = \tan^{-1}(1) = \pi/4 \text{ or } 5\pi/4$$

Since $a > 0$ and $b > 0$, the correct choice for $\theta$ is $\pi/4$. Similarly, for $1 - i$, we get

$$R = \sqrt{1^2 + 1^2} = \sqrt{2}$$

and

$$\theta = \tan^{-1}(-1/1) = \tan^{-1}(-1) = 3\pi/4 \text{ or } 7\pi/4$$

and since $a > 0$ and $b < 0$, we must choose $7\pi/4$. Hence, we have

$$2 + 2i = 2\sqrt{2}e^{i\pi/4}$$

and

$$1 - i = \sqrt{2}e^{i7\pi/4}$$

---

It is now straightforward to multiply or divide two complex numbers, represented as $z_1 = R_1 e^{i\theta_1}$ and $z_2 = R_2 e^{i\theta_2}$: For multiplication,

$$z_1 z_2 = (R_1 e^{i\theta_1})(R_2 e^{i\theta_2}) = R_1 R_2 e^{i(\theta_1 + \theta_2)}$$

Thus, when multiplying two complex numbers, the $R$'s are multiplied and the $\theta$'s are added. For example, multiplying a complex number by $i$ simply rotates it $90°$ in the complex plane without changing its distance from the origin, while multiplication by $-1$ is equivalent to rotation through $180°$. Similarly, for division,

$$z_1/z_2 = (R_1 e^{i\theta_1})/(R_2 e^{i\theta_2}) = (R_1/R_2)e^{i(\theta_1 - \theta_2)}$$

---

**Example 2.2. Multiplication of Complex Numbers**
What is $(2 + 2i)(1 - i)$?

This can be solved in two different ways. Using the Cartesian form for these numbers,

$$(2 + 2i)(1 - i) = (2)(1) + (2i)(1) + (2)(-i) + (2i)(-i)$$

$$= 2 + 2i - 2i + 2$$

$$= 4$$

In polar form, we use the results from Example 2.1:

$$2 + 2i = 2\sqrt{2}e^{i\pi/4}$$

and

$$1 - i = \sqrt{2}e^{i7\pi/4}$$

Then
$$(2 + 2i)(1 - i) = (2\sqrt{2}e^{i\pi/4})(\sqrt{2}e^{i7\pi/4})$$
$$= 4e^{i2\pi}$$
$$= 4$$

Of course, the final answer cannot depend on whether we perform the multiplication in Cartesian or polar form.

---

When a complex number is expressed in polar form, it is straightforward to take it to an arbitrary power:
$$(Re^{i\theta})^n = R^n e^{in\theta}$$
and the roots of a complex number can be determined in a similar way:
$$\sqrt[n]{Re^{i\theta}} = (Re^{i\theta})^{1/n} = \sqrt[n]{R}e^{i\theta/n}$$

---

### Example 2.3. The Cube Root of 1

What is the cube root of 1?

In terms of real numbers, the cube root of 1 is just 1. However, when we consider complex numbers, we discover that 1 actually has three different cube roots! We have in polar form:
$$1^3 = 1$$
$$\left(e^{i2\pi/3}\right)^3 = e^{i2\pi} = 1$$
$$\left(e^{i4\pi/3}\right)^3 = e^{i4\pi} = 1$$

In the complex plane, these three numbers all lie on a unit circle separated from each other by an angle of $2\pi/3$ (or $60°$).

---

Finally, we must deal with one operation that is unique to complex numbers. This is called *complex conjugation*. If $z = a + bi$ is an arbitrary complex number, then its complex conjugate, written as $\bar{z}$ or $z^*$ (we will use the latter notation), is given by
$$z^* = a - bi$$

This corresponds to reflection in the complex plane through the real axis (Figure 2.4).

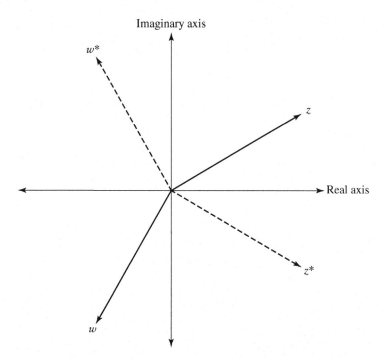

Fig. 2.4   If $z = a + bi$, then $z^* = a - bi$, corresponding to reflection through the real axis.

Some important properties of complex conjugation are

$$z^* z = (a - bi)(a + bi) = a^2 + b^2 = |z|^2$$
$$\left(z^*\right)^* = z$$
$$(w + z)^* = w^* + z^*$$ 
$$(wz)^* = w^* z^*$$

(2.3)

where $w$ and $z$ are any two complex numbers. In polar form, we have

$$\left(Re^{i\theta}\right)^* = Re^{-i\theta}$$

More generally, the complex conjugate of a complicated expression can be obtained simply by changing $i$ to $-i$ everywhere in the expression, e.g.,

$$\left(1 + i + e^{-ix}\right)^* = 1 - i + e^{ix}$$

## 2.2 Operators

### *Definition of an Operator*

The idea of a function is very familiar: a function is simply a fixed rule for taking a number and changing it into another number. A number is plugged into a function and out comes a different number. An *operator* is a rule for changing one *function* into another *function* (Figure 2.5).

A familiar example of an operator is the derivative operator $D$, which takes an input function $f(x)$ and produces as its output the derivative of that function, $df/dx$:

$$D[f(x)] = \frac{df}{dx}$$

e.g.,

$$D[x^2] = 2x$$
$$D[\sin x] = \cos x \tag{2.4}$$

and so on. We will be interested only in a special class of operators called *linear operators*. In order for an operator $L$ to be a linear operator, it must satisfy two properties: first, for every pair of functions $f(x)$ and $g(x)$,

$$L[f(x) + g(x)] = L[f(x)] + L[g(x)]$$

and second, for every function $f(x)$ and real or complex number $c$,

$$L[cf(x)] = cL[f(x)]$$

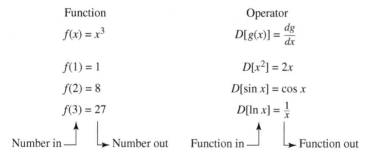

Function

$f(x) = x^3$

$f(1) = 1$

$f(2) = 8$

$f(3) = 27$

Number in ⌐ ⌐► Number out

Operator

$D[g(x)] = \frac{dg}{dx}$

$D[x^2] = 2x$

$D[\sin x] = \cos x$

$D[\ln x] = \frac{1}{x}$

Function in ⌐ ⌐► Function out

Fig. 2.5 A function is a rule for taking a number and turning it into a different number. An operator is a rule for taking a function and turning it into a different function.

## Example 2.4. Determining Whether Operators are Linear

Determine whether or not the following operators are linear:
(a) $A[g(x)] = g(x)^2$,
(b) the derivative operator, $D[g(x)] = dg/dx$.

(a) Note that

$$A[f(x) + g(x)] = [f(x) + g(x)]^2 = f(x)^2 + g(x)^2 + 2f(x)g(x)$$

and

$$A[f(x)] + A[g(x)] = f(x)^2 + g(x)^2$$

Thus, $A[f(x) + g(x)] \neq A[f(x)] + A[g(x)]$, so $A$ is *not* a linear operator.

(b) For the derivative operator, we have

$$D[f(x) + g(x)] = \frac{d}{dx}[f(x) + g(x)] = \frac{df}{dx} + \frac{dg}{dx} = D[f(x)] + D[g(x)]$$

and

$$D[cf(x)] = \frac{d}{dx}[cf(x)] = c\frac{df}{dx} = cD[f(x)]$$

So the derivative operator *is* a linear operator.

### Eigenfunctions and Eigenvalues

Suppose that for a particular linear operator $L$, we can find a function $f(x)$ which has the property

$$L[f(x)] = cf(x)$$

where $c$ can be a real or a complex number. In other words, applying $L$ to the function $f$ simply gives us $f$ back again multiplied by the number $c$. In this case, we say that $f$ is an *eigenfunction* of $L$ with *eigenvalue* $c$. The actual set of eigenfunctions, along with their corresponding eigenvalues, will depend on $L$.

## Example 2.5. The Eigenfunctions and Eigenvalues of the Derivative Operator

If $g$ is an eigenfunction of the derivative operator $D$, it must satisfy

$$D[g(x)] = \frac{dg}{dx} = cg(x) \tag{2.5}$$

This differential equation has the general solution $g(x) = Ae^{cx}$, where $A$ is an arbitrary constant, and $c$ is the eigenvalue appearing in Equation (2.5). Hence, $g(x) = Ae^{cx}$ is the most general possible eigenfunction of $D$ with eigenvalue $c$. Note that in this case, any complex number can be an eigenvalue of $D$.

---

Of course, most functions are *not* eigenfunctions of a given operator. For example, in the case of the derivative operator, it is obvious that

$$D[\sin(x)] = \cos(x) \neq c\sin(x)$$

$$D[x^2] = 2x \neq cx^2$$

$$D[\ln(x)] = 1/x \neq c\ln(x)$$

Thus, an eigenfunction is a very special kind of function.

---

## Example 2.6. The Eigenfunctions and Eigenvalues of the One-Dimensional Parity Operator

Find the eigenfunctions and eigenvalues of the one-dimensional parity operator, $\Pi$, defined by

$$\Pi[g(x)] = g(-x)$$

i.e., the parity operator reflects the function $g(x)$ through $x = 0$ (Figure 2.6).

Again, we must solve the equation $\Pi[g(x)] = cg(x)$. Using the definition of the parity operator, we get

$$\Pi[g(x)] = g(-x) = cg(x) \tag{2.6}$$

This equation has no obvious solution, so we apply $\Pi$ twice to the function $g(x)$:

$$\Pi^2[g(x)] = \Pi[g(-x)] = g(x) \tag{2.7}$$

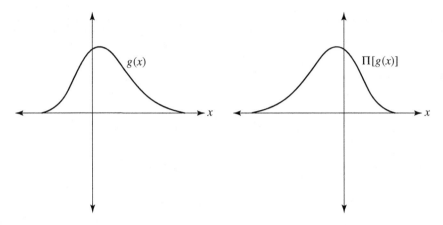

Fig. 2.6   The one-dimensional parity operator $\Pi$ reflects the function $g(x)$ through $x = 0$.

(Note that the notation $L^n[g(x)]$ means to apply the operator $n$ times to the function $g(x)$, e.g., $L^3[g(x)] = L[L[L[g(x)]]]$.) If we assume that $g(x)$ is an eigenfunction of $\Pi$ with eigenvalue $c$, we also have

$$\Pi^2[g(x)] = \Pi[cg(x)] = c\Pi[g(x)] = c^2 g(x) \tag{2.8}$$

Combining Equations (2.7) and (2.8), we get

$$g(x) = c^2 g(x)$$

which has the solutions $c = \pm 1$. Note that in contrast to Example 2.5, there are not an infinite number of eigenvalues but only two discrete eigenvalues. We can now determine what functions correspond to each eigenvalue. For $c = +1$, Equation (2.6) gives $g(-x) = g(x)$, so that $g(x)$ is an arbitrary even function. For $c = -1$, Equation (2.6) yields $g(-x) = -g(x)$, so $g(x)$ is an arbitrary odd function.

---

Note that when an eigenfunction of a linear operator is multiplied by an arbitrary constant, it remains an eigenfunction with the same eigenvalue (Problem 2.11).

## PROBLEMS

**2.1** Evaluate all of the following, and express all of your final answers in the form $a + bi$:

(a) $i(2 - 3i)(3 + 5i)$

(b) $i/(i-1)$

(c) $(1+i)^{30}$

**2.2** In the complex plane, there are 5 different fifth roots of 1. Determine the five values for $\sqrt[5]{1}$, and express them in polar form.

**2.3** Suppose that $z = 1 + e^{i\theta}$. Calculate $z^*$, $z^2$, and $|z|^2$. Your expression for $|z|^2$ should not contain any imaginary numbers.

**2.4** Suppose that a complex number $z$ has the property that $z^* = z$. What does this indicate about $z$?

**2.5** Reduce $i^i$ to a real number.

**2.6** What is wrong with the following argument?

$$\sqrt{\frac{1}{-1}} = \frac{\sqrt{1}}{\sqrt{-1}}$$

$$\sqrt{-1} = \frac{1}{i} \tag{2.9}$$

$$i = \frac{1}{i}$$

Therefore,

$$(i)(i) = 1$$
$$-1 = 1 \tag{2.10}$$

**2.7** Determine which of the following are linear operators, and which are not.

(i) The parity operator $\Pi[f(x)] = f(-x)$.

(ii) The translation operator $T[f(x)] = f(x+1)$.

(iii) The operator $L[f(x)] = f(x) + 1$.

**2.8** Consider the identity operator $I$, defined by $I[f(x)] = f(x)$.

(a) Show that $I$ is a linear operator.

(b) Find the eigenfunctions and corresponding eigenvalues of $I$.

**2.9** Suppose that the function $f(x)$ is an eigenfunction of the linear operator $P$ with eigenvalue $p$, and $f(x)$ is also an eigenfunction of the linear operator $Q$ with eigenvalue $q$. Show that $PQ[f(x)] =$

$QP[f(x)]$, where $PQ[f(x)]$ means to first apply the operator $Q$ to $f(x)$, and then apply $P$ to the result.

**2.10** Consider the square of the derivative operator $D^2$.

(a) Show that $D^2$ is a linear operator.

(b) Find the eigenfunctions and corresponding eigenvalues of $D^2$.

(c) Give an example of an eigenfunction of $D^2$ which is *not* an eigenfunction of $D$.

**2.11** Let $f(x)$ be an eigenfunction of a linear operator $L$ with eigenvalue $a$. Show that $cf(x)$ (where $c$ is a constant) is an eigenfunction of $L$ with eigenvalue $a$.

**2.12** Consider the following operator $L$:

$$L[f(x)] = \int_0^x f(s)\,ds$$

(a) Show that $L$ is a linear operator.

(b) Find the eigenfunctions of $L$, or show that $L$ has no eigenfunctions.

Chapter 3

# The Schrödinger equation

The Schrödinger equation, developed by Erwin Schrödinger and published in 1926, forms the basis of modern quantum mechanics. Indeed, it is one of the most important equations in all of physics, and much of the remainder of this book will be based on it.

The way in which the Schrödinger equation describes the behavior of particles is fundamentally different from the corresponding description in classical mechanics. In classical mechanics, a particle has a fixed position in three-dimensional space, given by the vector $\mathbf{r}$. This position is a function of time, $t$, so a complete description of the motion of the particle is given by its trajectory, $\mathbf{r}(t)$; the main problem in classical mechanics is to determine $\mathbf{r}(t)$.

In contrast, in quantum mechanics, a particle no longer has a definite trajectory $\mathbf{r}(t)$. Instead, we start with the idea from Chapter 1 that matter can be treated as a wave. In particular, we will assume that any particle can be described by a *wave function*, $\Psi(\mathbf{r}, t)$, which gives the amplitude of the wave as a function of the three-dimensional position in space, $\mathbf{r}$, and of the time, $t$.

This leads to an obvious question: what is the physical meaning of $\Psi(\mathbf{r}, t)$, i.e., what does it tell us about the particle? Although we must abandon the hope of determining the position of the particle as a function of time, what we can derive from the wave function is the *probability* that the particle will be found in a given region of space at a given time. Furthermore, all of the other observable physical characteristics of this particle (e.g., its momentum, energy, etc.) are related to $\Psi(\mathbf{r}, t)$. The relationship between observable quantities and $\Psi(\mathbf{r}, t)$ will be examined in more detail in Section 3.2. First, however, we will derive the equation which determines $\Psi(\mathbf{r}, t)$: the Schrödinger equation.

## 3.1    Derivation of the Schrödinger Equation

Unfortunately, it is no more possible to "derive" the Schrödinger equation from purely mathematical arguments than it is to derive Newton's law of gravitation or $F = ma$. The only basis for developing a physical theory is that it describes experimental reality. What we can do, however, is make some reasonable assumptions and show that these lead to the Schrödinger equation. We will then solve the Schrödinger equation and discover that it does, indeed, make numerous correct predictions. In particular, it provides an accurate description of the hydrogen atom and many other phenomena at the atomic and subatomic scales.

For simplicity, first consider a wave in one dimension, travelling in the $+x$ direction. A wave with frequency $\nu$ and wavelength $\lambda$ can be written in the form

$$\Psi(x,t) = A\cos(2\pi x/\lambda - 2\pi\nu t) \tag{3.1}$$

What does Equation (3.1) mean? The shape of the wave can be derived by fixing $t$ and treating $\Psi$ as a function only of $x$. Then Equation (3.1) represents a wave, frozen in time, which oscillates sinusoidally as a function of $x$ with wavelength $\lambda$ and amplitude $A$. Alternately, we can fix the position $x$ on the wave to be a constant and consider how $\Psi$ varies with $t$; in this case, we see that our fixed point on the wave oscillates up and down sinusoidally with frequency $\nu$ and amplitude $A$ (Figure 3.1). To simplify this wave equation, it is conventional to make the substitutions:

$$k = 2\pi/\lambda$$

and

$$\omega = 2\pi\nu$$

where $k$ is called the *wave number*, and $\omega$ is the *angular frequency*. In terms of $k$ and $\omega$, Equation (3.1) simplifies to

$$\Psi = A\cos(kx - \omega t) \tag{3.2}$$

We note one other property of the wave described by Equation (3.2) (or Equation (3.1)); it represents a wave travelling to the right with velocity $\omega/k$. To see this, consider, for example, the maximum $x_{max}$ in the amplitude of the wave at $x = 0$, $t = 0$. At some later time $t$, this will still be a maximum of the cosine function as long as $kx_{max} - \omega t = 0$, or

$$x_{max} = (\omega/k)\,t \tag{3.3}$$

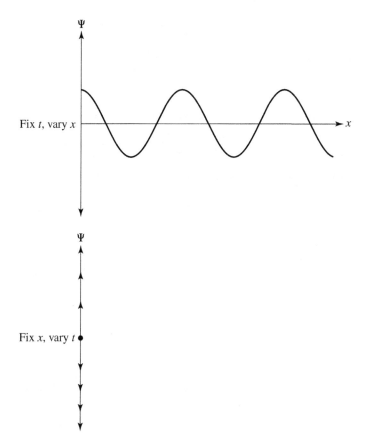

Fix $t$, vary $x$

Fix $x$, vary $t$

Fig. 3.1    The equation $\Psi(x,t) = A\cos(2\pi x/\lambda - 2\pi\nu t)$ represents a wave with amplitude $A$ and wavelength $\lambda$, oscillating in time with frequency $\nu$.

Equation (3.3) shows that the maximum moves in the $+x$ direction, i.e., to the right, with velocity $\omega/k$. This same argument can be made about any other point on the wave, which means that the entire wave moves in the $+x$ direction with velocity

$$v = \omega/k$$

This velocity is called the *phase velocity* (Figure 3.2).

So far, we have confined our attention to one dimension, but Equation (3.2) can be generalized to three dimensions by using a three-dimensional position vector $\mathbf{r}$ in place of the one-dimensional position $x$. In this case, we must also take $k$ to be a vector, $\mathbf{k}$, called the *wave vector*,

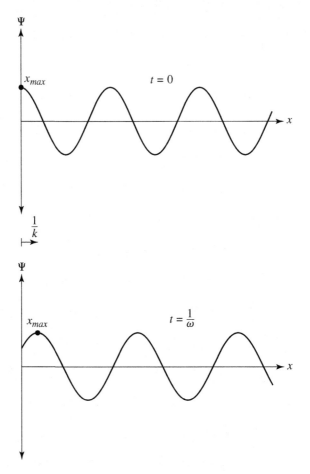

Fig. 3.2   The equation $\Psi(x, t) = A\cos(kx - \omega t)$ represents a wave moving in the $+x$ direction with phase velocity $\omega/k$.

and Equation (3.2) becomes

$$\Psi(\mathbf{r}, t) = A\cos(\mathbf{k}\cdot\mathbf{r} - \omega t)$$

From our previous argument, this represents a wave travelling in the $\mathbf{k}$ direction. However, this is not the most general possible form for such a wave. A sine function serves just as well as a cosine function to represent an oscillating wave, so the most general form for a wave moving in the $\mathbf{k}$ direction is

$$\Psi(\mathbf{r}, t) = A_1 \cos(\mathbf{k}\cdot\mathbf{r} - \omega t) + A_2 \sin(\mathbf{k}\cdot\mathbf{r} - \omega t) \tag{3.4}$$

where $A_1$ and $A_2$ are constants which determine both the amplitude and phase of the wave. Although this is a perfectly acceptable way to represent a general wave, it is somewhat awkward. As an alternative, we can write

$$\Psi(\mathbf{r}, t) = Be^{i(\mathbf{k}\cdot\mathbf{r} - \omega t)} \tag{3.5}$$

where $B$ is now a complex number, and in classical mechanics, it is understood that we take the *real* part of the right-hand side of Equation (3.5).

Equations (3.4) and (3.5) are completely equivalent. To see this expand Equation (3.5) out using Equation (2.2) and write $B = B_1 + iB_2$. Then Equation (3.5) becomes

$$\begin{aligned}\Psi(\mathbf{r}, t) = {}& B_1 \cos(\mathbf{k}\cdot\mathbf{r} - \omega t) - B_2 \sin(\mathbf{k}\cdot\mathbf{r} - \omega t) \\ & + iB_2 \cos(\mathbf{k}\cdot\mathbf{r} - \omega t) + iB_1 \sin(\mathbf{k}\cdot\mathbf{r} - \omega t)\end{aligned} \tag{3.6}$$

By choosing $B_1 = A_1$ and $B_2 = -A_2$ and taking the real part of the right-hand side of Equation (3.6), we get the same result as Equation (3.4). We will now use Equation (3.5) to represent our wave function.

In order to derive a plausible equation for the wave function, we need to make a connection between $\mathbf{k}$, $\omega$, and the momentum and energy of the particle. In Chapter 1 we noted two relations between the properties of waves and physically-measurable quantities. Specifically, for light, the energy $E$ and the frequency $\nu$ are related by

$$E = h\nu$$

and for matter waves, the momentum $p$ and the wavelength $\lambda$ are related by

$$p = h/\lambda$$

We will assume that both of these properties apply to matter waves. The equation for energy, $E = h\nu$, corresponds to

$$E = (h/2\pi)(2\pi\nu) = \hbar\omega$$

and the momentum equation, $p = h/\lambda$, becomes

$$p = (h/2\pi)(2\pi/\lambda) = \hbar k \tag{3.7}$$

We can generalize this equation from one dimension to three dimensions. In three dimensions, $\mathbf{k}$ points in the direction of motion of the particle, which is also the direction of the momentum vector, $\mathbf{p}$. Hence, Equation (3.7) becomes

$$\mathbf{p} = \hbar\mathbf{k}$$

For a non-relativistic particle with mass $m$ and with no potential energy, the relationship between $p$ and $E$ is given by

$$E = \frac{p^2}{2m} \tag{3.8}$$

It is now possible to generate expressions for $E$ and $p$ from Equation (3.5) by taking the appropriate derivatives. Taking the derivative of Equation (3.5) with respect to time, we get

$$\frac{\partial \Psi}{\partial t} = -i\omega B e^{i(\mathbf{k}\cdot\mathbf{r} - \omega t)}$$

Multiplying both sides by $i\hbar$ gives energy on the right-hand side:

$$i\hbar\frac{\partial \Psi}{\partial t} = \hbar\omega B e^{i(\mathbf{k}\cdot\mathbf{r} - \omega t)}$$
$$= E\Psi \tag{3.9}$$

Note that $i\hbar(\partial/\partial t)$ is a *linear operator*, and $\Psi$ is an *eigenfunction* of this operator with *eigenvalue* $E$.

Now we need to find a similar operator which gives the momentum $\mathbf{p}$. Expanding the dot product in Equation (3.5) gives

$$\Psi(\mathbf{r}, t) = B e^{i(k_x x + k_y y + k_z z - \omega t)}$$

so that

$$\frac{\partial \Psi}{\partial x} = i k_x B e^{i(k_x x + k_y y + k_z z - \omega t)}$$

$$\frac{\partial \Psi}{\partial y} = i k_y B e^{i(k_x x + k_y y + k_z z - \omega t)}$$

$$\frac{\partial \Psi}{\partial z} = i k_z B e^{i(k_x x + k_y y + k_z z - \omega t)}$$

Then

$$\nabla\Psi = \hat{x}\frac{\partial \Psi}{\partial x} + \hat{y}\frac{\partial \Psi}{\partial y} + \hat{z}\frac{\partial \Psi}{\partial z}$$
$$= (i k_x \hat{x} + i k_y \hat{y} + i k_z \hat{z}) B e^{i(k_x x + k_y y + k_z z - \omega t)}$$
$$= i\mathbf{k}\Psi$$

Since we want $\mathbf{p} = \hbar\mathbf{k}$ to multiply $\Psi$ on the right-hand side, we multiply both sides by $-i\hbar$:

$$-i\hbar\nabla\Psi = \hbar\mathbf{k}\Psi = \mathbf{p}\Psi$$

In summary, we now have two operators for which $\Psi$ is an eigenfunction; one gives the energy $E$ as the eigenvalue, and the other gives the momentum $\mathbf{p}$ as the eigenvalue:

$$i\hbar\frac{\partial\Psi}{\partial t} = E\Psi$$

$$-i\hbar\nabla\Psi = \mathbf{p}\Psi$$

We now derive an equation which corresponds to Equation (3.8). Applying the momentum operator $-i\hbar\nabla$ *twice* to $\Psi$ produces two factors of $\mathbf{p}$:

$$(-i\hbar\nabla)\cdot(-i\hbar\nabla)\Psi = \mathbf{p}\cdot\mathbf{p}\Psi$$

$$= p^2\Psi$$

Hence,

$$-\hbar^2\nabla^2\Psi = p^2\Psi$$

(Recall that $\nabla^2$ is shorthand for $\nabla\cdot\nabla = \partial^2/\partial x^2 + \partial^2/\partial y^2 + \partial^2/\partial z^2$.) Then

$$-\frac{\hbar^2}{2m}\nabla^2\Psi = \frac{p^2}{2m}\Psi \tag{3.10}$$

The right-hand sides of Equations (3.9) and (3.10) are manifestly equal, since $p^2/2m = E$. Then equating the left-hand sides of these equations gives us a differential equation satisfied by $\Psi$:

$$-\frac{\hbar^2}{2m}\nabla^2\Psi = i\hbar\frac{\partial\Psi}{\partial t} \tag{3.11}$$

So far, all we have shown is that the wave function given in Equation (3.5) satisfies Equation (3.11), and further, that it is an eigenfunction of the operators on the left-hand and right-hand sides of Equation (3.11) with eigenvalues $p^2/2m$ and $E$, respectively. It is at this point that we make an unjustified leap: what happens if we now add a potential energy $V$ to the system? In a classical system, the total energy is just the sum of the kinetic and potential energies:

$$\frac{p^2}{2m} + V = E$$

This suggests that we modify Equation (3.11) to read

$$-\frac{\hbar^2}{2m}\nabla^2\Psi + V\Psi = i\hbar\frac{\partial\Psi}{\partial t} \tag{3.12}$$

(Note that unlike the kinetic energy and total energy, which correspond to operators containing various derivatives, the operator corresponding to the potential energy is simply multiplication of $\Psi$ by $V$.)

Equation (3.12) is the *Schrödinger equation*, possibly the most important equation in all of 20th-century physics. Note that we have not *derived* this equation in a mathematical sense. We have constructed Equation (3.12) so that for $V = 0$, the equation will be satisfied by the wave function $\Psi = e^{i(\mathbf{k}\cdot\mathbf{r}-\omega t)}$. However, this particular wave function will *not* satisfy Equation (3.12) if $V \neq 0$. We simply postulate that Equation (3.12) will give the "correct" wave function $\Psi$ for *any* potential $V$. We will see that the predictions of the Schrödinger equation do agree with a wide variety of physical phenomena.

In its most general form, $\Psi$ is a *complex* function of three spatial coordinates $x$, $y$, and $z$, and of the time $t$. The potential $V$, in general, is also a function of $x$, $y$, $z$, and $t$. Hence, in the most general case, Equation (3.12) is really shorthand for

$$-\frac{\hbar^2}{2m}\left(\frac{\partial^2\Psi(x,y,z,t)}{\partial x^2} + \frac{\partial^2\Psi(x,y,z,t)}{\partial y^2} + \frac{\partial^2\Psi(x,y,z,t)}{\partial z^2}\right)$$
$$+ V(x,y,z,t)\Psi(x,y,z,t) = i\hbar\frac{\partial\Psi(x,y,z,t)}{\partial t}$$

$$(3.13)$$

Most practical applications of the Schrödinger equation involve a simplification of the most general case, e.g., motion in only one dimension, potentials which are independent of time, etc. For these cases, the Schrödinger equation assumes a much simpler form than Equation (3.13). For example, for a particle moving in one dimension, the Schrödinger equation becomes

$$-\frac{\hbar^2}{2m}\frac{\partial^2\Psi(x,t)}{\partial x^2} + V(x,t)\Psi(x,t) = i\hbar\frac{\partial\Psi(x,t)}{\partial t}$$

**Example 3.1. A Solution of the One-Dimensional Schrödinger Equation**

Consider the one-dimensional infinite square-well potential of width $a$, shown in Figure 3.3. The potential $V(x)$ is given by

$$V(x) = 0, \quad \text{for } 0 \leq x \leq a$$

$$V(x) = \infty, \quad \text{for } x < 0 \text{ or } x > a$$

Of course, no physical potential can be truly infinite, but this potential will be a good approximation for any system with sharp potential barriers such that $V \gg E$. The infinite potential barriers force $\Psi(x,t)$ to be zero outside of the potential well, and give the boundary conditions $\Psi(0,t) = 0$

and $\Psi(a,t) = 0$. (This will be discussed in more detail in Chapter 4.) Note further that the potential in this case is independent of time.

For $0 \leq x \leq a$, the potential is zero, so the Schrödinger equation can be written as

$$-\frac{\hbar^2}{2m}\frac{\partial^2 \Psi(x,t)}{\partial x^2} = i\hbar \frac{\partial \Psi(x,t)}{\partial t} \tag{3.14}$$

To solve this equation, we assume that the solution has the form

$$\Psi(x,t) = \psi(x)\chi(t) \tag{3.15}$$

where $\psi$ is a function only of $x$ and is independent of $t$, while $\chi$ is a function only of $t$ and is independent of $x$. How do we know that the solution of Equation (3.14) can be written in the form of Equation (3.15)? In fact, we don't. The only way to determine if the solution has this form is to see if we can find functions $\psi(x)$ and $\chi(t)$ for which $\Psi(x,t) = \psi(x)\chi(t)$ satisfies Equation (3.14). However, many partial differential equations in physics do, in fact, yield solutions of the form given in Equation (3.15). (Of course, we wouldn't be introducing this solution if it didn't work in this case!) This general method of solution is called *separation of variables*.

Substituting Equation (3.15) into Equation (3.14) gives

$$-\frac{\hbar^2}{2m}\frac{\partial^2 \psi(x)}{\partial x^2}\chi(t) = i\hbar\psi(x)\frac{\partial \chi(t)}{\partial t}$$

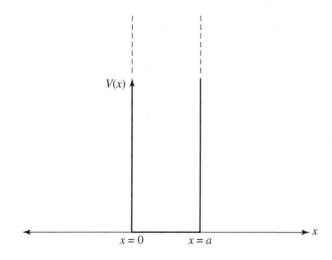

Fig. 3.3   The infinite square-well potential.

and dividing both sides by $(-\hbar^2/2m)\psi(x)\chi(t)$ yields

$$\frac{1}{\psi(x)}\frac{\partial^2\psi(x)}{\partial x^2} = -\frac{2mi}{\hbar}\frac{1}{\chi(t)}\frac{\partial\chi(t)}{\partial t} \tag{3.16}$$

Note that the left-hand side of Equation (3.16) is a function only of $x$ and is independent of $t$, while the right-hand side is a function only of $t$ and is independent of $x$. This apparent contradiction can be resolved by noting that there is only one function that satisfies both of these requirements: a constant, which is independent of *both* $x$ and $t$. Setting both sides of Equation (3.16) equal to the constant $C$, we get

$$\frac{d^2\psi(x)}{dx^2} = C\psi(x) \tag{3.17}$$

and

$$-\frac{2mi}{\hbar}\frac{d\chi(t)}{dt} = C\chi(t) \tag{3.18}$$

where the partial derivatives have become total derivatives, since each equation now contains only a single independent variable.

Consider first the equation for $\psi(x)$. The general solution to Equation (3.17) is either a sum of two exponentials with real arguments (for $C > 0$) or a sum of two trigonometric functions with real arguments (for $C < 0$). However, the boundary conditions give an additional constraint on the solution. The infinite barriers produce $\Psi(0, t) = 0$ and $\Psi(a, t) = 0$, which means that $\psi(0) = \psi(a) = 0$. However, a sum of two exponentials with real arguments has at most one value of $x$ for which $\psi(x) = 0$. Hence, $\psi(x)$ must be a sum of trigonometric functions, namely

$$\psi(x) = A_1 \sin\sqrt{-C}x + A_2 \cos\sqrt{-C}x \tag{3.19}$$

where $A_1$ and $A_2$ are constants, and $C$ must be negative, so $-C$ is positive. The condition that $\psi(0) = 0$ means that $A_2 = 0$ in Equation (3.19), giving

$$\psi(x) = A_1 \sin\sqrt{-C}x$$

while the condition $\psi(a) = 0$ can be satisfied for

$$\sqrt{-C}a = n\pi, \quad n = 1, 2, 3, \ldots \tag{3.20}$$

We will examine the general solution in more detail in Chapter 4; here we will simply make use of a single solution, $n = 1$, which gives

$$C = -\pi^2/a^2$$

and

$$\psi(x) = A_1 \sin\left(\frac{\pi x}{a}\right)$$

Now we can solve for $\chi(t)$ using Equation (3.18). Taking $C = -\pi^2/a^2$ in this equation gives

$$\frac{d\chi}{dt} = -\frac{i\hbar\pi^2}{2ma^2}\chi$$

which has the solution

$$\chi = A_3 e^{-i\hbar\pi^2 t/2ma^2}$$

Combining the solutions for $\psi(x)$ and $\chi(t)$ into an expression for $\Psi(x,t)$ gives

$$\Psi(x,t) = A\sin\left(\frac{\pi x}{a}\right) e^{-i\hbar\pi^2 t/2ma^2}, \quad \text{for } 0 \le x \le a$$

$$= 0, \quad \text{for } x < 0 \text{ or } x > a$$

In this solution, $A = A_1 A_3$ remains an arbitrary constant. However, the value of $A$ will be determined in the next section, when we examine the physical interpretation of the wave function.

Note that we have found only a single solution corresponding to $n = 1$ in Equation (3.20). The general solution, as well as the physical significance of the values for $n$, will be examined in the next chapter.

---

## 3.2 The Meaning of the Wave Function

In the previous section, we derived the Schrödinger equation, which describes the behavior of the wave function $\Psi$ associated with a particle moving in a potential $V$. This leaves an obvious question: what does $\Psi$ tell us about the physical behavior of the particle?

The interpretation of $\Psi$ was provided by Max Born. Recall that $\Psi(\mathbf{r},t)$ is complex, so it cannot by itself represent a physically-measurable quantity. Born argued that the square of the absolute value of the wave function, $|\Psi(\mathbf{r},t)|^2 = \Psi^*(\mathbf{r},t)\Psi(\mathbf{r},t)$, which is always a real number, gives the *probability* per unit volume of finding the particle at the position $\mathbf{r}$ at the time $t$. Hence $\Psi^*\Psi$ is called the *probability density*. Since probability is a pure number, the probability density must have units of 1/volume in three dimensions and 1/length in one dimension. Note that this represents the abandonment of one of the basic ideas of classical mechanics: a particle no longer has a definite position in space that can be described as a known function of time. Instead, the wave function is used to calculate the *probability* of finding the particle in a given region of space.

As an example, consider the wave function derived in Example 3.1 for a particle in an infinite square well:

$$\Psi(x,t) = A \sin\left(\frac{\pi x}{a}\right) e^{-i\hbar\pi^2 t/2ma^2}, \quad \text{for } 0 \le x \le a$$
$$= 0, \quad \text{for } x < 0 \text{ or } x > a \tag{3.21}$$

This solution satisfies the Schrödinger equation for an *arbitrary* value of $A$. This is a result of the fact that if $\Psi$ is any solution of the Schrödinger equation, then $c\Psi$ is also a solution for any complex number $c$:

$$-\frac{\hbar^2}{2m}\nabla^2\Psi + V\Psi = i\hbar\frac{\partial\Psi}{\partial t}$$

$$-c\frac{\hbar^2}{2m}\nabla^2\Psi + cV\Psi = ci\hbar\frac{\partial\Psi}{\partial t}$$

$$-\frac{\hbar^2}{2m}\nabla^2(c\Psi) + V(c\Psi) = i\hbar\frac{\partial(c\Psi)}{\partial t}$$

Hence, the value of $A$ cannot be determined from the Schrödinger equation alone. But it can be determined from the definition of the probability density.

According to the Born prescription, the probability density is

$$\Psi^*\Psi = \left(A^* \sin\left(\frac{\pi x}{a}\right) e^{i\hbar\pi^2 t/2ma^2}\right)\left(A \sin\left(\frac{\pi x}{a}\right) e^{-i\hbar\pi^2 t/2ma^2}\right)$$
$$= |A|^2 \sin^2\left(\frac{\pi x}{a}\right) \tag{3.22}$$

for $0 \le x \le a$, and $\Psi^*\Psi = 0$ for $x$ outside of this range. (Note also that the time $t$ has dropped out of the expression for the probability density; for this wave function the probability density is independent of time.) Now we require that the probability density satisfy an additional requirement, namely

$$\int \Psi^*\Psi \, d^3\mathbf{r} = 1 \tag{3.23}$$

where the integral is taken over all of space. The justification for Equation (3.23) follows from the definition of the probability density; if $\Psi^*(\mathbf{r},t)\Psi(\mathbf{r},t)$ gives the probability per unit volume of finding the particle at position $\mathbf{r}$, then the integral of this quantity over all of space gives the probability of finding it somewhere, which must be 1. A wave function which satisfies Equation (3.23) is said to be *normalized*.

For the infinite square-well wave function given by Equation (3.21), we can substitute $\Psi^*\Psi$ from Equation (3.22) into Equation (3.23) to find the

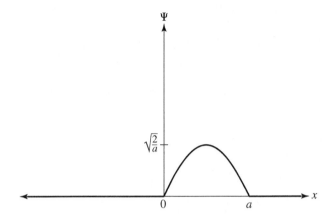

Fig. 3.4   The normalized wave function given by Equation (3.25) for a particle in an infinite square-well potential at time $t = 0$.

value of the constant $A$:

$$\int_0^a |A|^2 \sin^2\left(\frac{\pi x}{a}\right) dx = 1$$

$$|A|^2 \frac{1}{2} a = 1$$

so that

$$|A|^2 = \frac{2}{a} \tag{3.24}$$

If we take $A$ to be a positive real number, then Equation (3.24) has a unique solution: $A = \sqrt{2/a}$, and our normalized wave function is then

$$\Psi(x, t) = \sqrt{\frac{2}{a}} \sin\left(\frac{\pi x}{a}\right) e^{-i\hbar \pi^2 t/2ma^2}, \quad \text{for } 0 \leq x \leq a$$

$$= 0, \quad \text{for } x < 0 \text{ or } x > a \tag{3.25}$$

This wave function is shown in Figure 3.4 at $t = 0$.

Note, however, that Equation (3.24) has an infinite number of other solutions. We could just as easily have taken $A = -\sqrt{2/a}$. More generally, it is easy to show (see Exercise 3.7) that $A$ can be any complex number of the form $A = e^{i\theta}\sqrt{2/a}$, where $\theta$ is an arbitrary (real) constant. This leads to an obvious question: which value of $A$ do we choose? The answer is that it doesn't matter. We will see that all *observable* quantities depend only on $|A|^2$, not on $A$ itself. Hence, any value of $A$ which satisfies Equation (3.24) will give the same predictions for any measurements that we make. In practice, it is conventional to take (as we have) $A$ to be real and positive.

The Born interpretation tells us that the probability per unit length that the particle will be found at a point $x$ is given by $\Psi^*(x,t)\Psi(x,t)$. Therefore, the probability of finding the particle in some region is simply the integral of $\Psi^*\Psi$ over that region.

---

**Example 3.2. The Probability of Finding a Particle at a Particular Location**

Consider the particle described by the wave function of Equation (3.25). What is the probability $P(x < \epsilon)$ that the particle lies within a distance $\epsilon$ of the left-hand wall of the square well with $\epsilon \ll a$?

The desired probability is given by

$$
\begin{aligned}
P(x < \epsilon) &= \int_{x=0}^{\epsilon} \Psi^*\Psi \, dx \\
&= \int_{x=0}^{\epsilon} \frac{2}{a} \sin^2 \frac{\pi x}{a} \, dx \\
&= \frac{\epsilon}{a} - \frac{1}{2\pi} \sin \frac{2\pi\epsilon}{a}
\end{aligned}
\tag{3.26}
$$

In the limit where $\epsilon \ll a$, the second term in Equation (3.26) can be expanded as $\sin(x) \approx x - x^3/6 + \cdots$, giving

$$
P(x < \epsilon) \approx \frac{2}{3}\pi^2 \left(\frac{\epsilon}{a}\right)^3
$$

---

Although $\Psi^*\Psi$ can be used to derive the probability of finding the particle at a particular location, it contains more information than this. We can treat $\Psi^*\Psi$ as the probability distribution function for the position of the particle. To see how this works, consider first a much more practical example: the calculation of a grade point average. If $f_A$ is the fraction of a student's total grades which are A's, $f_B$ is the fraction which are B's, and so on, then the student's grade point average is simply

$$
\text{grade point average} = f_A 4 + f_B 3 + f_C 2 + f_D 1 + f_F 0
$$

Similarly, for any discrete distribution $p(j)$, the average value of $j$ (denoted $\langle j \rangle$) is just

$$
\langle j \rangle = \sum_j P(j) j
\tag{3.27}
$$

If the quantity of interest is continuous rather than discrete, then Equation (3.27) changes from a sum to an integral, giving

$$
\langle x \rangle = \int P(x) x \, dx
$$

where the integral is taken over all allowed values of $x$.

Now consider a large number of identical particles all described by the wave function $\Psi$, and suppose we measure the position of every one of these particles. Since $\Psi^*\Psi$ is the distribution function for each of the individual particle positions, it follows that the *average* position of the particles, $\langle x \rangle$, will be given by

$$\langle x \rangle = \int_{-\infty}^{\infty} \Psi^*(x,t)x\Psi(x,t)\,dx \tag{3.28}$$

(Of course, Equation (3.28) gives the theoretical average position; the actual measured average will tend toward this value as the number of measurements goes to infinity.) Note that we have written the right-hand side in a peculiar way, inserting $x$ between $\Psi^*$ and $\Psi$. The reason for this will become apparent shortly.

---

**Example 3.3. The Average Position of a Particle**
Consider the wave function for the infinite square well given in Equation (3.25). The mean value for the position of the particle is

$$\langle x \rangle = \int_{-\infty}^{\infty} \Psi^*(x,t)x\Psi(x,t)\,dx$$

$$= \int_{0}^{a} \frac{2}{a}\sin^2\left(\frac{\pi x}{a}\right)x\,dx$$

$$= \frac{a}{2}$$

This result, that $\langle x \rangle = a/2$, is exactly what we would expect, since the wave function is symmetric and is centered at $a/2$.

---

What about other observable quantities? Consider, for instance, the momentum. First consider the special case where $\Psi$ is an eigenfunction of the momentum operator, so that

$$-i\hbar\nabla\Psi = \mathbf{p}\Psi$$

or, in the one-dimensional case,

$$-i\hbar\frac{\partial\Psi}{\partial x} = p\Psi$$

For this special case, we have

$$\int_{-\infty}^{\infty} \Psi^*(x,t)\left(-i\hbar\frac{\partial}{\partial x}\right)\Psi(x,t)\,dx = \int_{-\infty}^{\infty} \Psi^*(x,t)p\Psi(x,t)\,dx$$

$$= p \tag{3.29}$$

assuming that $\Psi$ is normalized. In this case, the integral in Equation (3.29) gives the momentum $p$. Now we *postulate* that even if $\Psi$ is not an eigenfunction of the momentum operator, the average value of the momentum will be given by

$$\langle p \rangle = \int_{-\infty}^{\infty} \Psi^*(x,t) \left( -i\hbar \frac{\partial}{\partial x} \right) \Psi(x,t)\, dx$$

This brings us to one of the central ideas in quantum mechanics: every quantity we can measure, such as position, momentum, energy, angular momentum, etc., is associated with a corresponding *linear operator*. Measurable quantities are called, in quantum mechanics, *observables*, and we have already noted the correspondence between several observables and their operators:

$$\text{OBSERVABLE} \leftrightarrow \text{OPERATOR}$$

$$\text{position} \quad x \leftrightarrow x$$

$$\text{momentum} \quad \mathbf{p} \leftrightarrow -i\hbar \nabla$$

$$\text{energy} \quad E \leftrightarrow i\hbar \frac{\partial}{\partial t}$$

When a wave function $\Psi$ is an *eigenfunction* of an operator, then it represents a state in which the particle has a definite, fixed value for that observable. For example, if

$$\Psi = e^{-iky - i\omega t} \tag{3.30}$$

then we have

$$-i\hbar \nabla \Psi = -i\hbar \hat{y} \frac{\partial \Psi}{\partial y}$$

$$= -\hbar k \hat{y} \Psi$$

Thus, the wave function given by Equation (3.30) represents a particle in a state of definite momentum: it has momentum $\hbar k$ in the $-y$ direction. If the momentum of this particle is measured, the result will be a single, definite answer: the momentum will be $-\hbar k \hat{y}$.

In classical physics, of course, *all* measurements behave this way. A particle is always assumed, for instance, to have a definite momentum. While experimental error may limit the precision with which the momentum of a particle can be measured, it is always taken to be a well-defined, fixed quantity. This assumption is not true in quantum mechanics. Consider, for example, the square well wave function in Equation (3.25). If we apply the one-dimensional momentum operator to this wave function, we get

$$-i\hbar\frac{\partial}{\partial x}\Psi = -i\hbar\frac{\partial}{\partial x}\sqrt{\frac{2}{a}}\sin\left(\frac{\pi x}{a}\right)e^{-i\hbar\pi^2 t/2ma^2}$$

$$= -i\hbar\frac{\pi}{a}\sqrt{\frac{2}{a}}\cos\left(\frac{\pi x}{a}\right)e^{-i\hbar\pi^2 t/2ma^2}$$

which is manifestly not equal to a constant times $\Psi$. Thus, the square-well wave function is *not* an eigenfunction of the momentum operator, and it represents a particle which is *not* in a state of definite momentum. Of course, if the momentum of the particle is measured, a definite value is always obtained, but there is no way to predict ahead of time what this value will be. This is one of the places where quantum mechanics differs radically from classical physics: particles can be in states in which the momentum, position, or energy of the particle is not well-defined.

However, what *is* well-defined is the mean value of any observable, called the *expectation value*. The expectation value can be calculated for any observable, using the corresponding operator. If the operator $O$ corresponds to an observable $o$, then the expectation value of $o$ is just

$$\langle o \rangle = \int \Psi^*(\mathbf{r}, t)O\Psi(\mathbf{r}, t)\, d^3\mathbf{r}$$

For example, for a one-dimensional wave function $\Psi(x, t)$, we get

$$\langle p \rangle = \int_{-\infty}^{\infty} \Psi^* \left(-i\hbar\frac{\partial}{\partial x}\right)\Psi\, dx$$

$$\langle x \rangle = \int_{-\infty}^{\infty} \Psi^* x\Psi\, dx$$

$$\langle E \rangle = \int_{-\infty}^{\infty} \Psi^* \left(i\hbar\frac{\partial}{\partial t}\right)\Psi\, dx$$

(IMPORTANT: Note that the wave function *must* be normalized before the expectation value is computed!)

---

## Example 3.4. Expectation Value of the Momentum

Consider the expectation value of the momentum for a particle with the wave function given by Equation (3.25):

$$
\begin{aligned}
\langle p \rangle &= \int_{-\infty}^{\infty} \Psi^* \left( -i\hbar \frac{\partial}{\partial x} \right) \Psi \, dx \\
&= \int_0^a \sqrt{\frac{2}{a}} \sin\left( \frac{\pi x}{a} \right) e^{i\hbar \pi^2 t / 2ma^2} \left( -i\hbar \frac{\partial}{\partial x} \right) \sqrt{\frac{2}{a}} \sin\left( \frac{\pi x}{a} \right) e^{-i\hbar \pi^2 t / 2ma^2} \, dx \\
&= \int_0^a \frac{2}{a} \frac{\pi}{a} (-i\hbar) \sin\left( \frac{\pi x}{a} \right) \cos\left( \frac{\pi x}{a} \right) dx \\
&= 0
\end{aligned}
$$

Thus, the expectation value of the momentum is zero. This means that if $p$ is measured for a set of particles, all described by Equation (3.25), there is no way to predict in advance the outcome of an individual measurement, but the mean value of $p$, calculated by averaging over many measurements, will tend toward zero. This result for the expectation value makes sense, since the corresponding classical system is a particle bouncing back and forth inside the potential well; its momentum averaged over many back-and-forth trips is zero.

---

Using the same technique, the expectation value of functions of an observable can also be calculated. For a particle moving in one dimension, for instance, we have

$$
\langle x^2 \rangle = \int_{-\infty}^{\infty} \Psi^* x^2 \Psi \, dx
$$

$$
\begin{aligned}
\langle p^2 \rangle &= \int_{-\infty}^{\infty} \Psi^* \left( -i\hbar \frac{\partial}{\partial x} \right) \left( -i\hbar \frac{\partial}{\partial x} \right) \Psi \, dx \\
&= \int_{-\infty}^{\infty} \Psi^* \left( -\hbar^2 \frac{\partial^2}{\partial x^2} \right) \Psi \, dx
\end{aligned}
$$

## 3.3   The Time-Independent Schrödinger Equation: Qualitative Solutions and the Origin of Quantization

### *Derivation of the Time-Independent Schrödinger Equation*

In this section, we will consider a special case of the Schrödinger equation, but this special case will occupy much of the remainder of this book. For the most general version of the Schrödinger equation,

$$
-\frac{\hbar^2}{2m} \nabla^2 \Psi + V(\mathbf{r}, t) \Psi = i\hbar \frac{\partial \Psi}{\partial t}
$$

the various cases of interest are all embodied in the choice of the potential $V(\mathbf{r}, t)$. Now consider a special class of potentials: those for which $V$ is a function only of position $\mathbf{r}$, and is independent of time $t$. This is a very commonly encountered case; for example, the potential that an electron feels in an atom or in a crystal lattice can be treated this way. For this case, the Schrödinger equation becomes

$$-\frac{\hbar^2}{2m}\nabla^2\Psi + V(\mathbf{r})\Psi = i\hbar\frac{\partial\Psi}{\partial t} \tag{3.31}$$

It is possible to make a further simplification by considering only a particular class of solutions. We will require that the wave function $\Psi$ be an eigenfunction of the energy operator, $i\hbar(\partial/\partial t)$, so that this wave function corresponds to a state of definite energy $E$. This means that

$$i\hbar\frac{\partial\Psi(\mathbf{r}, t)}{\partial t} = E\Psi(\mathbf{r}, t) \tag{3.32}$$

Substituting this expression for the time derivative into Equation (3.31), we obtain

$$-\frac{\hbar^2}{2m}\nabla^2\Psi(\mathbf{r}, t) + V(\mathbf{r})\Psi(\mathbf{r}, t) = E\Psi(\mathbf{r}, t) \tag{3.33}$$

Equations (3.32) and (3.33) can be solved using separation of variables. Assume a solution of the form

$$\Psi(\mathbf{r}, t) = \psi(\mathbf{r})\chi(t) \tag{3.34}$$

where $\psi$ is a function only of position, and $\chi$ is a function only of time. Substituting this expression into Equation (3.32) gives

$$i\hbar\psi(\mathbf{r})\frac{\partial\chi(t)}{\partial t} = E\psi(\mathbf{r})\chi(t)$$

The factors of $\psi(\mathbf{r})$ cancel, and the resulting equation for $\chi(t)$ reduces to

$$\frac{d\chi(t)}{dt} = -\frac{iE}{\hbar}\chi(t)$$

which has the solution

$$\chi(t) = e^{-iEt/\hbar} \tag{3.35}$$

Now we substitute the expression for $\Psi(\mathbf{r}, t)$ from Equation (3.34) into Equation (3.33) to yield

$$-\frac{\hbar^2}{2m}\nabla^2\psi(\mathbf{r})\chi(t) + V(\mathbf{r})\psi(\mathbf{r})\chi(t) = E\psi(\mathbf{r})\chi(t)$$

Here the factor $\chi(t)$ cancels to give

$$-\frac{\hbar^2}{2m}\nabla^2\psi(\mathbf{r}) + V(\mathbf{r})\psi(\mathbf{r}) = E\psi(\mathbf{r}) \tag{3.36}$$

Equation (3.36) is an extremely important version of the Schrödinger equation, called the *time-independent* Schrödinger equation. It can be used whenever the potential $V$ is independent of time to find wave functions that are states of definite energy.

Once a solution $\psi(\mathbf{r})$ to the time-independent Schrödinger equation is derived, it can be used to recover the full wave function using the time-dependent piece of the solution in Equation (3.35):

$$\Psi(\mathbf{r}, t) = \psi(\mathbf{r})e^{-iEt/\hbar}$$

Often, however, the quantity of interest will be the time-independent wave function, $\psi(\mathbf{r})$, because many physical quantities can be derived from $\psi(\mathbf{r})$ alone, without reference to the full wave function $\Psi(\mathbf{r}, t)$. For instance, the probability density for the particle, $\Psi^*(\mathbf{r}, t)\Psi(\mathbf{r}, t)$, is just

$$\Psi^*(\mathbf{r}, t)\Psi(\mathbf{r}, t) = \psi(\mathbf{r})^* e^{iEt/\hbar}\psi(\mathbf{r})e^{-iEt/\hbar}$$
$$= \psi^*(\mathbf{r})\psi(\mathbf{r})$$

We can also write expectation values in terms of $\psi(\mathbf{r})$. The expectation value of any observable $o$ which corresponds to an operator $O$ that does not depend on time is

$$\langle o \rangle = \int \psi(\mathbf{r})^* e^{iEt/\hbar} O\psi(\mathbf{r})e^{-iEt/\hbar} \, d^3\mathbf{r}$$
$$= \int \psi(\mathbf{r})^* O\psi(\mathbf{r}) \, d^3\mathbf{r}$$

Clearly, both the probability density and the expectation values are independent of time. Of course, these arguments apply *only* to states of definite energy, which satisfy the time-independent Schrödinger equation. For this reason, such states are also called *stationary* states. Of course, the wave function itself still has a time dependence, given by the factor $e^{-iEt/\hbar}$, but when the observable $o$ is actually measured, the time dependent factors cancel out in the calculation of the expectation value: $e^{iEt/\hbar}e^{-iEt/\hbar} = 1$.

The time-independent Schrödinger equation suggests the definition of a new operator, $H$, given by

$$H = -\frac{\hbar^2}{2m}\nabla^2 + V \tag{3.37}$$

Then the time-independent Schrödinger equation can be written in the compact form

$$H\psi = E\psi$$

so that $E$ is the eigenvalue of $H$. The operator $H$ defined by Equation (3.37) is called the *Hamiltonian* operator, or just the Hamiltonian. (The Hamiltonian is named after William Hamilton, an Irish mathematician who died in 1865, long before the birth of quantum mechanics. The name originated in classical mechanics, where $H$ corresponds to the total energy of a particle.)

## Qualitative Solutions and the Origin of Quantization

In attempting to solve the time-independent Schrödinger equation, the first point to note (see Problem 3.8) is that if $\psi_1(\mathbf{r})$ and $\psi_2(\mathbf{r})$ are two different solutions of the time-independent Schrödinger equation with the same value of $E$, then $\psi_1(\mathbf{r}) + \psi_2(\mathbf{r})$ is also a solution with energy $E$. Furthermore, if $\psi(\mathbf{r})$ is a solution with energy $E$, then $c\psi(\mathbf{r})$ (where $c$ is any complex number) is also a solution with energy $E$. Thus, any solution of Equation (3.36) can always be multiplied by an arbitrary constant to obtain another solution, so we always have the freedom to normalize the wave function by the appropriate choice of the multiplicative constant.

In Equation (3.36), the potential $V(\mathbf{r})$ will generally be determined by whatever system is under consideration. But how do we determine $E$, the total energy of the system? This is generally *not* something that is specified in advance. However, for many potentials, it is not possible to find solutions for arbitrary choices of $E$; rather, solutions exist only for certain *discrete* choices for $E$. This is the origin of energy quantization.

To see how this arises, consider the one-dimensional version of Equation (3.36):

$$-\frac{\hbar^2}{2m}\frac{d^2\psi}{dx^2} + V(x)\psi = E\psi$$

where $\psi$ is a function of $x$. This equation can be rewritten as

$$\frac{d^2\psi}{dx^2} = \frac{2m}{\hbar^2}[V(x) - E]\psi \tag{3.38}$$

Assume that we have a particular potential $V(x)$ and $E$, and we want to solve for $\psi$.

Equation (3.38) can be used to determine the sign of $d^2\psi/dx^2$, which gives the direction in which the wave function curves: if $d^2\psi/dx^2 > 0$, then $\psi(x)$ is concave up, while if $d^2\psi/dx^2 < 0$, then $\psi(x)$ is concave down.

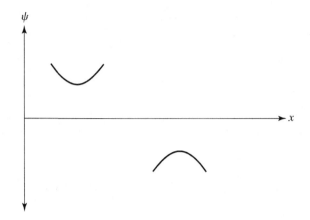

Fig. 3.5   If $d^2\psi/dx^2 > 0$ for $\psi > 0$, or $d^2\psi/dx^2 < 0$ for $\psi < 0$, then $\psi$ curves away from the $x$-axis.

Then for $E < V(x)$, the factor multiplying $\psi$ on the right-hand side of Equation (3.38) will be positive, so that $d^2\psi/dx^2$ and $\psi$ will have the same sign, either both positive or both negative. If $\psi > 0$ and $d^2\psi/dx^2 > 0$, then $\psi$ lies above the $x$-axis and is concave upward, while if $\psi < 0$ and $d^2\psi/dx^2 < 0$, then $\psi$ lies below the $x$-axis and is concave downward. In either case, $\psi(x)$ will curve away from the horizontal axis (Figure 3.5).

On the other hand, if $E > V(x)$ we can draw the opposite conclusion. In this case, the factor multiplying $\psi$ on the right-hand side of Equation (3.38) will be negative, so that $d^2\psi/dx^2$ and $\psi$ will have opposite signs. Then when $\psi$ lies above the horizontal axis, the function $\psi(x)$ will be concave downward, while if $\psi$ lies below the horizontal axis, it will be concave upward. In either case, $\psi(x)$ will curve toward the $x$-axis (Figure 3.6).

These different possibilities have a clear interpretation in the classical case. If $E < V(x)$ outside of a finite region of space, then the particle is bound in that potential, while $E > V(x)$ corresponds to an unbound particle. Consider a bound particle first with the potential and energy shown in Figure 3.7.

A classical particle with the indicated energy will oscillate back and forth in the region between $-x_1$ and $x_1$, i.e., in the region for which $E > V(x)$. The velocity of this particle is given by conservation of energy: $mv^2/2 = E - V(x)$. This velocity reaches zero at $x = -x_1$ and $x = x_1$, and the particle moves back in the opposite direction. Note that the particle can never enter the regions $x > x_1$ or $x < -x_1$, because in these regions it would

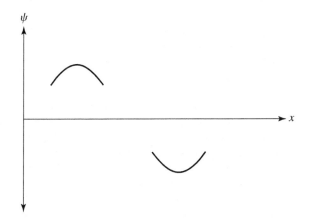

Fig. 3.6 If $d^2\psi/dx^2 < 0$ for $\psi > 0$, or $d^2\psi/dx^2 > 0$ for $\psi < 0$, then $\psi$ curves toward the $x$-axis.

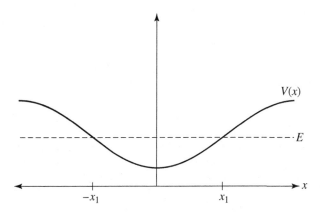

Fig. 3.7 A particle with energy $E$ moves in the potential shown here. Classically, the particle is bound in the region $-x_1 < x < x_1$.

have negative kinetic energy, which is impossible. Hence, this is called the classically forbidden region.

Now we can sketch a qualitative solution to the time-independent Schrödinger equation using the behavior of $\psi$ illustrated in Figures 3.5 and 3.6. Taking $\psi > 0$, we note that $\psi(x)$ will be concave down in the region $-x_1 < x < x_1$ and concave up outside this region. One solution, therefore, will behave as in Figure 3.8. This solution corresponds to the lowest-energy bound state, so it is called the *ground-state* wave function.

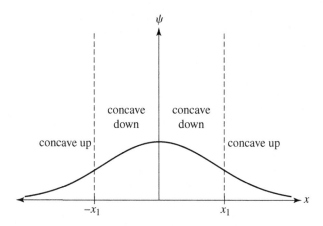

Fig. 3.8 A function $\psi(x)$ corresponding to the energy and potential shown in Figure 3.7. $\psi(x)$ is concave down in the region $-x_1 < x < x_1$ and concave up on either side of this region. This function does not cross the $x$-axis, and it corresponds to the ground state.

Now we see another stark contrast between the predictions of quantum mechanics and classical mechanics. The function $\psi$ is nonzero in the classically forbidden region, so there is a nonzero probability to find the particle in this region. The particle can penetrate into a region that classical mechanics predicts it does not have enough energy to reach! This effect has far-reaching consequences, some of which will be discussed in the next chapter.

Note further that $\psi$ must be very finely tuned in the classically forbidden region. When $E < V(x)$, so that $\psi(x)$ curves away from the $x$-axis, the function has a tendency to "blow up" to positive or negative infinity. This is bad because then the wave function cannot be normalized to give a meaningful probability density. Only by choosing the amplitude just right will $\psi(x)$ decline smoothly to zero, always with positive curvature, as $x$ goes to infinity. This "dangerous" behavior of the wave function in the classically forbidden region has an important consequence. Suppose we change the energy by a tiny amount, from $E$ to $E'$. Then, if we choose $\psi$ to go to zero on the left-hand side, when $x \to -\infty$, it will have the "wrong" slope when it enters the classically forbidden region on the right-hand side. Depending on the actual solution we choose, the function will either blow up toward $+\infty$ or cross the $x$-axis and fall away to $-\infty$ (Figure 3.9).

Does this mean that the time-independent Schrödinger equation has only a single solution for this particular potential? No. If the energy is increased by a large enough amount, a second acceptable solution for $\psi(x)$

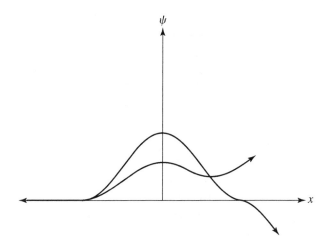

Fig. 3.9  If the energy in Figure 3.7 is changed slightly, there is no longer a well-behaved solution to the Schrödinger equation; the wave function blows up to either $+\infty$ or $-\infty$.

is encountered, which is shown in Figure 3.10. This state is called the *first excited state*, since it has the lowest energy of any state beyond the ground state. (The next state is called the *second excited state*, and so on.) Note further that $\psi(x)$ for the first excited state crosses the $x$-axis exactly one time, while the ground state function does not cross the $x$-axis at all. This corresponds to a general result for bound states: $\psi(x)$ for the ground state crosses the $x$-axis zero times, $\psi(x)$ for the first excited state crosses the $x$-axis one time, $\psi(x)$ for the second excited state crosses the $x$-axis two times, and so on.

Thus, for this potential, the bound-state solutions of the Schrödinger equation occur only at fixed, *discrete* values of the energy $E$. This is the origin of energy quantization: for states which are classically bound, the time-independent Schrödinger equation has solutions only at discrete values of the energy. So a particle in a bound state is no longer free to have any energy at all; it can take on only energies which correspond to solutions of the Schrödinger equation.

Now consider what happens for an unbound particle. Using the same potential, we consider the behavior of the particle when its energy is larger than the potential everywhere (Figure 3.11). In this case, $V(x) - E < 0$, so $\psi(x)$ curves toward the $x$-axis everywhere. Now $\psi$ can no longer "blow up," since it never curves away from the $x$-axis. Therefore, if $E$ is changed

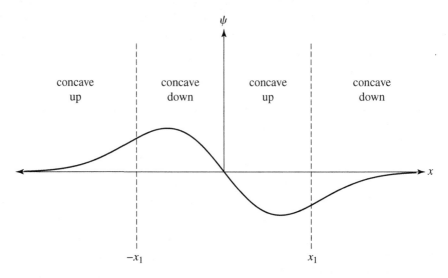

Fig. 3.10　A second function $\psi(x)$ corresponding to the potential shown in Figure 3.7, with a larger energy than displayed in that figure. $\psi(x)$ is concave up for $x < -x_1$ and $0 < x < x_1$, and concave down for $-x_1 < x < 0$ and $x > x_1$. This function crosses the $x$-axis once and corresponds to the first excited state.

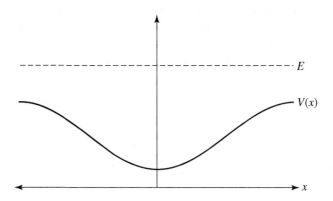

Fig. 3.11　A particle with energy $E$ moves in the potential shown here. Since $E > V(x)$ everywhere, the particle is not bound in this potential.

slightly, the Schrödinger equation still yields an acceptable solution for $\psi(x)$. So while the bound states are quantized, the unbound states are not; they can have any energy (larger than $V(x)$) and are not restricted to a set of discrete energies. Note also that a single potential, such as the one discussed here, can have both bound and unbound states.

## PROBLEMS

**3.1** A particle of mass $m$ is moving in one dimension in a potential $V(x,t)$. The wave function for the particle is

$$\Psi(x,t) = Axe^{-(\sqrt{km}/2\hbar)x^2}e^{-i\sqrt{k/m}(3/2)t}$$

for $-\infty < x < +\infty$, where $k$ and $A$ are constants.
(a) Show that $V$ is independent of $t$, and determine $V(x)$.
(b) Normalize this wave function.
(c) Using the normalized wave function, calculate $\langle x \rangle$, $\langle x^2 \rangle$, $\langle p \rangle$, and $\langle p^2 \rangle$.

**3.2** Determine which of the following one-dimensional wave functions represent states of definite momentum. For each wave function that does correspond to a state of definite momentum, determine the momentum.
(a) $\psi(x) = e^{ikx}$
(b) $\psi(x) = xe^{ikx}$
(c) $\psi(x) = \sin(kx) + i\cos(kx)$
(d) $\psi(x) = e^{ikx} + e^{-ikx}$

**3.3** The wave function for a particle is $\Psi(x,t) = \sin(kx)[i\cos(\omega t/2) + \sin(\omega t/2)]$, where $k$ and $\omega$ are constants.
(a) Is this particle in a state of definite momentum? If so, determine the momentum.
(b) Is this particle in a state of definite energy? If so, determine the energy.

**3.4** A particle with mass $m$ is moving in one dimension near the speed of light so that the relation

$$E = p^2/2m$$

for the kinetic energy is no longer valid. Instead, the total energy is given by

$$E^2 = p^2c^2 + m^2c^4$$

Hence, we can no longer use the Schrödinger equation. Suppose the wave function $\Psi(x,t)$ for the particle is an eigenfunction of the energy operator and an eigenfunction of the momentum operator, and also assume that there is no potential energy $V$. Derive a linear differential equation for $\Psi(x,t)$.

**3.5** A particle with mass $m$ is moving in one dimension in the potential $V(x)$. The particle is in a state of definite energy $E$, but it is *not* in a state of definite momentum $p$. Show that $\langle p^2 \rangle$, $\langle V(x) \rangle$, and $E$ are related by

$$\frac{\langle p^2 \rangle}{2m} + \langle V(x) \rangle = E$$

**3.6** Consider the solution to the Schrödinger equation for the infinite square well with $n = 2$ rather than $n = 1$ in Equation (3.20). Derive $\Psi(x, t)$ for this case, and normalize this wave function.

**3.7** Suppose that a wave function $\Psi(\mathbf{r}, t)$ is normalized. Show that the wave function $e^{i\theta}\Psi(\mathbf{r}, t)$, where $\theta$ is an arbitrary real number, is also normalized.

**3.8** Suppose that $\psi_1$ and $\psi_2$ are two different solutions of the time-independent Schrödinger equation with the same energy $E$.
(a) Show that $\psi_1 + \psi_2$ is also a solution with energy $E$.
(b) Show that $c\psi_1$ is also a solution of the Schrödinger equation with energy $E$.

**3.9** A particle moves in one dimension in the potential shown here. The energy $E$ is shown on the graph, and the particle is in its ground state.

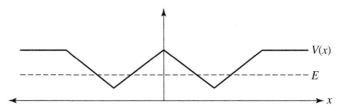

(a) Sketch $\psi(x)$ for this particle.
(b) You make a measurement to find the particle. Indicate on your graph the point or points at which you are most likely to find it.

**3.10** A particle moves in one dimension in the potential shown here. The energy $E$ is shown on the graph, and the particle is in its first excited state.

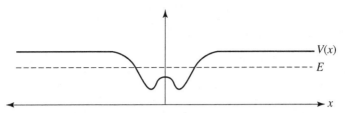

(a) Sketch $\psi(x)$ for this particle.

(b) You make a measurement to find the particle. Indicate on your graph the point or points at which you are most likely to find it.

**3.11** A particle moving in one dimension is described by the function $\psi(x)$ shown here:

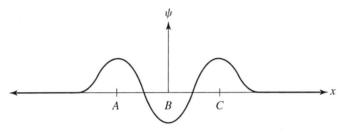

(a) You make a measurement to locate the particle. Which one of the following is true?

(i) You will always find the particle at point $B$.

(ii) You are most likely to find the particle at points $A$ or $C$ and least likely to find the particle at point $B$.

(iii) You are most likely to find the particle at points $A$, $B$, or $C$.

Explain your answer.

(b) Which one of the following potentials $V(x)$ could give rise to this $\psi(x)$?

(i)

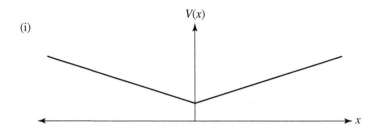

(ii)

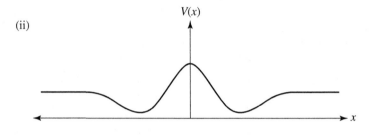

(iii)

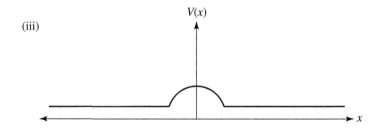

Chapter 4

# Solutions of the one-dimensional time-independent Schrödinger equation

In this chapter we will examine some exact solutions of the one-dimensional time-independent Schrödinger equation,

$$-\frac{\hbar^2}{2m}\frac{d^2\psi}{dx^2} + V(x)\psi = E\psi \tag{4.1}$$

Of course, the real world is three-dimensional, but Equation (4.1) can be applied whenever a particle moves only in a single direction. Consider, for example, an electron travelling through an evacuated tube (Figure 4.1) with an electrostatic potential $\Phi(x)$. Since the electron has charge $-e$, it experiences the potential $V(x) = -e\Phi(x)$, and this system can be treated as effectively one-dimensional. Another example is the motion of electrons in *semiconductor heterostructures*. These are materials formed by joining together two or more different semiconductors. At the junction between the materials, the electron experiences a change in the potential, and it is possible to construct materials with potentials that mimic some of those considered in this chapter.

Classically, there are two possible types of states: *bound states*, in which the particle is confined to move in a finite region, and *unbound states*, in which the particle can escape to infinity. Because the solutions for unbound states are quite different from those of bound states, we will treat the two cases separately. For unbound states, we will calculate the probability that an incident particle will reflect off of or transmit through a particular potential. For bound states, we will calculate the wave functions and energies. As noted in the previous chapter, the bound state energy levels will, in general, be discrete rather than continuous.

$V(x) = -e\Phi(x)$

$e$

$x$

Fig. 4.1   An electron moving in one dimension in the potential $V(x) = -e\Phi(x)$.

## 4.1   Unbound States: Scattering and Tunneling

In this section we will consider only the case of piecewise constant potentials, i.e., potentials with step-like discontinuities (Figure 4.2). Consider a range in $x$ over which $V(x)$ is a constant, $V_0$. Then Equation (4.1) becomes

$$-\frac{\hbar^2}{2m}\frac{d^2\psi}{dx^2} + V_0\psi = E\psi$$

which can be rewritten as

$$\frac{d^2\psi}{dx^2} + \frac{2m(E - V_0)}{\hbar^2}\psi = 0 \tag{4.2}$$

This is the simplest possible version of the time-independent Schrödinger equation. To find a solution, consider the general equation of the form

$$\frac{d^2y}{dx^2} + Ay = 0 \tag{4.3}$$

where $A$ is an arbitrary constant. We try a solution of the form

$$y = Ce^{mx}$$

where both $C$ and $m$ are constants. Then

$$\frac{d^2y}{dx^2} = Cm^2e^{mx}$$

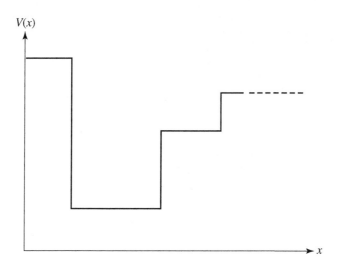

Fig. 4.2   An example of a piecewise constant potential.

Substituting these expressions for $d^2y/dx^2$ and $y$ into Equation (4.3), we obtain

$$m^2 C e^{mx} + A C e^{mx} = 0$$

which is satisfied as long as $m$ is chosen so that $m^2 + A = 0$, while $C$ can have any value at all. Thus, $m = \pm\sqrt{-A}$, and the general solution is

$$y = C_1 e^{\sqrt{-A}\,x} + C_2 e^{-\sqrt{-A}\,x} \tag{4.4}$$

where $C_1$ and $C_2$ are arbitrary constants. (Note that $C_1$ and $C_2$ need not be real; they can also be complex numbers.)

This solution will have quite different behavior depending on whether $A$ is positive or negative. If $A$ is negative, then the quantities under the square roots will be positive, and the solution given by Equation (4.4) will be the sum of a positive and a negative exponential. On the other hand, if $A$ is positive, then $\sqrt{-A}$ will be imaginary. In this case we take $\sqrt{-A} = i\sqrt{A}$, and Equation (4.4) becomes

$$y = C_1 e^{i\sqrt{A}\,x} + C_2 e^{-i\sqrt{A}\,x} \tag{4.5}$$

Note that this solution can also be written as a sum of trigonometric functions (see Problem 4.1):

$$y = D_1 \cos(\sqrt{A}\,x) + D_2 \sin(\sqrt{A}\,x) \tag{4.6}$$

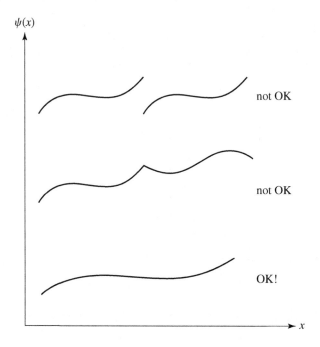

Fig. 4.3　Both $\psi(x)$ and $d\psi(x)/dx$ must be continuous.

The solutions given by Equations (4.5) and (4.6) are completely equivalent, so the question of which one to use is a matter of convenience; it will usually be easier to use one form or the other depending on the boundary conditions of the problem at hand.

For a particle with a given energy $E$ moving in a piecewise constant potential with step discontinuities such as in Figure 4.2, we will obtain different solutions in the regions with different values of $V_0$. The constants appearing in the solutions must then be chosen so that the wave functions "join up" at each step, i.e., both $\psi(x)$ and $d\psi(x)/dx$ must be continuous (Figure 4.3).

We can now apply this set of solutions to the Schrödinger equation in the form of Equation (4.2). Consider first the simplest case, where $V_0 = 0$, so we are dealing with a "free particle" and no potential. In this case we clearly have $E - V_0 > 0$, so we choose a solution of the form given in Equation (4.5) or (4.6). The former will be more convenient for our current purposes; we obtain

$$\psi = C_1 e^{i\left(\sqrt{2mE}/\hbar\right)x} + C_2 e^{-i\left(\sqrt{2mE}/\hbar\right)x}$$

The physical interpretation of this solution is clearer if we use it to derive the time-dependent wave function,

$$\Psi(x,t) = \psi(x)e^{-iEt/\hbar}$$
$$= C_1 e^{i[(\sqrt{2mE}/\hbar)x - Et/\hbar]} + C_2 e^{i[-(\sqrt{2mE}/\hbar)x - Et/\hbar]} \qquad (4.7)$$

Now recall that energy is related to momentum via $E = p^2/2m$, so that $\sqrt{2mE} = p$. Furthermore, in the previous chapter we derived relations between momentum and wavenumber ($p = \hbar k$) and between energy and frequency ($E = \hbar\omega$). Substituting all of these into Equation (4.7), we get

$$\Psi(x,t) = C_1 e^{i(kx-\omega t)} + C_2 e^{i(-kx-\omega t)} \qquad (4.8)$$

We have already seen wave functions of this form in Chapter 3; the function $e^{i(kx-\omega t)}$ represents a wave moving in the $+x$ direction (i.e., to the right in a conventional coordinate system) with energy $\hbar\omega$ and momentum $\hbar k$, while the second term represents a particle moving to the left with energy $\hbar\omega$ and momentum $-\hbar k$. We know that Equation (4.8) is an eigenfunction of the energy operator, since it was derived using the time-independent Schrödinger equation. Now consider what happens if we apply the momentum operator to each term separately:

$$-i\hbar\frac{\partial}{\partial x}C_1 e^{i(kx-\omega t)} = \hbar k C_1 e^{i(kx-\omega t)}$$

$$-i\hbar\frac{\partial}{\partial x}C_2 e^{i(-kx-\omega t)} = -\hbar k C_2 e^{i(-kx-\omega t)}$$

As expected, the first term in our wave function, which represents a rightward moving particle, is an eigenfunction of the momentum operator with momentum $\hbar k$. Similarly, the second term is an eigenfunction of the momentum operator with momentum $-\hbar k$. On the other hand, it is easy to verify that the total wave function (if $C_1 \neq 0$ and $C_2 \neq 0$) is *not* an eigenfunction of the momentum operator (see Problem 4.2).

It might seem puzzling that the general form for the wave function of a free particle moving in one dimension consists of pieces representing particles simultaneously moving both to the left and to the right. Consider, however, a particle reflecting off of a boundary. The wave function in this case simultaneously represents both the rightward-moving incident particle and the leftward-moving reflected particle. Unlike classical mechanics where the motion of the particle is specified by its position as a function of time $x(t)$, and $x$ simply tracks the particle as it first moves to the right and later moves to the left, the quantum mechanical wave function $\Psi$ *simultaneously* encodes the particle moving to the right and reflecting back to the left.

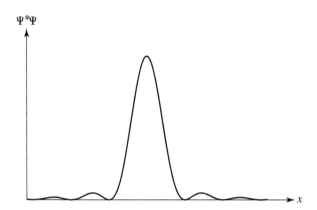

Fig. 4.4   $\Psi^*\Psi$ for a wave packet generated by summing over all waves in the interval $k = 0$ to $k = k_0$, where each wave is given equal amplitude.

The constants $C_1$ and $C_2$ in Equation (4.8) are free parameters which can be chosen to match the physical system. Hence, a free particle moving purely to the right can be expressed in terms of Equation (4.8) by setting $C_2 = 0$. Similarly, a free particle moving to the left has a wave function given by Equation (4.8) with $C_1 = 0$.

What do we know about the position of the particle? Consider the wave function for a purely rightward-moving particle,

$$\Psi(x, t) = C_1 e^{i(kx - \omega t)} \tag{4.9}$$

Calculating $\Psi^*\Psi$ gives

$$\Psi^*\Psi = C_1^* e^{-i(kx - \omega t)} C_1 e^{i(kx - \omega t)}$$

$$= C_1^* C_1$$

which is independent of position. This means that the particle is equally likely to be found anywhere in space! In fact, Equation (4.9) represents an idealization. In theory, a particle which is in an exact momentum state will be spread out over an arbitrarily large distance, but in practice, a physical particle will correspond to a *sum* of waves like those in Equation (4.9), each with a slightly different momentum and energy (or, equivalently, a slightly different $k$ and $\omega$). These waves can be summed to produce a value for $\Psi^*\Psi$ which is localized in space. Such a sum of waves is called a *wave packet*, and it can be normalized. For example, in Figure 4.4, $\Psi^*\Psi$ is shown for a wave packet generated by summing over all waves in the interval $k = 0$ to $k = k_0$, where each wave is given equal amplitude.

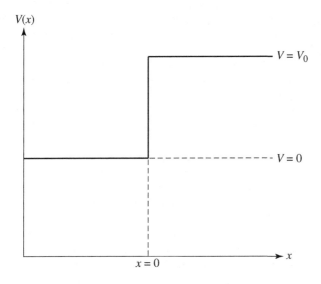

Fig. 4.5   A step-function potential.

## Scattering From Step-Function Potentials

In solving the Schrödinger equation for scattering in one dimension, we will idealize the situation and treat the particles as eigenfunctions of momentum; we will still be able to derive physically-interesting results for this case. Consider first the "step-function" potential shown in Figure 4.5. The physical location of the step in the potential is arbitrary, so we will take it to lie at $x = 0$. Similarly, as in classical mechanics, we are free to choose the zero of the potential anywhere, so we will take $V = 0$ for $x < 0$, while $V = V_0$ for $x \geq 0$.

We can distinguish two different cases here: either $E < V_0$, or $E > V_0$. In a classical system, the behavior of the particle is easy to calculate. Consider first a classical particle moving from left to right with energy $E < V_0$. The particle will not be able to enter the region $x > 0$, and it will simply bounce back to the left with the magnitude of its velocity unchanged. A classical particle with $E > V_0$ will penetrate into the region $x > 0$, but the magnitude of its velocity will decrease as its momentum changes from $p_1 = \sqrt{2mE}$ on the left-hand side of the barrier to $p_2 = \sqrt{2m(E - V_0)}$ on the right-hand side. Quantum mechanics predicts a very different behavior as we shall now see.

## Step-Function Potential With $E > V_0$

Consider first the high-energy case for which a classical particle will simply travel into the $x > 0$ region with reduced momentum. For this case, the Schrödinger equation for a constant potential, Equation (4.2), has the form

$$\frac{d^2\psi}{dx^2} + \frac{2mE}{\hbar^2}\psi = 0 \qquad (4.10)$$

for $x < 0$, and

$$\frac{d^2\psi}{dx^2} + \frac{2m(E - V_0)}{\hbar^2}\psi = 0 \qquad (4.11)$$

for $x \geq 0$. Since both $E$ and $E - V_0$ are positive, our solution in both regions can be expressed in the form of either Equation (4.5) or (4.6). Either form can be used to solve the problem, but the solution will turn out to be easier to interpret using exponentials (Equation (4.5)) rather than trigonometric functions. The solution to the Schrödinger equation in these two regions is

$$\psi_1(x) = A_1 e^{i\left(\sqrt{2mE}/\hbar\right)x} + B_1 e^{-i\left(\sqrt{2mE}/\hbar\right)x}, \qquad x < 0$$

$$\psi_2(x) = A_2 e^{i\left(\sqrt{2m(E-V_0)}/\hbar\right)x} + B_2 e^{-i\left(\sqrt{2m(E-V_0)}/\hbar\right)x}, \; x \geq 0$$

where $A_1$, $B_1$, $A_2$, and $B_2$ are constants which need to be determined from the boundary conditions. To simplify the calculation, we make the substitutions

$$k_1 = \frac{\sqrt{2mE}}{\hbar}$$

$$k_2 = \frac{\sqrt{2m(E - V_0)}}{\hbar}$$

so that $\hbar k_1$ and $\hbar k_2$ give the magnitude of the momentum of the particle on the left and right sides of the step, respectively. Then the solution becomes

$$\psi_1(x) = A_1 e^{ik_1 x} + B_1 e^{-ik_1 x}, \quad x < 0 \qquad (4.12)$$

$$\psi_2(x) = A_2 e^{ik_2 x} + B_2 e^{-ik_2 x}, \quad x \geq 0 \qquad (4.13)$$

As noted previously, the constants $A_1$, $B_1$, $A_2$, and $B_2$ must be chosen so that the wave function and its derivative are continuous at the boundary between the two solutions:

$$\psi_1(0) = \psi_2(0)$$

and

$$\frac{d\psi_1}{dx}(0) = \frac{d\psi_2}{dx}(0)$$

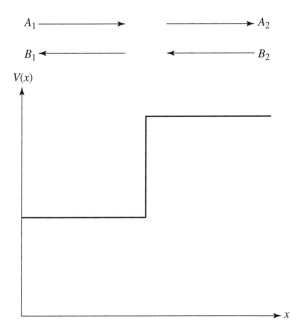

Fig. 4.6   In Equations (4.14) and (4.15), the "$A_1$" term represents a rightward-moving particle on the left side of the step, the "$B_1$" term represents a leftward-moving particle on the left side of the step, the "$A_2$" term represents a rightward-moving particle on the right side of the step, and the "$B_2$" term represents a leftward-moving particle on the right side of the step.

However, before applying these boundary conditions, it is useful to understand the physical significance of the terms in Equations (4.12) and (4.13). The time-dependent wave functions for the particle, derived from Equations (4.12) and (4.13), are

$$\Psi_1(x,t) = A_1 e^{i(k_1 x - Et/\hbar)} + B_1 e^{i(-k_1 x - Et/\hbar)}, \quad x < 0 \qquad (4.14)$$

$$\Psi_2(x,t) = A_2 e^{i(k_2 x - Et/\hbar)} + B_2 e^{i(-k_2 x - Et/\hbar)}, \quad x \geq 0 \qquad (4.15)$$

Thus, the "$A_1$" term represents a rightward-moving particle on the left side of the step, the "$B_1$" term represents a leftward-moving particle on the left side of the step, the "$A_2$" term represents a rightward-moving particle on the right side of the step, and the "$B_2$" term represents a leftward-moving particle on the right side of the step (Figure 4.6). However, not all of these terms make physical sense. Assuming that the particle is initially in the region $x < 0$, travelling to the right, we expect $A_1 \neq 0$. Classically the particle will travel across the step and continue travelling to the right, so we expect $A_2 \neq 0$. We also want to allow for the possibility that the particle can scatter backwards off of the step; although this cannot happen classically, it cannot be ruled out in a quantum system, so we must allow for $B_1 \neq 0$. However, one term makes no sense at all: the "$B_2$" term represents a particle originating at $x = +\infty$ and moving to the left. There is no way to produce such a particle trajectory from a particle initially moving to the right on the left-hand side of the step. Therefore, on physical grounds, we set $B_2 = 0$.

Then our wave functions simplify to

$$\psi_1(x) = A_1 e^{ik_1 x} + B_1 e^{-ik_1 x}, \qquad\qquad x < 0$$
$$\psi_2(x) = A_2 e^{ik_2 x}, \qquad\qquad\qquad\qquad x \geq 0 \qquad (4.16)$$

Using our two boundary conditions at $x = 0$, we find the requirement that $\psi_1(0) = \psi_2(0)$ gives

$$A_1 + B_1 = A_2 \qquad (4.17)$$

while the requirement that $d\psi_1/dx = d\psi_2/dx$ at $x = 0$ yields

$$ik_1 A_1 - ik_1 B_1 = ik_2 A_2 \qquad (4.18)$$

Since $k_1$ and $k_2$ are specified by the values of $E$ and $V_0$, we have three unknowns, $A_1$, $B_1$, and $A_2$, and only two equations constraining them, Equations (4.17) and (4.18). Hence, we cannot solve for $A_1$, $B_1$, and $A_2$, but we can express two of the unknowns in terms of the third. Keeping $A_2$ as our only unknown, we get the solution

$$A_1 = \frac{A_2}{2} \left( 1 + \frac{k_2}{k_1} \right) \qquad (4.19)$$

and

$$B_1 = \frac{A_2}{2} \left( 1 - \frac{k_2}{k_1} \right) \qquad (4.20)$$

so that the wave functions become

$$\psi_1(x) = \frac{A_2}{2} \left( 1 + \frac{k_2}{k_1} \right) e^{ik_1 x} + \frac{A_2}{2} \left( 1 - \frac{k_2}{k_1} \right) e^{-ik_1 x}, \, x < 0$$
$$\psi_2(x) = A_2 e^{ik_2 x}, \qquad\qquad\qquad\qquad\qquad\qquad x \geq 0$$

This is the complete solution, but what does it mean physically? Recall that for a normalized wave function, the probability of finding a particle at a given location $x$ is given by $\psi(x)^* \psi(x)$. Our wave functions are not so well behaved, since they represent an idealized, single-momentum state. However, it is still possible to derive useful results from them. In particular, we can define the probabilities that the particle will be reflected at the step $(R)$ and that it will be transmitted across the step $(T)$. In analogy to results from classical electromagnetism, we take

$$R = \frac{\text{(reflected amplitude)}^2}{\text{(incident amplitude)}^2}$$

and

$$T = \frac{\text{(transmitted amplitude)}^2}{\text{(incident amplitude)}^2}$$

We require that $R + T = 1$, since the particle must be either transmitted or reflected. Since we are dealing with complex quantities, the square of the incident amplitude is $A_1^* A_1$, and the square of the reflected amplitude is $B_1^* B_1$, so

$$R = \frac{B_1^* B_1}{A_1^* A_1} \tag{4.21}$$

(Note that in general, the terms giving the transmitted and reflected amplitudes will also include a factor depending on the velocity of the particle, but these factors cancel in Equation (4.21), since they are evaluated in the same region.) The transmission probability is then simply

$$T = 1 - R$$

Substituting our amplitudes from Equations (4.19) and (4.20) into Equation (4.21) gives us the reflection probability

$$R = \frac{(k_1 - k_2)^2}{(k_1 + k_2)^2} \tag{4.22}$$

while

$$T = 1 - R = 1 - \frac{(k_1 - k_2)^2}{(k_1 + k_2)^2}$$

This is a truly remarkable result; it tells us that there is a nonzero probability that the particle will not be transmitted across the step but will actually reflect backwards! The probability of reflection (Equation (4.22)) can be rewritten in terms of $E$ and $V_0$:

$$R = \left( \frac{\sqrt{E} - \sqrt{E - V_0}}{\sqrt{E} + \sqrt{E - V_0}} \right)^2$$

It is reasonable to ask why such reflection is not observed in classical systems. For example, suppose we have an electron moving through a region of space in which the electric potential changes abruptly. If we take the electron to have an energy equal to twice the potential step, $E = 2V_0$, we obtain $R = 0.17$, a nonnegligible reflection probability. The answer lies in our assumption that the potential is an infinitely sharp step. In any real physical system, the step-function potential will have a nonzero width in the $x$ direction. As long as this width is much larger than the de Broglie wavelength of the scattering particle, the system will lie in the classical regime, and the calculation we have just performed will be invalid.

---

**Example 4.1. When Is a Step-Function Potential in the Quantum Regime?**
As an example, consider an electron accelerated through a 100 V potential difference. How narrow would the step-function potential need to be in order to observe quantum effects?

The energy of the electron in this case is

$$E = e\Phi$$

$$= (1.6 \times 10^{-19} \text{ C})(100 \text{ V})$$

$$= 1.6 \times 10^{-17} \text{ J}$$

Its de Broglie wavelength is then

$$\lambda = h/p$$

$$= h/\sqrt{2mE}$$

$$= (6.6 \times 10^{-34} \text{ J s})/\sqrt{(2)(9.1 \times 10^{-31} \text{ kg})(1.6 \times 10^{-17} \text{ J})}$$

$$= 1.2 \times 10^{-10} \text{ m}$$

Thus, the step in the potential would need to rise from 0 to $V_0$ in a length less than $10^{-10}$ m, about an atomic radius. In practice, the step width would need to be much larger than this in order for the system to lie in the purely classical regime.

---

In general, quantum effects in step potentials can only be seen at the atomic or nuclear level. An important example ($\alpha$ decay) will be discussed below.

*Step-Function Potential With $E < V_0$*

Now consider scattering from a step-function potential when the energy of the incident particle is less than the height of the step. Again, the

Schrödinger equation will be given by Equations (4.10) and (4.11), but now $E - V_0$ will be negative. Hence, in the region $x < 0$, the solution to the Schrödinger equation will be identical to what we obtained in the previous section:

$$\psi_1(x) = A_1 e^{i\left(\sqrt{2mE}/\hbar\right)x} + B_1 e^{-i\left(\sqrt{2mE}/\hbar\right)x}$$

However, in the region $x \geq 0$, we expect a solution of the form given by Equation (4.4), i.e., a solution which looks like

$$\psi_2(x) = A_2 e^{\left(\sqrt{2m(V_0-E)}/\hbar\right)x} + B_2 e^{-\left(\sqrt{2m(V_0-E)}/\hbar\right)x}$$

Note that the quantity under the square root in the exponential is positive (since $V_0 - E > 0$), so the exponentials are real. As in the previous section, we define

$$k_1 = \frac{\sqrt{2mE}}{\hbar}$$

$$k_2 = \frac{\sqrt{2m(V_0 - E)}}{\hbar}$$

As before, $\hbar k_1$ is the magnitude of the momentum of the particle on the left side of the step, but $\hbar k_2$ does not have a similar physical significance. Now we write the solution as

$$\psi_1(x) = A_1 e^{ik_1 x} + B_1 e^{-ik_1 x}, \quad x < 0$$
$$\psi_2(x) = A_2 e^{k_2 x} + B_2 e^{-k_2 x}, \quad x \geq 0$$

As in the previous section, we can make a physical argument to eliminate one of the unknown constants. Note that if $A_2 \neq 0$, the wave function $\psi_2(x)$ will "blow up" in the limit where $x \to \infty$. This problem can only be avoided by taking $A_2 = 0$, yielding

$$\psi_2(x) = B_2 e^{-k_2 x}, \quad x \geq 0$$

The requirement that $\psi_1 = \psi_2$ at $x = 0$ gives

$$A_1 + B_1 = B_2$$

while the requirement that $d\psi_1/dx = d\psi_2/dx$ at $x = 0$ gives

$$ik_1 A_1 - ik_1 B_1 = -k_2 B_2$$

These two equations can be solved to express $A_1$ and $B_1$ as functions of $B_2$:

$$A_1 = \frac{B_2}{2}\left(1 + i\frac{k_2}{k_1}\right)$$

$$B_1 = \frac{B_2}{2}\left(1 - i\frac{k_2}{k_1}\right)$$

and the full wave function becomes

$$\psi_1(x) = \frac{B_2}{2}\left(1 + i\frac{k_2}{k_1}\right)e^{ik_1 x} + \frac{B_2}{2}\left(1 - i\frac{k_2}{k_1}\right)e^{-ik_1 x}, \; x < 0$$

$$\psi_2(x) = B_2 e^{-k_2 x}, \qquad\qquad\qquad\qquad x \geq 0$$

As in the previous section, we can calculate the transmission and reflection probabilities. The probability of reflection is

$$R = \frac{B_1^* B_1}{A_1^* A_1}$$

$$= \frac{(B_2/2)^2\left(1 + i\dfrac{k_2}{k_1}\right)\left(1 - i\dfrac{k_2}{k_1}\right)}{(B_2/2)^2\left(1 - i\dfrac{k_2}{k_1}\right)\left(1 + i\dfrac{k_2}{k_1}\right)}$$

$$= 1$$

and the transmission probability is $T = 1 - R = 0$. In this case, the quantum mechanical calculation produces a result in agreement with the classical result: the particle will always reflect back at the step boundary.

However, there is one strange result here that does not agree with the classical calculation. In the classical case, the particle stops exactly at the step and reflects back. However, in our quantum mechanical calculation, the wave function is nonzero for $x > 0$. This indicates that the probability of finding the particle in the classically forbidden region ($x > 0$) is nonzero. This penetration into the classically forbidden region is a purely quantum mechanical effect that has no analog in classical mechanics. It is so small as to be unobservable at the macroscopic scale, but it can have important consequences at the atomic and nuclear scales as shown in the next section.

An important limiting case of our solution occurs in the case of an "infinitely high" potential barrier. Of course, no physical potential can be infinite, but this simply means that $V_0$ is much larger than any typical particle energies: $E \ll V_0$. In this limit, $k_2 \to \infty$, and $\psi_2(x)$ becomes negligible for any $x > 0$. In this limit, therefore, it is a good approximation to take as a boundary condition, $\psi(x) = 0$ at $x = 0$.

*Tunneling*

In the previous section, we saw that the wave function can penetrate into the classically forbidden region of a step-function potential. Now we examine what happens if the step has finite width. Consider the step potential shown in Figure 4.7. This potential has $V = 0$ for $x < 0$ and $x \geq a$, and

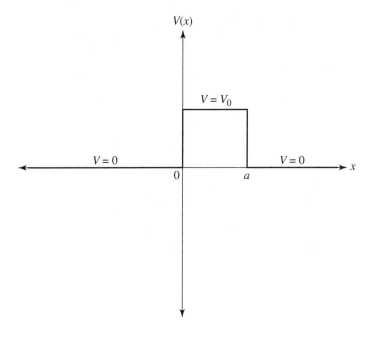

Fig. 4.7   A potential with a step of finite width.

$V = V_0$ for $0 \leq x < a$. We assume that a particle is incident from the left with energy $E < V_0$. In classical physics, such a particle will always bounce off of the barrier and reflect back to the left. However, quantum mechanics makes a very different prediction: the particle will sometimes be able to tunnel completely through the barrier and emerge on the other side!

To see why this happens, recall our qualitative solutions to the Schrödinger equation from the previous chapter. We expect the wave function to oscillate in the regions $x < 0$ and $x > a$ but to curve away from the horizontal axis for $0 < x < a$. We saw that for an infinitely wide step, the particle penetrates into the classically forbidden region. If the potential is reduced to zero after a finite distance, as in Figure 4.7, the wave function spills out onto the other side, where it oscillates. The magnitude of this effect can be calculated using the Schrödinger equation.

This potential yields three distinct solutions to the Schrödinger equation, corresponding to the three regions $x < 0$, $0 \leq x < a$, and $x \geq a$. As in the previous section, we get

$$\psi_1(x) = A_1 e^{ik_1 x} + B_1 e^{-ik_1 x}, \ x < 0$$

$$\psi_2(x) = A_2 e^{k_2 x} + B_2 e^{-k_2 x}, \ 0 \le x < a$$

with

$$k_1 = \frac{\sqrt{2mE}}{\hbar}$$

$$k_2 = \frac{\sqrt{2m(V_0 - E)}}{\hbar}$$

while the solution for $x \ge a$ is similar to the solution for $x < 0$, with the leftward-moving part of the wave function deleted (as in Equation (4.16)):

$$\psi_3(x) = B_3 e^{ik_1 x}$$

Now we have two pairs of boundary conditions:

$$\psi_1(0) = \psi_2(0) \tag{4.23}$$

$$\frac{d\psi_1}{dx}(0) = \frac{d\psi_2}{dx}(0) \tag{4.24}$$

and

$$\psi_2(a) = \psi_3(a) \tag{4.25}$$

$$\frac{d\psi_2}{dx}(a) = \frac{d\psi_3}{dx}(a) \tag{4.26}$$

Equations (4.23)–(4.24) give

$$A_1 + B_1 = A_2 + B_2$$

$$ik_1 A_1 - ik_1 B_1 = k_2 A_2 - k_2 B_2$$

and Equations (4.25)–(4.26) give

$$A_2 e^{k_2 a} + B_2 e^{-k_2 a} = B_3 e^{ik_1 a}$$

$$k_2 A_2 e^{k_2 a} - k_2 B_2 e^{-k_2 a} = ik_1 B_3 e^{ik_1 a}$$

These equations can be solved to derive four of the unknown coefficients in terms of the fifth. We are interested primarily in the transmission probability, given here by

$$T = \frac{B_3^* B_3}{A_1^* A_1}$$

After some tedious algebra, we obtain

$$T = \frac{1}{1 + [V_0^2/4E(V_0 - E)]\sinh^2(k_2 a)} \tag{4.27}$$

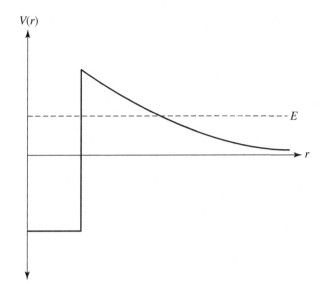

Fig. 4.8    The potential experienced by an $\alpha$ particle in a nucleus.

Clearly, $T$ is nonzero even when $E < V_0$, so a particle can tunnel through the barrier and emerge on the other side! The classical regime corresponds to the limit $k_2 a \gg 1$, so that $T \to 0$; this is why tunneling is never observed in classical systems.

Tunneling *is* seen, however, on the nuclear scale. An important example is $\alpha$ decay, in which a heavy nucleus such as $U^{238}$ emits an $\alpha$ particle (consisting of two protons and two neutrons). The potential seen by the $\alpha$ particle consists of an attractive nuclear force at short distances plus the Coulomb repulsion from the remainder of the nucleus at large distances (see Figure 4.8). The total energy $E$ can be measured for the emitted $\alpha$ particle, and it is found to be lower than the height of the potential barrier. The only way for an $\alpha$ particle to escape, therefore, is by quantum-mechanical tunneling. As a consequence, the typical lifetimes of $\alpha$-emitting nuclei are enormous. For example, $U^{238}$ has a lifetime of $\tau = 6.5 \times 10^9$ years: half the age of the universe!

Another application of tunneling is seen in the *tunnel diode*. This is a solid-state device in which, over a certain range in applied voltage, electrons tunnel through a potential barrier. This tunneling current increases with applied voltage up to a maximum value and then decreases with voltage. This results in an interesting property for the tunnel diode: over some

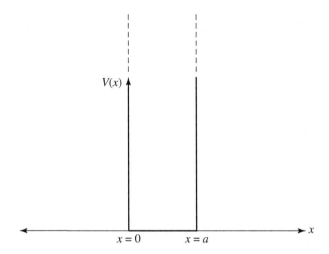

Fig. 4.9   The infinite square-well potential.

range in applied voltage, the current *decreases* as the voltage *increases*. Thus, over this range in voltage, the resistance is negative!

## 4.2   Bound Systems

We now move from unbound systems to bound systems. We will consider two important examples of bound systems in one dimension: the infinite square-well potential and the harmonic oscillator. The infinite square well has less physical importance than the harmonic oscillator, but it is simpler to solve and will illustrate concepts that can be applied elsewhere. Both of these one-dimensional problems will serve as a "warm-up" for the more physically-relevant (but also more complicated) three-dimensional systems examined in Chapter 6.

### The Infinite Square Well

Consider the one-dimensional infinite square-well potential of width $a$, shown in Figure 4.9. This potential has $V(x) = 0$ for $0 \leq x \leq a$, and $V(x) = \infty$ for $x < 0$ and $x > a$. Of course, no physical potential can be truly infinite, but this potential will be a good approximation for any system with sharp potential barriers such that $V \gg E$.

For our idealized system, it is clear that a particle with *any* energy $E$ will be in a bound state. The Schrödinger equation for $0 \leq x \leq a$ can be written

$$\frac{d^2\psi}{dx^2} + \frac{2mE}{\hbar^2}\psi = 0 \tag{4.28}$$

Since $E > 0$, the general solution will have the form of either Equation (4.5) or Equation (4.6), which are completely equivalent. It will be more convenient, however, to use the trigonometric form of the solution (Equation (4.6)), which gives, as the general solution

$$\psi(x) = C_1 \cos\left(\sqrt{2mE/\hbar^2}\,x\right) + C_2 \sin\left(\sqrt{2mE/\hbar^2}\,x\right)$$

The boundary conditions in the problem imply that $\psi(0) = 0$ and $\psi(a) = 0$. The first of these gives

$$\psi(0) = C_1 \cos(0) + C_2 \sin(0) = C_1 = 0$$

Hence, $C_1 = 0$, and the wave function is simply

$$\psi(x) = C_2 \sin\left(\sqrt{2mE/\hbar^2}\,x\right) \tag{4.29}$$

The second boundary condition, $\psi(a) = 0$, implies

$$\psi(a) = C_2 \sin\left(\sqrt{2mE/\hbar^2}\,a\right) = 0 \tag{4.30}$$

We cannot take $C_2 = 0$ (or the wave function would vanish everywhere!), so we must assume that

$$\sin\left(\sqrt{2mE/\hbar^2}\,a\right) = 0$$

The function $\sin(x)$ is zero for $x = 0, \pi, 2\pi, \ldots, n\pi$. Hence, Equation (4.30) can only be satisfied if the argument of the sine function is an integer multiple of $\pi$:

$$\sqrt{2mE/\hbar^2}\,a = n\pi, \quad n = 1, 2, 3, \ldots \tag{4.31}$$

Note that we exclude the case $n = 0$, which would produce a wave function that vanishes everywhere. The only free parameter in Equation (4.31) is the energy $E$. This equation tells us that the Schrödinger equation has a solution only for certain discrete values of $E$ which satisfy Equation (4.31). Solving Equation (4.31) for $E$, we get

$$E_n = \frac{\hbar^2\pi^2}{2ma^2}n^2, \quad n = 1, 2, 3, \ldots \tag{4.32}$$

where we have labeled the energy with a subscript corresponding to the value of $n$ on the right-hand side.

It is clear that the energy of this system is quantized. The energy $E$ cannot have an arbitrary value as in classical physics. Instead, only the set of discrete values given by Equation (4.32) is allowed. Although we showed qualitatively in the previous chapter that the Schrödinger equation for bound systems leads to energy quantization, this is our first explicit calculation with the Schrödinger equation that demonstrates such quantization.

Equation (4.32) has an interesting corollary. Since the smallest allowed value for $n$ is $n = 1$, the smallest possible energy for the particle is

$$E_1 = \frac{\hbar^2 \pi^2}{2ma^2} \tag{4.33}$$

Hence, the particle cannot have zero energy; it must have at least the minimum energy specified by Equation (4.33). This energy is called the *zero-point energy* of the system.

The wave function $\psi(x)$ can now be simplified in Equation (4.29) by substituting Equation (4.32) for the energy $E$. This gives

$$\psi_n(x) = C_2 \sin \frac{n\pi x}{a}$$

where $\psi_n(x)$ is the form of the wave function corresponding to a particle with energy $E_n$. The constant $C_2$ is calculated by normalizing the wave function:

$$\int_0^a \psi_n(x)^* \psi_n(x)\, dx = |C_2|^2 \int_0^a \sin^2 \frac{n\pi x}{a}\, dx$$
$$= |C_2|^2 \frac{a}{2}$$
$$= 1$$

Taking, for convenience, $C_2$ to be real and positive, we get

$$C_2 = \sqrt{\frac{2}{a}}$$

Note that for this case, the coefficient $C_2$ does not depend on $n$. This will not necessarily be the case for solutions with other potentials. The normalized wave functions and corresponding energies are then

$$\psi_n(x) = \sqrt{\frac{2}{a}} \sin \frac{n\pi x}{a}, \quad 0 \le x \le a$$

$$\psi_n(x) = 0, \quad x < 0 \text{ or } x > a$$

$$E_n = \frac{\hbar^2 \pi^2}{2ma^2} n^2$$

$$n = 1, 2, 3, \dots$$

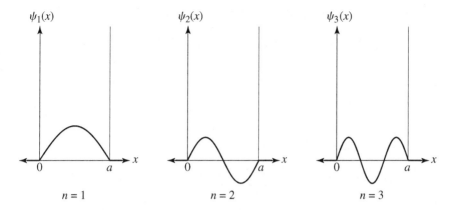

Fig. 4.10    The infinite square-well wave functions, $\psi_n(x)$, for $n = 1$, 2, and 3.

The first few wave functions (for $n = 1, 2, 3$) are shown in Figure 4.10. Note the resemblance to standing waves in a pipe. The wave functions $\psi_n(x)$ are alternately symmetric (for odd $n$) and antisymmetric (for even $n$) about $x = a/2$. In fact, it is possible to show (see Problem 4.10) that for *any* symmetric potential, the solutions of the Schrödinger equation will be either symmetric or antisymmetric. Note also that $\psi_n$ crosses the $x$-axis $n - 1$ times.

The probability of finding the particle in a small interval $dx$ at the location $x$ is just

$$\psi_n(x)^* \psi_n(x)\, dx = \frac{2}{a} \sin^2 \frac{n\pi x}{a} dx. \tag{4.34}$$

As $n \to \infty$, we expect to approach the classical regime. For large $n$, Equation (4.34) gives a function that resembles a fine tooth comb, giving a nearly uniform probability of finding the particle anywhere inside the square well. However, this is exactly what would be expected for the classical case of a particle bouncing continuously between the two walls of a closed container; the particle spends an equal amount of time at every point in the container.

A physical system of this sort can be constructed by sandwiching a thin layer of semiconductor between thicker layers of a different semiconductor (an example of a semiconductor heterostructure mentioned at the beginning of this chapter). The electrons are then free to move along the thin layer, but are confined in a potential well perpendicular to this layer. When such a structure is extended to three dimensions (so that the electrons are effectively confined to a single point), it produces a *quantum dot*.

### The Harmonic Oscillator Potential

We now examine a potential with more physical significance: the harmonic oscillator potential. The one-dimensional harmonic oscillator potential is

$$V(x) = \frac{1}{2}Kx^2 \tag{4.35}$$

This potential is familiar from classical mechanics, where it represents the potential energy of a mass attached to an ideal spring with spring constant $K$. Before solving the quantum harmonic oscillator, we first review the behavior of the classical harmonic oscillator. For a classical spring, the force is given by

$$F(x) = -Kx$$

so the equation of motion for the mass is

$$m\frac{d^2x}{dt^2} = -Kx$$

One example of a solution to this equation of motion is

$$x = A\cos(\omega t)$$

for the position of the mass, and

$$v = -A\omega\sin(\omega t)$$

for the velocity of the mass, where $\omega = \sqrt{K/m}$. Thus, the motion of the mass is sinusoidal. This solution gives a constant value for the total energy:

$$\begin{aligned} E &= \frac{1}{2}mv^2 + \frac{1}{2}Kx^2 \\ &= \frac{1}{2}mA^2\omega^2\sin^2(\omega t) + \frac{1}{2}KA^2\cos^2(\omega t) \\ &= \frac{1}{2}KA^2 \end{aligned} \tag{4.36}$$

Before proceeding to solve the Schrödinger equation with this potential, it is reasonable to ask why this potential would be of any interest at all. We certainly do not expect, for example, to see particles attached to atomic-scale springs! The answer lies in the fact that this potential is an excellent approximation to the motion of a particle undergoing small oscillations about the minimum of *any* potential. Consider an arbitrary potential $V(x)$, and choose the origin of the $x$-axis to lie at the minimum of this potential. Now consider a particle trapped near the minimum of the potential at $x = 0$. The potential can be approximated as a Taylor series near the origin:

$$V(x) = V(0) + V'(0)x + \frac{1}{2}V''(0)x^2 + \frac{1}{6}V'''(0)x^3 + \cdots$$

The first term in this equation is just a constant and can be ignored, since we are always free to redefine the zero of a potential. The minimum of the potential lies at $x = 0$, so $V'(0) = 0$, and the second term is zero. This leaves only terms proportional to $x^2$, $x^3$, $x^4$, and higher powers of $x$. But if $x$ is sufficiently small, then $x^2 \gg x^3 \gg x^4 \gg \cdots$, and we need worry only about the $x^2$ term. Thus, a particle undergoing small oscillations about the minimum of the potential will experience the approximate potential

$$V(x) = \frac{1}{2}V''(0)x^2$$

which has the same form as Equation (4.35) with $K = V''(0)$. For example, the motion of ions in a crystal lattice is often approximated using a harmonic oscillator potential.

We now proceed to solve the quantum harmonic oscillator. The Schrödinger equation with the one-dimensional harmonic oscillator potential is

$$-\frac{\hbar^2}{2m}\frac{d^2\psi}{dx^2} + \frac{1}{2}Kx^2\psi = E\psi \qquad (4.37)$$

In order to simplify the algebra involved in solving this equation, it is convenient to define a new independent variable $s$ given by

$$s = \frac{(Km)^{1/4}}{\hbar^{1/2}}x \qquad (4.38)$$

and a new constant $\lambda$ proportional to the energy:

$$\lambda = \frac{2}{\hbar}\sqrt{\frac{m}{K}}E \qquad (4.39)$$

Both $s$ and $\lambda$ are dimensionless; otherwise they have no special significance beyond simplifying the calculation.

In terms of $s$ and $\lambda$, Equation (4.37) simplifies to

$$\frac{d^2\psi}{ds^2} + (\lambda - s^2)\psi = 0 \qquad (4.40)$$

To derive a solution, consider what happens for $s^2 \gg \lambda$. In this limit, Equation (4.40) looks like

$$\frac{d^2\psi}{ds^2} - s^2\psi = 0 \qquad (4.41)$$

Even this simplified version of the equation cannot be solved exactly, but we can find an approximate solution valid for large $s$:

$$\psi = Ae^{s^2/2} + Be^{-s^2/2}$$

For this solution,

$$\frac{d^2\psi}{ds^2} = A(1+s^2)e^{s^2/2} - B(1-s^2)e^{-s^2/2}$$

and for $s \gg 1$

$$\frac{d^2\psi}{ds^2} \approx As^2 e^{s^2/2} + Bs^2 e^{-s^2/2} = s^2\psi$$

satisfying Equation (4.41). However, we also want our solution to be well behaved in the limit where $s \to \pm\infty$. This means that $A = 0$ (otherwise, the solution "blows up" at $\pm\infty$). Thus, we expect the solution to resemble, in the limit of large $s^2$,

$$\psi(s) \sim e^{-s^2/2} \tag{4.42}$$

This is clearly not an exact solution of Equation (4.40), but it suggests that we look for an exact solution of the form

$$\psi(s) = f(s)e^{-s^2/2} \tag{4.43}$$

Substituting this form for $\psi$ into Equation (4.40) and simplifying, we get a differential equation for $f(s)$:

$$\frac{d^2 f}{ds^2} - 2s\frac{df}{ds} + (\lambda - 1)f = 0 \tag{4.44}$$

To find a function $f(s)$ which satisfies this equation, we expand $f(s)$ out in a power series:

$$f(s) = \sum_{n=0}^{\infty} a_n s^n \tag{4.45}$$

where the coefficients $a_n$ must be chosen to satisfy Equation (4.44). Substituting this form for $f(s)$ into Equation (4.44) gives

$$\sum_{n=2}^{\infty} n(n-1)a_n s^{n-2} - 2\sum_{n=0}^{\infty} na_n s^n + (\lambda - 1)\sum_{n=0}^{\infty} a_n s^n = 0$$

Rewriting the first term in this equation in terms of $m = n - 2$ and combining the last two terms gives

$$\sum_{m=0}^{\infty} (m+2)(m+1)a_{m+2} s^m + (\lambda - 2n - 1)\sum_{n=0}^{\infty} a_n s^n = 0$$

(Note that $m$ is just a summation label, so that it can be changed back to $n$ in the first term of this equation.) In order for this equation to be satisfied, the left-hand side must be identically zero. This can only be achieved if the factor multiplying *every* power of $s$ is zero. This requirement gives

$$(n+2)(n+1)a_{n+2} + (\lambda - 2n - 1)a_n = 0$$

which can be used to fix $a_{n+2}$ in terms of $a_n$:

$$a_{n+2} = \frac{2n + 1 - \lambda}{(n+2)(n+1)}a_n \qquad (4.46)$$

A relationship of this sort is called a *recursion* relation. Given $a_0$ and $a_1$, Equation (4.46) can be iterated to calculate all of the other terms in the power series. However, an arbitrary choice for $a_0$ and $a_1$ will not, in general, give an acceptable solution. The reason is the solution should look like Equation (4.42) when $s$ is large. If the factor $f(s)$ multiplying the exponential in Equation (4.43) is a *finite* polynomial, then the exponential factor will dominate the polynomial at large $s$, giving the desired asymptotic behavior for $\psi(s)$ at large $s$ (Equation (4.42)). An *infinite* polynomial, on the other hand, can dominate the exponential at large $s$, leading to an unacceptable solution.

We therefore require that the power series must terminate at some finite value of $n$. This can be achieved by choosing an appropriate value of $\lambda$ in Equation (4.46). Recall that $\lambda$ is simply proportional to the energy $E$, and we expect the energy to be quantized, so that the Schrödinger equation will have solutions for only a set of discrete values of $E$. Thus, fixing the value of $\lambda$ in Equation (4.46) is simply equivalent to fixing the value of $E$ to give a solution for the Schrödinger equation. Consider what happens if we set

$$\lambda = 2n + 1 \qquad (4.47)$$

for some fixed value of $n$. As an example, suppose that we choose $\lambda = 13$, so that $\lambda = 2n + 1$ for $n = 6$. Then we fix a value for $a_0$ which determines $a_2$ through Equation (4.46). Then $a_2$ determines $a_4$, and $a_4$ determines $a_6$, at which point Equation (4.46) with $n = 6$ and $\lambda = 13$ gives $a_8 = 0$. Then $a_{10} = 0$, $a_{12} = 0$, and so on. However, this still leaves the odd values of $n$: $a_1, a_3, \ldots$. There is nothing to force this sequence of terms to terminate at a finite value of $n$ (since we chose $n$ to be even in Equation (4.47)). Therefore, to obtain a finite polynomial in Equation (4.45), we must choose $a_1 = 0$, so that *all* of the odd-numbered terms are zero. Conversely, if we take $n$ in Equation (4.47) to be odd, then $a_0$ and all of the other even-numbered terms must vanish.

This procedure produces a set of solutions $\psi_n(s)$, corresponding to our choice of $\lambda$ in Equation (4.47), and $\lambda$, in turn, gives the corresponding energies $E_n$ in Equation (4.39). Here is what the first few solutions look like:

$$n = 0, \quad \psi_0(s) = C_0 e^{-s^2/2} \tag{4.48}$$

$$n = 1, \quad \psi_1(s) = C_1(2s)e^{-s^2/2} \tag{4.49}$$

$$n = 2, \quad \psi_2(s) = C_2(4s^2 - 2)e^{-s^2/2} \tag{4.50}$$

$$n = 3, \quad \psi_3(s) = C_3(8s^3 - 12s)e^{-s^2/2} \tag{4.51}$$

$$n = 4, \quad \psi_4(s) = C_4(16s^4 - 48s^2 + 12)e^{-s^2/2} \tag{4.52}$$

where the constants $C_n$ must be fixed to normalize the wave functions. The normalization condition gives

$$C_n = \frac{1}{\pi^{1/4}\sqrt{2^n n!}}$$

These polynomials are called *Hermite polynomials*; they crop up in other areas of physics and mathematics. The harmonic oscillator wave functions for $n = 0, 1, 2, 3$ are shown in Figure 4.11. As expected (Problem 4.10), these wave functions are alternately even and odd, since the potential is symmetric about $x = 0$.

The energies corresponding to these wave functions can be derived from Equations (4.39) and (4.47):

$$2n + 1 = \lambda = \frac{2}{\hbar}\sqrt{\frac{m}{K}}E$$

so

$$E_n = \left(n + \frac{1}{2}\right)\hbar\sqrt{\frac{K}{m}}$$

Recall that the frequency of a classical harmonic oscillator is $\omega = \sqrt{K/m}$, so that the energies can be written

$$E_n = \left(n + \frac{1}{2}\right)\hbar\omega \tag{4.53}$$

Once again, we see the phenomenon of a zero-point energy: the lowest-energy state, $n = 0$, does not have zero energy. Instead, its energy is

$$E_0 = \frac{1}{2}\hbar\omega$$

Note further that the energy levels for the harmonic oscillator are evenly spaced; adjacent energy levels differ by $\hbar\omega$.

The behavior of $\psi^*\psi$ is shown in Figure 4.12 for the states $n = 0, 1, 2, 3$. It is clear that $\psi^*\psi$ for these wave functions is nonzero in the classically

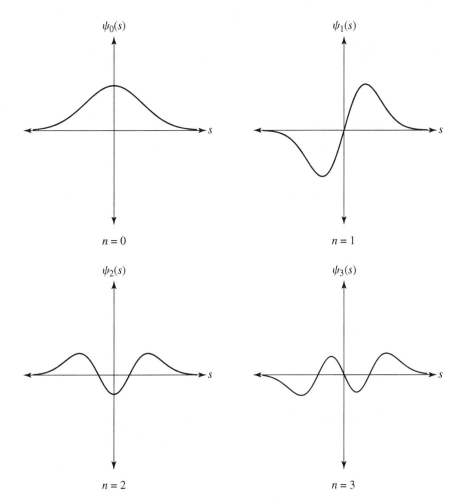

Fig. 4.11   The harmonic oscillator wave functions $\psi_n(s)$ for the lowest four energy states: $n = 0, 1, 2, 3$, where $s = [(Km)^{1/4}/\hbar^{1/2}]x$.

forbidden region $V > E$ (in fact, it is nonzero for all of the harmonic oscillator wave functions over all space). Once again, we see that the particle can penetrate into a region which, classically, it should not be able to reach.

These quantum probabilities can be compared with the expected result in the classical limit. Consider a classical harmonic oscillator with total energy given by Equation (4.36). The oscillating mass moves between the limits $x_- = -\sqrt{2E/K}$ and $x_+ = \sqrt{2E/K}$ with its largest velocity at the

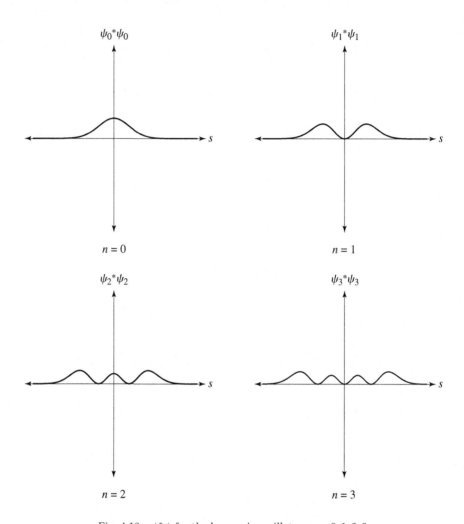

Fig. 4.12   $\psi^*\psi$ for the harmonic oscillator, $n = 0, 1, 2, 3$.

center of the oscillator and the smallest velocity near $x_-$ and $x_+$. Thus, the mass spends more time near $x_-$ and $x_+$ and less time near the middle of the oscillator, so if we take a random snapshot in time, we are most likely to find the particle near $x_-$ and $x_+$, and least likely to find it in the middle.

More quantitatively, if we pick a random time, the probability $P$ of finding the mass in a small interval $dx$ is proportional to the time $dt$ that it spends in that interval. But $dt = dx/v$, and $v$ can be determined from

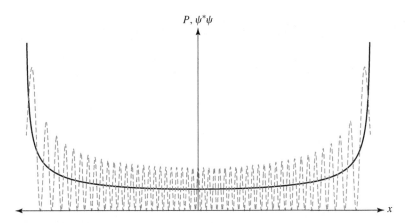

Fig. 4.13    The classical probability $P$ of finding the harmonic oscillator mass in a small interval $dx$ at a position $x$ is shown as a solid curve. The quantum probability density, $\psi_n(x)^*\psi(x)$ for $n = 50$, is shown as a dashed curve.

Equation (4.36):

$$v = \sqrt{(2E - Kx^2)/m}$$

Thus, we have

$$P \propto \sqrt{\frac{m}{2E - Kx^2}}\, dx$$

This classical probability, suitably normalized, is shown in Figure 4.13. This distribution clearly shows no relation to the low-$n$ quantum-mechanical probabilities in Figure 4.12. However, if we take a large value for $n$, corresponding to the classical limit, we obtain a result which begins to resemble the classical probability. In Figure 4.13, we show $\psi^*\psi$ for $n = 50$. It is clear that in this limit, the quantum mechanical probability approaches the classical result.

One application of the quantum harmonic oscillator is in the physics of diatomic molecules. The two nuclei in a diatomic molecule can vibrate about their equilibrium separation, and the potential for these nuclei can be approximated by a harmonic oscillator potential. It is found that the vibrational energy levels are well described by Equation (4.53). (See also Problem 4.13.)

### The Heisenberg Uncertainty Principle

Consider the ground state harmonic oscillator wave function derived in the previous section (Equation (4.48)). Suppose that we make repeated measurements of the position $x$. We will get a different answer each time we measure $x$, although the mean value of all of our measurements will converge to $\langle x \rangle$. Then we can define the "uncertainty" $\Delta x$ in our measurement of $x$ as the standard deviation of our measurements relative to $\langle x \rangle$:

$$(\Delta x)^2 = \langle x^2 \rangle - \langle x \rangle^2.$$

For the ground state harmonic oscillator wave function, $\langle x \rangle = 0$, so $(\Delta x)^2$ is given by

$$(\Delta x)^2 = \int_{-\infty}^{\infty} \frac{1}{\pi^{1/4}} e^{-s^2/2} \left( \frac{\hbar^{1/2}}{(Km)^{1/4}} s \right)^2 \left( \frac{1}{\pi^{1/4}} e^{-s^2/2} \right) ds$$
$$= \frac{1}{2} \frac{\hbar}{(Km)^{1/2}}$$

where we have used the conversion factor between $s$ and $x$ given in Equation (4.38). Similarly, we can make repeated measurements of the momentum, and define the uncertainty in the momentum to be $\Delta p$, where $(\Delta p)^2 = \langle p^2 \rangle - \langle p \rangle^2$. As in the case of the position, the mean value of $p$ is zero, so we have

$$(\Delta p)^2 = \int_{-\infty}^{\infty} \frac{1}{\pi^{1/4}} e^{-s^2/2} \left( -i\hbar \frac{(Km)^{1/4}}{\hbar^{1/2}} \right)^2 \frac{d^2}{ds^2} \left( \frac{1}{\pi^{1/4}} e^{-s^2/2} \right) ds$$
$$= \frac{1}{2} \hbar (Km)^{1/2}$$

Combining these two expressions gives

$$(\Delta x)^2 (\Delta p)^2 = \frac{\hbar^2}{4}.$$

However, it turns out that the ground state harmonic oscillator wave function gives the smallest possible value of $(\Delta x)^2 (\Delta p)^2$ (a result that we will not prove here). Therefore, for any wave function, we have

$$\Delta x \Delta p \geq \frac{\hbar}{2}$$

which is called the *Heisenberg uncertainty principle*. (As an example, see Problem 4.17).

The Heisenberg uncertainty principle gives a fundamental limit on the accuracy with which the position and momentum of a particle can be measured. The more precisely you measure the position of a particle, the less accurately you will be able to determine the momentum, and vice-versa. Note that it applies only to measurements made in the same direction. For example, there is nothing to prevent an arbitrarily precise measurement of the position along the $x$ axis and of the component of the momentum along the $y$ axis.

## PROBLEMS

**4.1** Show that the differential equation solution given in Equation (4.5),

$$y = C_1 e^{i\sqrt{A}x} + C_2 e^{-i\sqrt{A}x},$$

is completely equivalent to the solution in Equation (4.6),

$$y = D_1 \cos(\sqrt{A}x) + D_2 \sin(\sqrt{A}x),$$

and express $C_1$ and $C_2$ in terms of $D_1$ and $D_2$. If both $D_1$ and $D_2$ are real and nonzero, is it possible for both $C_1$ and $C_2$ to be real?

**4.2** Show that the general expression for the wave function for a free particle, given by Equation (4.8) as

$$\Psi(x,t) = C_1 e^{i(kx-\omega t)} + C_2 e^{i(-kx-\omega t)},$$

is not an eigenfunction of momentum unless $C_1 = 0$ or $C_2 = 0$.

**4.3** A particle with mass $m$ and energy $E$ is moving in one dimension from *right to left*. It is incident on the step potential $V(x) = 0$ for $x < 0$ and $V(x) = V_0$ for $x \geq 0$, where $V_0 > 0$, as shown on the diagram. The energy of the particle is $E > V_0$.

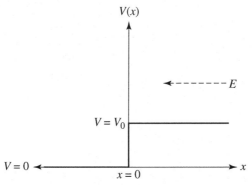

(a) Solve the Schrödinger equation to derive $\psi(x)$ for $x < 0$ and $x \geq 0$. Express the solution in terms of a single unknown constant.
(b) Calculate the value of the reflection coefficient $R$ for the particle.

**4.4** A particle with mass $m$ and energy $E$ is moving in one dimension from *left to right*. It is incident on the step potential $V(x) = 0$ for $x < 0$ and $V(x) = V_0$ for $x \geq 0$, where $V_0 > 0$, as shown on the diagram. The energy of the particle is exactly equal to $V_0$, i.e., $E = V_0$.

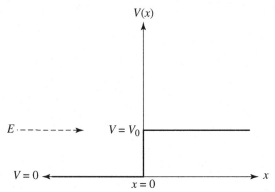

(a) Solve the Schrödinger equation to derive $\psi(x)$ for $x < 0$ and $x \geq 0$. Express the solution in terms of a single unknown constant.
(b) Calculate the value of the reflection coefficient $R$ for the particle.

**4.5** Consider reflection from a step potential of height $V_0$ with $E > V_0$, but now with an infinitely high wall added at a distance $a$ from the step (see diagram):

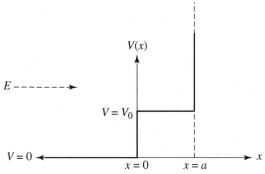

(a) Solve the Schrödinger equation to find $\psi(x)$ for $x < 0$ and $0 \leq x \leq a$. Your solution should contain only one unknown constant.
(b) Show that the reflection coefficient at $x = 0$ is $R = 1$. This is different from the value of $R$ previously derived without the infinite wall. What is the physical reason that $R = 1$ in this case?
(c) Which part of the wave function represents a leftward-moving particle at $x \leq 0$? Show that this part of the wave function is an

eigenfunction of the momentum operator, and calculate the eigen-value. Is the total wave function for $x \leq 0$ an eigenfunction of the momentum operator?

**4.6** An electron is accelerated through a potential difference of 3.0 eV and is incident on a finite potential barrier of height 5.0 eV and thickness $5.0 \times 10^{-10}$ m. What is the probability that the electron will tunnel through the barrier?

**4.7** Consider an infinite square-well potential of width $a$, but with the coordinate system shifted so that the infinite potential barriers lie at $x = -a/2$ and $x = a/2$ (see diagram):

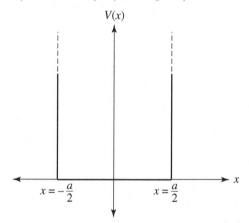

(a) Solve the Schrödinger equation for this case to calculate the normalized wave functions $\psi_n(x)$ and the corresponding energies $E_n$.

(b) Explain why you get the same energies as for the square well between $x = 0$ and $x = a$, but a different set of wave functions.

**4.8** A baseball of mass 0.14 kg is confined between two thick walls a distance 0.5 m apart. Calculate the zero-point energy of the baseball.

**4.9** A particle is trapped inside an infinite one-dimensional square well of width $a$ in the first excited state ($n = 2$).

(a) You make a measurement to locate the particle. At what positions are you most likely to find the particle? At what positions are you least likely to find it?

(b) Calculate $\langle p^2 \rangle$ for this particle.

**4.10** A particle is bound in a one-dimensional potential $V(x)$, where $V(x)$ is symmetric, i.e., $V(x) = V(-x)$.

(a) Suppose that $\psi(x)$ is a solution of the Schrödinger equation with energy $E$. Make the change of variables $y = -x$, and show that $\psi(y)$ is also a solution of the Schrödinger equation with energy $E$.

(b) Since the solutions of the Schrödinger equation for a fixed value of $E$ are unique (up to multiplication by a constant), the result from part (a) implies that $\psi(x) = c\psi(-x)$, where $c$ is an unknown constant. Use this result to show that $\psi(x)$ must be either even $[\psi(-x) = \psi(x)]$ or odd $[\psi(-x) = -\psi(x)]$.

(c) For a particle bound in a one-dimensional symmetric potential, so that $V(-x) = V(x)$, show that all of the following are true:

(i) $\psi^*\psi$ is a symmetric function,

(ii) $\langle x \rangle = 0$,

(iii) $\langle p \rangle = 0$.

**4.11** Consider the semi-infinite square well given by $V(x) = -V_0 < 0$ for $0 \le x \le a$ and $V(x) = 0$ for $x > a$. There is an infinite barrier at $x = 0$ (hence the name "semi-infinite"). A particle with mass $m$ is in a bound state in this potential with energy $E \le 0$.

(a) Solve the Schrödinger equation to derive $\psi(x)$ for $x \ge 0$. Use the appropriate boundary conditions and normalize the wave function so that the final answer does not contain any arbitrary constants.

(b) Show that the allowed energy levels $E$ must satisfy the equation

$$\tan\left(\frac{\sqrt{2m(E + V_0)}}{\hbar}a\right) + \sqrt{\frac{-(E + V_0)}{E}} = 0$$

(c) The equation in part (b) cannot be solved analytically to give the allowed energy levels, but simple solutions exist in certain special cases. Determine the conditions on $V_0$ and $a$ so that a bound state exists with $E = 0$.

**4.12** A particle of mass $m$ moves in a harmonic oscillator potential. The particle is in the first excited state.

(a) Calculate $\langle x \rangle$ for this particle.

(b) Calculate $\langle p \rangle$ for this particle.

(c) Calculate $\langle p^2 \rangle$ for this particle.

(d) At what positions are you most likely to find the particle? At what positions are you least likely to find it?

**4.13** The oscillation frequencies of a diatomic molecule are typically $10^{12}$ Hz–$10^{14}$ Hz. Derive an order of magnitude estimate for the harmonic oscillator constant $K$ for such molecules.

**4.14** A particle of mass $m$ is bound in a one-dimensional power law potential $V(x) = Kx^{\beta}$, where $\beta$ is an even positive integer. Show that the allowed energy levels are proportional to $m^{-\beta/(2+\beta)}$.

**4.15** A particle is moving in a simple harmonic oscillator potential $V(x) = \frac{1}{2}Kx^2$ for $x \geq 0$, but with an infinite potential barrier at $x = 0$ (the paddle ball potential). Calculate the allowed wave functions and corresponding energies. Do not worry about normalizing the wave functions.

**4.16** A particle moves in one dimension in the potential $V(x) = V_0 \ln(x/x_0)$ for $x > 0$, where $x_0$ and $V_0$ are constants with units of length and energy, respectively. There is an infinite potential barrier at $x = 0$. The particle drops from the first excited state with energy $E_1$ into the ground state with energy $E_0$, by emitting a photon with energy $E_1-E_0$. Show that the frequency of the photon emitted by this particle is independent of the mass of the particle.

**4.17** Calculate $\Delta x \Delta p$ for the ground state wave function in an infinite square well and show that it satisfies the Heisenberg uncertainty principle.

Chapter 5

# Math interlude B: Linear algebra

We have already seen that linear operators occupy a central place in quantum mechanics. In particular, in Chapter 3 we noted that for any observable quantity $o$, we can find a linear operator $O$ with the following properties:

1. If a particle is in a state with a definite value of $o$, then the wave function $\psi$ for the particle is an eigenfunction of $O$ with eigenvalue $o$:

$$O\psi = o\psi \qquad (5.1)$$

Note that $O$ is an *operator* and $o$ is a *number*, but $O$ and $o$ must have the same physical units (joules, meters, etc.). If Equation (5.1) is satisfied, then if a measurement of the observable is made, the result is guaranteed to be $o$.

2. If the particle is not in a state with a definite value of $o$, then $\psi$ will not be an eigenfunction of $O$, and Equation (5.1) will not be satisfied. If a measurement of the observable is made, there is no way to predict the result. However, the expectation value of $o$ is still a well-defined quantity given by

$$\langle o \rangle = \int \psi^* O \psi \; d^3\mathbf{r}$$

The properties of linear operators are part of a more general branch of mathematics called *linear algebra*, and it is this subject which we examine in more detail here.

## 5.1 Properties of Linear Operators

Recall that in order for an operator $O$ to be a linear operator, must have two properties:

$$O[cf(x)] = cOf(x)$$

and

$$O[f(x) + g(x)] = Of(x) + Og(x)$$

for all functions $f(x)$ and $g(x)$, and all complex numbers $c$.

We can go further and define addition or subtraction of linear operators. If $P$ and $Q$ are two linear operators, then their sum, $R = P + Q$, is defined by

$$Rf(x) = (P + Q)f(x) = Pf(x) + Qf(x)$$

A similar result holds for subtraction: if $R = P - Q$, then

$$Rf(x) = Pf(x) - Qf(x)$$

In fact, we have already implicitly used this definition when we introduced the Hamiltonian operator $H$. The Hamiltonian operator is defined as the sum of two operators:

$$H = -\frac{\hbar^2}{2m}\nabla^2 + V$$

so that

$$H\psi = -\frac{\hbar^2}{2m}\nabla^2\psi + V\psi$$

We can also define the product of two linear operators. If $P$ and $Q$ are linear operators, then $R = PQ$ is defined by

$$R\psi = (PQ)\psi$$
$$= P(Q\psi) \tag{5.2}$$

where Equation (5.2) means that we first apply the operator $Q$ to $\psi$, and then we apply the operator $P$ to the result. The operator $PQ$ is also called the *composition* of $P$ and $Q$.

---

## Example 5.1. Multiplication of Two Operators

Let $X$ be the one-dimensional position operator, $X\psi = x\psi$, and let $D$ be the derivative operator: $D\psi = d\psi/dx$. Calculate $DX$.

Applying $DX$ to an arbitrary function $\psi(x)$ gives

$$DX\psi = \frac{d}{dx}(x\psi)$$

$$= \psi + x\frac{d\psi}{dx}$$

$$= (1 + XD)\psi$$

so that

$$DX = 1 + XD \tag{5.3}$$

---

Example 5.1 demonstrates that the multiplication of operators differs from the multiplication of ordinary numbers in one important respect: it is not commutative! Clearly, Equation (5.3) implies that $DX \neq XD$.

In fact, the question of whether or not operators commute is of central importance in quantum mechanics. Hence, it is customary to define a special quantity called the *commutator*. For any two operators $A$ and $B$, the commutator is denoted by the symbol $[A, B]$, and it is defined as

$$[A, B] = AB - BA$$

When two operators do commute with each other, their commutator is, obviously, zero.

---

**Example 5.2. The Commutator of $H$ and $P$**

As an example, we calculate the commutator of the one-dimensional Hamiltonian operator $H$ and the one-dimensional momentum operator $P$.

Applying $[H, P]$ to an arbitrary wave function, $\psi(x)$, gives

$$[H, P]\psi = HP\psi - PH\psi$$

$$= \left[ -\frac{\hbar^2}{2m}\frac{\partial^2}{\partial x^2} + V(x) \right]\left[ -i\hbar\frac{d\psi}{dx} \right] - \left[ -i\hbar\frac{d}{dx} \right]\left[ -\frac{\hbar^2}{2m}\frac{\partial^2\psi}{\partial x^2} + V(x)\psi \right]$$

$$= i\frac{\hbar^3}{2m}\frac{\partial^3\psi}{\partial x^3} - i\hbar V(x)\frac{\partial\psi}{\partial x} - i\frac{\hbar^3}{2m}\frac{\partial^3\psi}{\partial x^3} + i\hbar\psi\frac{\partial V}{\partial x} + i\hbar V(x)\frac{\partial\psi}{\partial x}$$

$$= i\hbar\frac{\partial V}{\partial x}\psi$$

Hence,

$$[H, P] = i\hbar\frac{\partial V}{\partial x}$$

---

Some important general properties of commutators are (for any operators $A$, $B$, and $C$)

$$[A, B] = -[B, A] \tag{5.4}$$

$$[A, A] = 0 \tag{5.5}$$

$$[A + B, C] = [A, C] + [B, C] \tag{5.6}$$

$$[A, BC] = [A, B]C + B[A, C] \tag{5.7}$$

(See Problem 5.1.)

The importance of commutators for quantum mechanics arises in the following way. Suppose that I have a particle for which I want to measure two different observables, $a$ and $b$. (For instance, I might want to measure the position and momentum of the particle.) Is it possible for the particle to be in a state of definite $a$ and definite $b$ at the same time? The answer is "yes" but only if the corresponding operators $A$ and $B$ commute.

To see why this is the case, recall that in order for the particle to be in a state of definite $a$, it must be an eigenfunction of $A$ with eigenvalue $a$:

$$A\psi = a\psi \qquad (5.8)$$

Now we make an additional assumption: there are no other wave functions $\psi$ (other than multiples of $\psi$) that are eigenfunctions of $A$ with the same eigenvalue $a$. (If two different eigenfunctions have the same eigenvalue, and one eigenfunction is not a multiple of the other, then the eigenfunctions are said to be *degenerate*. The argument presented here can be extended to the case of degenerate states, but it is somewhat more complicated.) Now suppose that $A$ and $B$ do commute so that $[A, B] = 0$. Operating on both sides of Equation (5.8) with the operator $B$ gives

$$BA\psi = Ba\psi = aB\psi$$

But $A$ commutes with $B$, so we can rewrite the left-hand side of this equation to get

$$A(B\psi) = a(B\psi)$$

So we have shown that $B\psi$ is an eigenfunction of $A$ with eigenvalue $a$. However, we assumed that the eigenfunctions of $A$ were all nondegenerate, so that any eigenfunction of $A$ with eigenvalue $a$ must simply be a multiple of $\psi$. Thus, $B\psi$ must be a multiple of $\psi$, e.g., $b\psi$, and

$$B\psi = b\psi$$

So $\psi$ is simultaneously an eigenfunction of both $A$ and $B$. Physically, this means that the particle is in a state of definite $a$ and $b$, and both quantities can be measured simultaneously.

---

## Example 5.3. Simultaneous Eigenfunctions

For which potentials $V(x)$ is it possible to find solutions of the one-dimensional time-independent Schrödinger equation which are also states of definite momentum?

We calculated $[H, P]$ in Example 5.2, finding

$$[H, P] = i\hbar \frac{\partial V}{\partial x}$$

Solutions of the one-dimensional time-independent Schrödinger equation are eigenfunctions of $H$; in order for them also to be eigenfunctions of $P$, we must have $[H, P] = 0$ which implies $V = $ constant (and the constant can be set to zero). Hence, free particles are the only solutions of the time-independent Schrödinger equation that can be in states of definite momentum; it is precisely these states which we examined in the previous chapter.

---

Another very famous commutator is provided by the momentum and position operators. Consider the one-dimensional case, $[P, X]$:

$$[P, X]\psi = \left(-i\hbar \frac{\partial}{\partial x}\right) x\psi - x \left(-i\hbar \frac{\partial \psi}{\partial x}\right)$$
$$= -i\hbar \psi$$

which implies

$$[P, X] = -i\hbar$$

Thus, a particle can *never* be in a state which is simultaneously a state of definite momentum and a state of definite position. Of course, if we could measure both the position and momentum exactly, it would violate the Heisenberg uncertainly principle discussed in the previous chapter.

These results regarding commutators provide a useful blueprint for measuring quantities of interest. We will usually want our system to satisfy the time-independent Schrödinger equation, which implies that the wave function is an eigenfunction of $H$ and represents a state of definite energy $E$. We will then want to find a set of operators $A, B, C, \ldots$, which all commute with $H$ and also with each other:

$$[H, A] = 0$$
$$[H, B] = 0$$
$$[H, C] = 0$$
$$\vdots$$
$$[A, B] = 0$$
$$[A, C] = 0$$
$$[B, C] = 0$$
$$\vdots$$

The fact that all of these operators commute with each other means that we can find a wavefunction $\psi$ for which

$$H\psi = E\psi$$
$$A\psi = a\psi$$
$$B\psi = b\psi$$

and so on. The set of eigenvalues $E$, $a$, $b$, etc. can be used to specify $\psi$ and are called *good quantum numbers*. When a measurement of the corresponding observables is made, the results will be $E$, $a$, $b$, . . . .

## 5.2 Vector Spaces

In this section we introduce the concept of an abstract "vector space." First, consider a familiar three-dimensional vector. It can be represented in one of two equivalent ways: either as a quantity with a given magnitude and a direction in three dimensions, or as a set of components, $(x, y, z)$. Similarly, a two-dimensional vector can be represented as a quantity with a given magnitude and direction in the $x$-$y$ plane, or as a two-component quantity, $(x, y)$. It is the component representation which we will generalize. One can specify a vector in $n$ dimensions as a set of $n$ components: $(r_1, r_2, \ldots, r_n)$. Obviously, an $n$-dimensional vector cannot be represented in ordinary three-dimensional space as a quantity with a magnitude and direction, but any of the normal vector operations can be performed on it by using its components. For instance, the sum and dot product of two three-dimensional vectors $\mathbf{r} = (r_1, r_2, r_3)$ and $\mathbf{s} = (s_1, s_2, s_3)$ are just

$$\mathbf{r} + \mathbf{s} = (r_1 + s_1, r_2 + s_2, r_3 + s_3)$$

and

$$\mathbf{r} \cdot \mathbf{s} = r_1 s_1 + r_2 s_2 + r_3 s_3$$

In the same way, for two $n$-dimensional vectors, $\mathbf{r} = (r_1, r_2, \ldots, r_n)$ and $\mathbf{s} = (s_1, s_2, \ldots, s_n)$, the sum and dot product are

$$\mathbf{r} + \mathbf{s} = (r_1 + s_1, r_2 + s_2, \ldots, r_n + s_n)$$

and

$$\mathbf{r} \cdot \mathbf{s} = r_1 s_1 + r_2 s_2 + \cdots + r_n s_n$$

A *vector space* is simply a collection of objects (called vectors) which act, in a general way, like familiar three-dimensional vectors. For instance, the sum of two vectors $\mathbf{r}$ and $\mathbf{s}$, must also be a vector:

$$\mathbf{r} + \mathbf{s} = \mathbf{t} \tag{5.9}$$

while the product of a number $c$ and a vector $\mathbf{r}$ must also be a vector:

$$c\mathbf{r} = \mathbf{u} \qquad (5.10)$$

Note that for our ordinary three-dimensional vectors, $c$ can only be a real number, but for some vector spaces $c$ is assumed to be a complex number. Thus, there are two different kinds of vector spaces: real vector spaces (for which $c$ in Equation (5.10) is restricted to be a real number) and complex vector spaces (for which $c$ can be complex). *In quantum mechanics, we will deal exclusively with complex vector spaces.*

The properties given in Equations (5.9) and (5.10) may seem almost trivial, but they allow the notion of a vector to be generalized to a wide variety of other systems. For example, consider the set of real numbers. Clearly, they obey Equations (5.9) and (5.10) (as long as we restrict $c$ to be a real number). Hence, the set of all real numbers is a vector space with each real number acting as a vector. This result becomes obvious when we realize that the real numbers are just equivalent to one-component vectors, $(r_1)$; the set of real numbers forms a one-dimensional vector space.

A less obvious result is that the set of functions $f(x)$ can also be treated as a vector space. Clearly, the sum of two functions $f(x)$ and $g(x)$ is also a function:

$$f(x) + g(x) = h(x)$$

and we can multiply a function by a real or complex number to get another function. In fact, the set of functions behaves like an infinite-dimensional vector space! To see this, consider first the three-dimensional vector $\mathbf{r} = (1, 4, 9)$. We can display this vector in a rather strange way, plotting $r_i$ as a function of $i$ (see Figure 5.1). Similarly, an $n$-dimensional vector $(r_1, r_2, \ldots, r_n)$ can be displayed in a plot of $r_i$ as a function of $i$. An example for $n = 7$ is shown in Figure 5.2. Consider what happens to Figure 5.2 as we take the limit $n \to \infty$. The points become denser, merging into a continuous curve: a function! (See Figure 5.3.) This is a plausibility argument rather than a rigorous proof, but it can be shown rigorously that a suitably defined set of functions is, in fact, an infinite-dimensional vector space.

## Inner Products

As we have seen, one of the standard operations on a set of three-dimensional vectors is the dot product: $\mathbf{r} \cdot \mathbf{s} = r_1 s_1 + r_2 s_2 + r_3 s_3$. It is

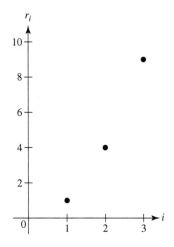

Fig. 5.1 A plot of $r_i$ as a function of $i$ for the three-dimensional vector $\mathbf{r} = (1, 4, 9)$.

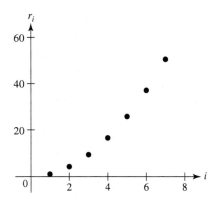

Fig. 5.2 A plot of $r_i$ as a function of $i$ for a 7-dimensional vector $\mathbf{r} = (r_1, r_2, \ldots, r_7)$.

trivial to generalize this to other finite-dimensional vectors, but what happens when we have an infinite-dimensional vector space, such as a set of functions?

In this case we need to define a more abstract concept called an *inner product*. Note that the dot product is a function which takes two vectors, $\mathbf{r}$ and $\mathbf{s}$, and produces a real number. Now suppose we have two vectors, $\psi$ and $\phi$, in an abstract vector space. In analogy with the dot product, we take the inner product to be a function which takes $\psi$ and $\phi$ as inputs and produces a real or complex number as the output. (As already noted,

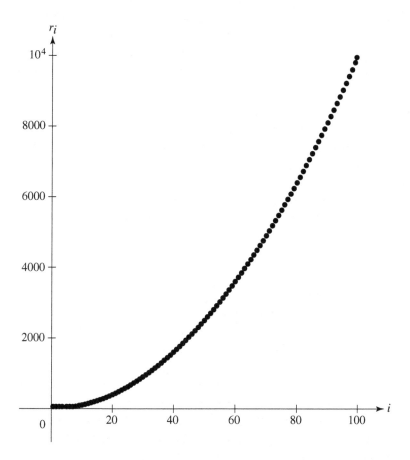

Fig. 5.3   A function is the limiting case of an $n$-dimensional vector when $n \to \infty$.

quantum mechanics makes use of *complex* vector spaces, so our inner products will produce complex numbers.) The inner product of $\psi$ and $\phi$ will be denoted $(\psi|\phi)$, such that

$$(\psi|\phi) = c$$

where $c$ is a complex number. Note that, in the world of mathematics, there is no standard notation for the inner product. In addition to the notation introduced here, $(\psi, \phi)$, $\langle \psi, \phi \rangle$, and $\langle \psi | \phi \rangle$ are also used. Later, we will encounter yet another notation for inner products, called the *Dirac notation*, which diverges in some respects from the standard mathematical notation for inner products, but which is ultimately equivalent to it.

If $\psi$, $\phi$, and $\theta$ are vectors in an arbitrary complex vector space, and $c$ is a complex number, then inner products have the following properties:

$$(\psi + \phi | \theta) = (\psi | \theta) + (\phi | \theta) \tag{5.11}$$

$$(\psi | c\phi) = c(\psi | \phi) \tag{5.12}$$

$$(\psi | \phi) = (\phi | \psi)^* \tag{5.13}$$

$$(\psi | \psi) \geq 0 \tag{5.14}$$

where, as usual, the $*$ denotes complex conjugation. Note that Equations (5.12) and (5.13) together imply that

$$(c\psi | \phi) = c^*(\psi | \phi) \tag{5.15}$$

In physics, the standard convention is given by Equations (5.12) and (5.15); mathematicians, on the other hand, use the reverse convention: $(c\psi | \phi) = c(\psi | \phi)$ and $(\psi | c\phi) = c^*(\psi | \phi)$. Needless to say, we will use the physics convention throughout.

For ordinary real three-dimensional vectors, the dot product satisfies Equations (5.11)–(5.14) and is therefore an inner product (see Problem 5.3). However, we are most interested in the vector space of functions, so we need to define an inner product for this case. We argued at the beginning of this section that a function resembles an $n$-dimensional vector in the limit where $n$ goes to infinity. We can use this analogy to derive a reasonable inner product for functions. For two $n$-dimensional vectors, $\mathbf{r} = (r_1, r_2, \ldots, r_n)$ and $\mathbf{s} = (s_1, s_2, \ldots, s_n)$, the $n$-dimensional inner product is $\mathbf{r} \cdot \mathbf{s} = r_1 s_1 + r_2 s_2 + \cdots + r_n s_n$. Now consider two real-valued functions, $f(x)$ and $g(x)$. The quantity analogous to the $n$-dimensional inner product is $f(x_1)g(x_1) + f(x_2)g(x_2) + \cdots + f(x_n)g(x_n)$, in the limit where $n \to \infty$. Taking the continuum limit, the inner product should be $\int f(x)g(x)\, dx$. For complex-valued functions, in order to satisfy Equations (5.13) and (5.14), the inner product becomes $\int f(x)^* g(x)\, dx$.

More rigorously, consider the set of complex-valued functions in three dimensions. For any two such functions, $f(\mathbf{r})$ and $g(\mathbf{r})$, the inner product will be defined as

$$(f | g) = \int f(\mathbf{r})^* g(\mathbf{r})\, d^3\mathbf{r} \tag{5.16}$$

where the integral is taken over all of three-dimensional space. For the frequently encountered case of functions in one dimension, this reduces to

$$(f | g) = \int_{x=-\infty}^{\infty} f(x)^* g(x)\, dx \tag{5.17}$$

Although we used a rough analogy to finite-dimensional vectors to motivate these expressions, these definitions do not depend on this argument.

All that is necessary for our expressions in Equations (5.16) and (5.17) to represent valid inner products is that they satisfy Equations (5.11)–(5.14). This is indeed the case (see Problem 5.3).

---

**Example 5.4. An Inner Product**

Consider two functions, $\psi(x) = e^{-x^2/2}$ and $\phi(x) = xe^{-x^2/2}$. What is $(\psi|\phi)$? We have

$$(\psi|\phi) = \int_{x=-\infty}^{\infty} \psi(x)^* \phi(x)\, dx$$

$$= \int_{x=-\infty}^{\infty} (e^{-x^2/2})^* xe^{-x^2/2}\, dx$$

$$= \int_{x=-\infty}^{\infty} xe^{-x^2}\, dx$$

but note that $xe^{-x^2}$ is an odd function [i.e., a function for which $f(-x) = -f(x)$], so the integral from $-\infty$ to 0 cancels the integral from 0 to $\infty$. Hence $(\psi|\phi) = 0$.

---

Some of the results from Chapter 3 can now be cast in a much more compact form, using inner product notation. For instance, the requirement that the wave function $\psi$ be normalized can now be written as

$$(\psi|\psi) = 1$$

In analogy with three-dimensional vectors, normalization means that the "length" of an infinite-dimensional vector is 1.

Similarly, the expectation value of an operator $O$ can now be expressed in the compact form:

$$\langle o \rangle = (\psi|O\psi)$$

*Adjoint and Hermitian Operators*

The inner product derived in the previous section can be used to pair up every operator with a second operator called its *adjoint operator*. If $A$ is an operator, then the adjoint operator of $A$ is written as $A^\dagger$, and it satisfies the equation

$$(\phi|A\psi) = (A^\dagger\phi|\psi) \tag{5.18}$$

for *all* $\phi$ and $\psi$.

## Example 5.5. The Adjoint of the Derivative Operator

Consider the one-dimensional derivative operator $D$; what is its adjoint? We can write, for arbitrary $\phi$ and $\psi$

$$(\phi|D\psi) = \int_{x=-\infty}^{\infty} \phi(x)^* \frac{d\psi}{dx} \, dx$$

Integration by parts gives

$$(\phi|D\psi) = [\phi(x)^*\psi(x)]_{-\infty}^{\infty} - \int_{x=-\infty}^{\infty} \frac{d\phi}{dx}^* \psi(x) \, dx \qquad (5.19)$$

If $\phi$ and $\psi$ represent wave functions for physical particles, we can assume that $\phi \to 0$ and $\psi \to 0$ as $x \to \pm\infty$, so the first term in Equation (5.19) vanishes, and the second term simplifies to

$$(\phi|D\psi) = \int_{x=-\infty}^{\infty} (-D\phi)^* \psi$$
$$= (-D\phi|\psi)$$

Hence, from the definition of the adjoint operator, $D^\dagger = -D$.

---

From the definition of the adjoint operator (Equation (5.18)), the following general properties can be derived:

$$(cP)^\dagger = c^* P^\dagger \qquad (5.20)$$
$$(P+Q)^\dagger = P^\dagger + Q^\dagger$$
$$(PQ)^\dagger = Q^\dagger P^\dagger \qquad (5.21)$$
$$(P^\dagger)^\dagger = P$$

where $c$ is a complex number, and $P$ and $Q$ are arbitrary operators. The only nonintuitive result here is the reversal in the order of the operators $P$ and $Q$ in Equation (5.21); this arises because of the way that operators are "peeled away" to form the adjoint:

$$(\phi|P[Q\psi]) = (P^\dagger\phi|Q\psi)$$
$$= (Q^\dagger[P^\dagger\phi]|\psi)$$

so $(PQ)^\dagger = Q^\dagger P^\dagger$.

Note that for the operator corresponding to multiplication by the complex number $c$, Equation (5.20) implies that

$$c^\dagger = c^*$$

It is possible for an operator to be equal to its own adjoint, i.e., $O^\dagger = O$. Such operators are called *self-adjoint* or *Hermitian*, and they occupy a special place in quantum mechanics.

---

**Example 5.6. The Position Operator Is Hermitian**

As an example, we now show that the one-dimensional position operator $X$ is Hermitian.

We can write, for arbitrary $\phi$ and $\psi$,

$$
\begin{aligned}
(\phi|X\psi) &= \int_{x=-\infty}^{\infty} \phi(x)^* x\psi(x)\,dx \\
&= \int_{x=-\infty}^{\infty} [x\phi(x)]^* \psi(x)\,dx \\
&= (X\phi|\psi)
\end{aligned}
$$

so $X^\dagger = X$, and $X$ is Hermitian.

---

The reason that Hermitian operators are important in quantum mechanics is that both the expectation values and the eigenvalues of Hermitian operators are real. Since observable quantities are always real, *we require that the operators corresponding to observables be Hermitian.*

Consider first the expectation value of a Hermitian operator $Q$: $\langle q \rangle = (\psi|Q\psi)$. In order for $\langle q \rangle$ to be real, it must equal its own complex conjugate. From Equation (5.13), we have

$$\langle q \rangle^* = (Q\psi|\psi)$$

From the definition of the adjoint operator, this is equivalent to

$$\langle q \rangle^* = (\psi|Q^\dagger \psi)$$

and, since $Q$ is Hermitian, we get

$$\langle q \rangle^* = (\psi|Q\psi) = \langle q \rangle$$

Hence, $\langle q \rangle$ is real.

Now suppose that $\psi$ is an eigenfunction of a Hermitian operator $Q$ with eigenvalue $q$. Since $Q$ is Hermitian,

$$(\psi|Q\psi) = (Q\psi|\psi)$$

which implies

$$
\begin{aligned}
(\psi|q\psi) &= (q\psi|\psi) \\
q(\psi|\psi) &= q^*(\psi|\psi) \\
q &= q^*
\end{aligned}
$$

so $q$ must be real. Hence, Hermitian operators have both real expectation values and real eigenvalues. We have already shown that the position operator is Hermitian; Problem 5.7 will show that several other operators corresponding to observables are also Hermitian.

## Basis Sets

Among the collection of all three-dimensional vectors, there are three vectors that occupy a special place: the unit vectors in the $x$, $y$, and $z$ directions, denoted $\hat{x}$, $\hat{y}$, and $\hat{z}$. This collection of three vectors $\hat{x}$, $\hat{y}$, and $\hat{z}$ has several important properties. Any three-dimensional vector $\mathbf{r}$ can be expressed as a *linear combination* of $\hat{x}$, $\hat{y}$, and $\hat{z}$, i.e., as the sum of the product of each of these three vectors with a different real number:

$$\mathbf{r} = c_1 \hat{x} + c_2 \hat{y} + c_3 \hat{z} \qquad (5.22)$$

This result indicates that we have "enough" vectors to do the job of decomposing every three-dimensional vector. However, we also don't have "too many" vectors in the sense that no one vector in our set of three can be expressed in terms of the other two. One can never write, for example,

$$\hat{z} = c_1 \hat{x} + c_2 \hat{y} \qquad (5.23)$$

If $\hat{z}$ could be expressed in this way, then $\hat{z}$ would be irrelevant; Equation (5.23) could just be substituted into Equation (5.22) and $\hat{z}$ could be eliminated from Equation (5.22). In an abstract vector space, a subset of vectors with these two properties (every vector can be represented as a linear combination of the vectors in the subset, and no one vector in the subset can be expressed as a linear combination of the rest) is called a *basis*. In general, an $n$-dimensional vector space will have $n$ distinct basis vectors.

The basis set $\hat{x}$, $\hat{y}$, and $\hat{z}$ has two other desirable properties. First of all, each vector has unit length, and second, each vector is perpendicular to the other two: $\hat{x} \cdot \hat{y} = \hat{x} \cdot \hat{z} = \hat{y} \cdot \hat{z} = 0$. A basis with these two additional properties is called an *orthonormal* basis.

In general, we can find a basis set for any vector space, but the basis will not be unique. In three dimensions, for example, there are an infinite number of orthonormal bases, obtained by rotating the $\hat{x}$, $\hat{y}$, $\hat{z}$ basis. For instance, another perfectly acceptable orthonormal basis is $(1/\sqrt{2})(\hat{x} + \hat{y})$, $(1/\sqrt{2})(-\hat{x} + \hat{y})$, $\hat{z}$.

Even infinite-dimensional vector spaces have basis sets. As an example, consider the set of functions $f(x)$ which are periodic with period $2\pi$. Thus, for these functions, $f(x + 2\pi) = f(x)$. (The inner product for this vector space will be defined by integration from 0 to $2\pi$ rather than $-\infty$ to $\infty$.) Since this is an infinite-dimensional vector space, its basis set will contain an infinite set of functions.

Then a familiar example of a basis for this vector space is the set of trigonometric functions:

$$\frac{1}{\sqrt{\pi}}\sin x, \frac{1}{\sqrt{\pi}}\sin(2x), \ldots, \frac{1}{\sqrt{\pi}}\sin(nx), \ldots$$

$$\frac{1}{\sqrt{\pi}}\cos x, \frac{1}{\sqrt{\pi}}\cos(2x), \ldots, \frac{1}{\sqrt{\pi}}\cos(nx), \ldots$$

Any periodic function can be written as a sum of these functions:

$$f(x) = \sum_{n=0}^{\infty} \frac{A_n}{\sqrt{\pi}}\sin(nx) + \sum_{n=0}^{\infty} \frac{B_n}{\sqrt{\pi}}\cos(nx) \qquad (5.24)$$

called a *Fourier series*. Further, this set of trigonometric functions forms an orthonormal basis, since

$$\left(\frac{1}{\sqrt{\pi}}\sin mx \,\middle|\, \frac{1}{\sqrt{\pi}}\sin nx\right) = \left(\frac{1}{\sqrt{\pi}}\cos mx \,\middle|\, \frac{1}{\sqrt{\pi}}\cos nx\right) = 0, \ m \neq n$$

$$= 1, \ m = n$$

and

$$\left(\frac{1}{\sqrt{\pi}}\sin mx \,\middle|\, \frac{1}{\sqrt{\pi}}\cos nx\right) = 0$$

for all $m$ and $n$.

Although any periodic function can be written as the sum of trigonometric basis functions, as in Equation (5.24), we need a procedure for determining the constants $A_n$ and $B_n$ which multiply these trigonometric functions. Again, an analogy to three-dimensional vectors is instructive. Consider an arbitrary three-dimensional vector $\mathbf{r}$, expanded out as

$$\mathbf{r} = c_1\hat{x} + c_2\hat{y} + c_3\hat{z} \qquad (5.25)$$

and suppose that we need to determine $c_1$, $c_2$, and $c_3$. Since our basis is orthonormal, we can take the dot product of $\mathbf{r}$ with $\hat{x}$ to obtain

$$\mathbf{r}\cdot\hat{x} = c_1(\hat{x}\cdot\hat{x}) + c_2(\hat{y}\cdot\hat{x}) + c_3(\hat{z}\cdot\hat{x})$$

$$= c_1(1) + c_2(0) + c_3(0)$$

$$= c_1$$

So $c_1 = \mathbf{r}\cdot\hat{x}$. Similarly, $c_2 = \mathbf{r}\cdot\hat{y}$ and $c_3 = \mathbf{r}\cdot\hat{z}$. This result shows that $c_1$, $c_2$, and $c_3$ are the magnitudes of the projection of $\mathbf{r}$ onto the $x$, $y$, and $z$ axes, respectively. Therefore, an alternate way of writing Equation (5.25) is

$$\mathbf{r} = (\mathbf{r}\cdot\hat{x})\hat{x} + (\mathbf{r}\cdot\hat{y})\hat{y} + (\mathbf{r}\cdot\hat{z})\hat{z} \qquad (5.26)$$

Now suppose we take an arbitrary periodic function $f(x)$ and wish to expand it out in the form of Equation (5.24). In analogy to Equation (5.26), we write

$$f(x) = \left(f|\frac{1}{\sqrt{\pi}}\sin x\right)\frac{1}{\sqrt{\pi}}\sin x + \left(f|\frac{1}{\sqrt{\pi}}\sin 2x\right)\frac{1}{\sqrt{\pi}}\sin 2x + \cdots$$

$$+ \left(f|\frac{1}{\sqrt{\pi}}\sin nx\right)\frac{1}{\sqrt{\pi}}\sin nx + \cdots + \left(f|\frac{1}{\sqrt{\pi}}\cos x\right)\frac{1}{\sqrt{\pi}}\cos x$$

$$+ \left(f|\frac{1}{\sqrt{\pi}}\cos 2x\right)\frac{1}{\sqrt{\pi}}\cos 2x + \cdots + \left(f|\frac{1}{\sqrt{\pi}}\cos nx\right)\frac{1}{\sqrt{\pi}}\cos nx + \cdots$$

Thus, the Fourier coefficients $A_n$ and $B_n$ in Equation (5.24) are the inner products of $f(x)$ with the appropriate basis functions, e.g.,

$$A_n = \left(f|\frac{1}{\sqrt{\pi}}\sin nx\right)$$

$$= \frac{1}{\sqrt{\pi}}\int_{x=0}^{2\pi} f(x)\sin nx\, dx$$

and

$$B_n = \left(f|\frac{1}{\sqrt{\pi}}\cos nx\right)$$

$$= \frac{1}{\sqrt{\pi}}\int_{x=0}^{2\pi} f(x)\cos nx\, dx$$

Of course, the trigonometric functions are not the only possible basis set, and this is where the connection to quantum mechanics becomes relevant. Recall that for an arbitrary potential $V(x)$, we can find a set of solutions $\psi_n(x)$ for the Schrödinger equation. This set of solutions will itself form a basis set, so that an arbitrary function can be expressed as a linear combination of these solutions:

$$f(x) = \sum_n c_n \psi_n(x)$$

The usefulness of this sort of expansion will become apparent later.

**PROBLEMS**

**5.1** Verify the commutator properties given in Equations (5.4)–(5.7), i.e., for any operators $A$, $B$, and $C$, show that

$$[A, B] = -[B, A]$$

$$[A, A] = 0$$

$$[A + B, C] = [A, C] + [B, C]$$

$$[A, BC] = [A, B]C + B[A, C]$$

**5.2** Consider a particle moving in three dimensions. Is it possible for the particle to be in a state of definite $p_x$ and $y$, i.e., can both its $y$-coordinate and its momentum in the $x$ direction be known at the same time?

**5.3** (a) Verify that the ordinary dot product for three-dimensional vectors satisfies all of the properties of an inner product, given by Equations (5.11)–(5.14).
(b) Verify that the inner product for complex-valued, three-dimensional functions defined in Equation (5.16) satisfies Equations (5.11)–(5.14).

**5.4** (a) The operators $A$, $B$, and $C$ are all Hermitian with $[A, B] = C$. Show that $C = 0$.
(b) The operators $A$ and $B$ are both Hermitian with $[A, B] = i\hbar$. Determine whether or not $AB$ is a Hermitian operator.

**5.5** The one-dimensional parity operator $\Pi$ is defined by $\Pi\psi(x) = \psi(-x)$. In other words, $\Pi$ changes $x$ into $-x$ everywhere in the function.
(a) Is $\Pi$ a Hermitian operator?
(b) For what potentials $V(x)$ is it possible to find a set of wavefunctions which are eigenfunctions of the parity operator and solutions of the one-dimensional time-independent Schrödinger equation?

**5.6** (a) Let $Q$ be an operator which is not a function of time, and let $H$ be the Hamiltonian operator. Show that

$$i\hbar\frac{\partial}{\partial t}\langle q \rangle = \langle [Q, H] \rangle$$

Here $\langle q \rangle$ is the expectation value of $Q$ for an arbitrary time-dependent wave function $\Psi$, which is not necessarily an eigenfunction of $H$, and $\langle [Q, H] \rangle$ is the expectation value of the commutator of $Q$ and $H$ for the same wave function. This result is known as *Ehrenfest's theorem*.
(b) Use this result to show that

$$\frac{\partial}{\partial t}\langle p \rangle = \left\langle -\frac{\partial V}{\partial x} \right\rangle$$

What is the classical analog of this equation?

**5.7** (a) Show that the one-dimensional momentum operator is Hermitian.

(b) Use this result to show that the one-dimensional Hamiltonian operator $H$ with potential $V(x)$ is Hermitian. What (reasonable) assumption must be made about $V(x)$ to derive this result?

**5.8** Suppose that the operator $T$ is defined by $T = \alpha Q^\dagger Q$, where $\alpha$ is a real number, and $Q$ is an operator (not necessarily Hermitian). Show that $T$ is Hermitian.

**5.9** Determine all potentials $V(x)$ for which it is possible to find a set of solutions of the time-independent Schrödinger equation which are also eigenfunctions of the position operator $X$, or else show that no such potentials exist.

**5.10** Suppose that two operators $P$ and $Q$ satisfy the commutation relation

$$[P, Q] = Q$$

Suppose that $\psi$ is an eigenfunction of the operator $P$ with eigenvalue $p$. Show that $Q\psi$ is also an eigenfunction of $P$, and find its eigenvalue.

**5.11** The operator $F$ is defined by $F\psi(x) = \psi(x+a) + \psi(x-a)$, where $a$ is a nonzero constant. Determine whether or not $F$ is a Hermitian operator.

Chapter 6

# Solutions of the three-dimensional time-independent Schrödinger equation

In Chapter 4 we examined solutions of the one-dimensional time-independent Schrödinger equation,

$$-\frac{\hbar^2}{2m}\frac{d^2\psi}{dx^2} + V(x)\psi = E\psi$$

Of course, the real world is three-dimensional, so in this chapter we will solve the full three-dimensional, time-independent Schrödinger equation. Along the way, we will need to understand the behavior of angular momentum in quantum mechanics. The crowning achievement of the solution of the three-dimensional Schrödinger equation (and of this chapter) will be a description of the hydrogen atom.

The three-dimensional Schrödinger equation is:

$$-\frac{\hbar^2}{2m}\nabla^2\Psi(\mathbf{r},t) + V(\mathbf{r})\Psi(\mathbf{r},t) = i\hbar\frac{\partial\Psi(\mathbf{r},t)}{\partial t}$$

We will assume throughout this chapter that the wave function represents a state of definite energy $E$, so that the Schrödinger equation can be written in the time-independent form:

$$-\frac{\hbar^2}{2m}\nabla^2\psi(\mathbf{r}) + V(\mathbf{r})\psi(\mathbf{r}) = E\psi(\mathbf{r}) \qquad (6.1)$$

In order to solve Equation (6.1), we need to choose a particular coordinate system, e.g., rectangular, cylindrical, or spherical. Of course, the physical solution does not depend on the coordinate system we choose; instead, the correct choice of coordinates can simplify the form of the solution by taking advantage of the symmetries in the problem. For example, a central force, defined as a force which points radially inward or outward, corresponds to a potential that is a function only of radial distance $r$. (An example is the potential experienced by an electron in a hydrogen atom.) For such a

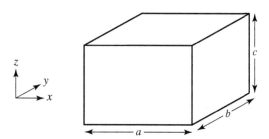

Fig. 6.1   A particle of mass $m$ is confined to a rectangular box with sides of length $a$, $b$, and $c$.

potential, the spherical coordinate system is most appropriate. However, as a warm-up, we will first examine a problem which can be solved in a simple way in rectangular coordinates.

## 6.1   Solution in Rectangular Coordinates

Consider a particle of mass $m$, confined to move in a rectangular box with sides of length $a$, $b$, and $c$ (Figure 6.1). We choose a rectangular coordinate system with $\hat{x}$ along the side of length $a$, $\hat{y}$ along the side of length $b$, and $\hat{z}$ along the side of length $c$, and one corner of the box at the origin. Then the potential is simply

$$V(\mathbf{r}) = 0, \, 0 < x < a$$
$$0 < y < b$$
$$0 < z < c$$

with infinite potential barriers at $x = 0$, $x = a$, $y = 0$, $y = b$, and $z = 0$, $z = c$. This is the three-dimensional analog of the infinite one-dimensional square-well potential discussed in Chapter 4.

Inside the box, where $V = 0$, Equation (6.1) becomes

$$-\frac{\hbar^2}{2m}\nabla^2\psi(\mathbf{r}) = E\psi(\mathbf{r})$$

which can be written as

$$\frac{\partial^2\psi}{\partial x^2} + \frac{\partial^2\psi}{\partial y^2} + \frac{\partial^2\psi}{\partial z^2} = -\frac{2mE}{\hbar^2}\psi \qquad (6.2)$$

As in Chapter 3, we use separation of variables. (The solution here is slightly more complicated than the separation of variables solutions in

Chapter 3, since we now have three independent variables.) We take a trial solution of the form

$$\psi(x, y, z) = \psi_1(x)\psi_2(y)\psi_3(z) \tag{6.3}$$

In this equation, $\psi_1(x)$ is an unknown function of $x$ and is independent of $y$ and $z$. Similarly, $\psi_2(y)$ is an unknown function of $y$ which is independent of $x$ and $z$, and so on. Our job is then to find the functions $\psi_1$, $\psi_2$, and $\psi_3$. Substituting Equation (6.3) into Equation (6.2) gives

$$\frac{\partial^2 \psi_1(x)}{\partial x^2}\psi_2(y)\psi_3(z) + \psi_1(x)\frac{\partial^2 \psi_2(y)}{\partial y^2}\psi_3(z) + \psi_1(x)\psi_2(y)\frac{\partial^2 \psi_3(z)}{\partial z^2}$$

$$= -\frac{2mE}{\hbar^2}\psi_1(x)\psi_2(y)\psi_3(z)$$

and dividing both sides by $\psi_1(x)\psi_2(y)\psi_3(z)$ yields

$$\frac{1}{\psi_1(x)}\frac{\partial^2 \psi_1(x)}{\partial x^2} + \frac{1}{\psi_2(y)}\frac{\partial^2 \psi_2(y)}{\partial y^2} + \frac{1}{\psi_3(z)}\frac{\partial^2 \psi_3(z)}{\partial z^2} = -\frac{2mE}{\hbar^2} \tag{6.4}$$

Now consider what happens if we move the second and third terms on the left-hand side over to the right-hand side:

$$\frac{1}{\psi_1(x)}\frac{\partial^2 \psi_1(x)}{\partial x^2} = -\frac{2mE}{\hbar^2} - \frac{1}{\psi_2(y)}\frac{\partial^2 \psi_2(y)}{\partial y^2} - \frac{1}{\psi_3(z)}\frac{\partial^2 \psi_3(z)}{\partial z^2} \tag{6.5}$$

The left-hand side of this equation is a function only of $x$ and is independent of $y$ and $z$. On the other hand, the right-hand side is a function only of $y$ and $z$ and is independent of $x$. There is only one function that satisfies both of these requirements: a constant, which is independent of $x$, $y$, and $z$. Hence, we can set both the left-hand side and the right-hand side of Equation (6.5) equal to some (still to be determined) constant, which we will call $C_x$:

$$\frac{1}{\psi_1(x)}\frac{d^2 \psi_1(x)}{dx^2} = C_x \tag{6.6}$$

But now note that we can again begin with Equation (6.4) and, instead of leaving the first term on the left-hand side, we can leave the second or third terms. Leaving the second term on the left-hand side, a similar argument gives

$$\frac{1}{\psi_2(y)}\frac{d^2 \psi_2(y)}{dy^2} = C_y \tag{6.7}$$

where $C_y$ is another undetermined constant, while leaving the third term on the left-hand side produces

$$\frac{1}{\psi_3(z)}\frac{d^2 \psi_3(z)}{dz^2} = C_z \tag{6.8}$$

Thus, we have transformed a single partial differential equation (Equation (6.4)) into three ordinary differential equations. Note that $C_x$, $C_y$, and $C_z$ are not completely independent. Adding Equations (6.6), (6.7), and (6.8), and comparing with Equation (6.4), we get

$$C_x + C_y + C_z = -\frac{2mE}{\hbar^2}$$

It is possible to put Equations (6.6)–(6.8) into a form that we have already seen. Define the new constants $E_x$, $E_y$, and $E_z$ to be given by $C_x = -2mE_x/\hbar^2$, $C_y = -2mE_y/\hbar^2$, and $C_z = -2mE_z/\hbar^2$. Note that $E_x$, $E_y$, and $E_z$ have no physical significance, but their sum does; it is the total energy:

$$E_x + E_y + E_z = E$$

If we rewrite Equations (6.6)–(6.8) in terms of $E_x$, $E_y$, and $E_z$, instead of $C_x$, $C_y$, and $C_z$, we get the three equations

$$\frac{d^2\psi_1(x)}{d^2x} + \frac{2mE_x}{\hbar^2}\psi_1(x) = 0 \qquad (6.9)$$

$$\frac{d^2\psi_2(y)}{d^2y} + \frac{2mE_y}{\hbar^2}\psi_2(y) = 0 \qquad (6.10)$$

$$\frac{d^2\psi_3(z)}{d^2z} + \frac{2mE_z}{\hbar^2}\psi_3(z) = 0 \qquad (6.11)$$

supplemented by the boundary condition that the wave function must vanish on the sides of the box, which gives $\psi_1(x) = 0$ at $x = 0$ and $x = a$, $\psi_2(y) = 0$ at $y = 0$ and $y = b$, and $\psi_3(z) = 0$ at $z = 0$ and $z = c$. But we have seen equations of this form before. Equations (6.9), (6.10), and (6.11) all have the form of the Schrödinger equation for a one-dimensional infinite square well (Equation (4.28)) with the same boundary conditions as the infinite one-dimensional square well. Hence, the solutions to these equations are the same as the solutions previously derived in Chapter 4, namely,

$$\psi_1(x) \propto \sin\left(\frac{n_x\pi x}{a}\right)$$

with corresponding energy

$$E_n = \frac{\hbar^2\pi^2}{2ma^2}n_x^2$$

and similarly for $\psi_2(y)$ and $\psi_3(z)$. Using Equation (6.3) to reassemble the wave function, we get

$$\psi(x,y,z) = A\sin\left(\frac{n_x\pi x}{a}\right)\sin\left(\frac{n_y\pi y}{b}\right)\sin\left(\frac{n_z\pi z}{c}\right)$$

where $n_x$, $n_y$, and $n_z$ can each take on positive integer values, and $A$ is the normalization constant. (This normalization constant is $A = \sqrt{8/V}$, where $V$ is the volume of the box; see Problem 6.1.) The energy corresponding to a given $n_x$, $n_y$, and $n_z$ is

$$E = E_x + E_y + E_z$$

$$= \left(\frac{\hbar^2 \pi^2}{2m}\right)\left(\frac{n_x^2}{a^2} + \frac{n_y^2}{b^2} + \frac{n_z^2}{c^2}\right)$$

Consider what happens for the special case of a cube of side $a$. For this case, the wave function with quantum numbers $n_x$, $n_y$, and $n_z$ is

$$\psi(x, y, z) = A \sin\left(\frac{n_x \pi x}{a}\right) \sin\left(\frac{n_y \pi y}{a}\right) \sin\left(\frac{n_z \pi z}{a}\right)$$

with corresponding energy levels

$$E = \left(\frac{\hbar^2 \pi^2}{2ma^2}\right)(n_x^2 + n_y^2 + n_z^2)$$

Now we see an interesting new phenomenon. Consider the two states $n_x = 1$, $n_y = 1$, $n_z = 2$ and $n_x = 1$, $n_y = 2$, and $n_z = 1$. These correspond to two different wave functions, but they have the same energy. This illustrates the phenomenon of *degeneracy*. Two different states are said to be degenerate if they have the same energy but different wave functions. (As noted in the previous chapter, degeneracy in linear algebra occurs when two different eigenvectors have the same eigenvalue. In this case the two different wave functions are both eigenfunctions of the Hamiltonian $H$, and they have the same eigenvalue $E$.) Note that our two wave functions in this case are related by an interchange of the $x$- and $y$-axes, which leaves the potential unchanged. Degeneracies often arise from this sort of symmetry.

## 6.2 Angular Momentum

Before moving on to examine quantum mechanical systems with spherical symmetry, it is necessary to derive a quantum mechanical treatment of angular momentum. Recall that for a classical particle with momentum $\mathbf{p}$ at position $\mathbf{r}$ relative to the origin, the angular momentum is a vector given by the cross product of $\mathbf{r}$ and $\mathbf{p}$ (Figure 6.2):

$$\mathbf{L} = \mathbf{r} \times \mathbf{p}$$

We now need to derive a quantum mechanical operator corresponding to $\mathbf{L}$. Note that we already have an operator $\mathbf{R}$ corresponding to the position

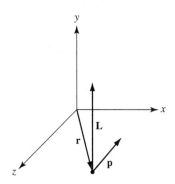

Fig. 6.2   In classical mechanics, the angular momentum **L** for a particle with momentum **p** at the position **r** relative to the origin is **L** = **r**×**p**.

**r**, namely multiplication by **r**, and we have an operator **P** corresponding to the momentum **p**, namely $-i\hbar\nabla$. (We will use the lowercase symbols **r** and **p** to refer to the physical position and momentum, and the uppercase symbols **R** and **P** to refer to the corresponding operators.) Hence, the operator corresponding to angular momentum should simply be

$$\mathbf{L} = \mathbf{R}\times\mathbf{P} \qquad (6.12)$$

In practice, however, it is easier to break **L** down into its components, and to calculate the operators corresponding to the $x$, $y$, and $z$ components of angular momentum, namely, $L_x$, $L_y$, and $L_z$. Then Equation (6.12) gives

$$L_x = YP_z - ZP_y$$
$$L_y = ZP_x - XP_z$$
$$L_z = XP_y - YP_x$$

All three of these operators are Hermitian.

---

**Example 6.1. Show that the Operator $L_z$ is Hermitian**

The adjoint of $L_z$ is

$$L_z^\dagger = (XP_y)^\dagger - (YP_x)^\dagger$$

The rules for taking adjoints of sums and products of operators (Chapter 5) give

$$L_z^\dagger = P_y^\dagger X^\dagger - P_x^\dagger Y^\dagger$$

But the momentum and position operators are Hermitian, so that

$$L_z^\dagger = P_y X - P_x Y$$

Now recall that $P_y$ commutes with $X$, and $P_x$ commutes with $Y$, so that

$$L_z^\dagger = X P_y - Y P_x = L_z$$

so $L_z$ is Hermitian. The argument is similar for the other components of **L** (Problem 6.5).

---

Now consider how well we can measure the angular momentum of a particle. Suppose that we have a particle for which we would like to measure the angular momentum exactly. For this to be possible, the particle must be in an eigenstate of each component of **L**, i.e., $L_x$, $L_y$, and $L_z$. However, this is possible only if $L_x$, $L_y$, and $L_z$ all commute with each other. Consider, for example, whether or not $L_x$ and $L_y$ commute with each other. We have

$$[L_x, L_y] = [Y P_z - Z P_y, Z P_x - X P_z] \tag{6.13}$$

To simplify this expression, we need to know all commutators of the form $[P_z, X]$, $[P_y, Z]$, and so on. In the previous chapter, we derived the commutator for the one-dimensional momentum and position operators, which corresponds to

$$[X, P_x] = i\hbar$$

and similarly for the $y$ and $z$ components. But what happens if the position and momentum in this equation correspond to different components, e.g., does $P_y$ commute with $X$? In general, all of the position and momentum operators commute with each other as long they correspond to different components. For instance,

$$[X, P_y]\psi = x(-i\hbar)\frac{\partial\psi}{\partial y} - (-i\hbar)\frac{\partial}{\partial y}(x\psi) = 0$$

A shorthand for this result that encapsulates all of these commutation relations is

$$[R_\alpha, P_\beta] = i\hbar\delta_{\alpha\beta}$$

where $x$, $y$, and $z$ correspond to $\alpha = 1, 2, 3$ and $\beta = 1, 2, 3$, and $\delta_{\alpha\beta}$ is called the *Kronecker delta*; it has the value

$$\delta_{\alpha\beta} = 1, \ \alpha = \beta$$
$$= 0, \ \alpha \neq \beta$$

Furthermore, the different components of position and momentum all commute with each other, i.e.,

$$[X, Y] = [X, Z] = [Y, Z] = 0$$

and

$$[P_x, P_y] = [P_x, P_z] = [P_y, P_z] = 0$$

(see Problem 6.6).

The simplification of Equation (6.13) also requires the use of the identities (from Chapter 5),

$$[A + B, C] = [A, C] + [B, C]$$

and

$$[AB, C] = A[B, C] + [A, C]B$$

These identities allow Equation (6.13) to be reduced to

$$[L_x, L_y] = [YP_z, ZP_x] - [ZP_y, ZP_x] - [YP_z, XP_z] + [ZP_y, XP_z]$$

for which the second and third terms are zero (since they contain only factors which commute with each other), while the first and fourth terms reduce to

$$\begin{aligned}[L_x, L_y] &= YP_zZP_x - ZP_xYP_z + ZP_yXP_z - XP_zZP_y \\ &= YP_x[P_z, Z] - XP_y[P_z, Z] \\ &= -i\hbar(YP_x - XP_y) \\ &= i\hbar L_z\end{aligned}$$

Similarly,

$$[L_y, L_z] = i\hbar L_x$$

and

$$[L_z, L_x] = i\hbar L_y$$

Thus, none of the three components of **L** commutes with any of the others. This means that we cannot measure the full angular momentum exactly; instead, we can measure only a single component of the angular momentum! We will normally take this component to be $L_z$, i.e., we will look for states which are eigenfunctions of $L_z$. This choice is, however, arbitrary. It is a convenient choice because, in spherical coordinates, the angular variable $\theta$ is normally measured relative to the $z$-axis. But this does not mean that there is anything special about $L_z$; we could just as easily have chosen to measure $L_x$ or $L_y$ instead.

Can we measure anything else about the angular momentum other than the value of a single component? In fact, we can also measure the square of the magnitude of the angular momentum. The corresponding operator is $L^2 = L_x^2 + L_y^2 + L_z^2$, and this operator commutes with $L_z$:

$$[L^2, L_z] = [L_x^2 + L_y^2 + L_z^2, L_z]$$

$$= [L_x^2, L_z] + [L_y^2, L_z] + [L_z^2, L_z]$$

The last term is zero, and the other terms give

$$[L^2, L_z] = L_x[L_x, L_z] + [L_x, L_z]L_x + L_y[L_y, L_z] + [L_y, L_z]L_y$$

$$= -i\hbar L_x L_y - i\hbar L_y L_x + i\hbar L_y L_x + i\hbar L_x L_y$$

$$= 0$$

This means that a particle can be in an eigenstate of $L^2$ and $L_z$ simultaneously, i.e., we can measure the total magnitude squared of the angular momentum and its component in the $z$ direction.

But what result will we get if we actually do measure these two quantities? The answer is given by the eigenvalues of $L^2$ and $L_z$. One might think that these eigenvalues are fairly arbitrary and would depend on the particular wave function (as is, for example, the case for the eigenvalues of $H$, which correspond to energy). However, this is not the case. If $\psi$ is an eigenfunction of both $L^2$ and $L_z$, then the eigenvalues of these operators are actually restricted to a small class of possible values, which we will now calculate.

Before beginning this calculation, we need to point out that there are actually two kinds of angular momentum that are observed at the atomic level. The first is the familiar *orbital angular momentum*, which is equivalent, classically, to the orbit of a particle. An example of this is the angular momentum of the electron as it orbits the nucleus of an atom. However, particles, such as the electron and proton, also have an internal angular momentum called *spin angular momentum*. Naively, one can imagine these particles behaving like rotating balls. However, spin differs so much from our intuitive ideas of rotational motion that this analogy is quite crude. We will be mostly concerned with orbital angular momentum in this chapter and will deal with spin angular momentum in Chapter 8. However, the discussion which follows in this section will be kept as general as possible. When dealing with general angular momentum (as opposed to orbital or spin angular momentum), we will use the symbol **J**. Then **L** will always refer to orbital angular momentum, and spin angular momentum will be denoted by the operator **S**.

Assume, then, that the angular momentum operator **J** obeys the commutation relations we derived above, namely,

$$[J_x, J_y] = i\hbar J_z \tag{6.14}$$
$$[J_z, J_x] = i\hbar J_y$$
$$[J_y, J_z] = i\hbar J_x$$

Furthermore, $J^2$ commutes with each of the individual operators, $J_x$, $J_y$, and $J_z$. We will now assume that $\psi$ is an eigenfunction of $J^2$ and $J_z$, and we will calculate the possible eigenvalues of these two operators.

To perform this calculation, we will make use of a special set of operators called *ladder operators*. This will allow us to find the allowed eigenvalues of $J^2$ and $J_z$ without even calculating what the wave function looks like (although we will, in fact, find explicit forms for the orbital angular momentum eigenfunctions in the next section). The ladder operators are designated $J_+$ and $J_-$, and they are defined by

$$J_+ = J_x + iJ_y$$
$$J_- = J_x - iJ_y$$

These operators, $J_+$ and $J_-$, are also called the raising and lowering operators, respectively. Note that $J_+$ and $J_-$ are not Hermition operators, e.g., $J_+^\dagger = J_x^\dagger - iJ_y^\dagger = J_-$. Hence, these two operators do not correspond to observable quantities. Rather, they can be used to turn one eigenfunction of $J^2$ and $J_z$ into another eigenfunction with a different set of eigenvalues.

Suppose, for instance, that $\psi$ is an eigenfunction of $J^2$ and $J_z$ with eigenvalues $\alpha, \beta$, so that

$$J^2\psi = \alpha\psi$$
$$J_z\psi = \beta\psi$$

and suppose that $J_+$ operates on this eigenfunction to give some new wave function $\phi$:

$$J_+\psi = \phi$$

It is now possible to find $J^2\phi$ and $J_z\phi$. Note that $J^2$ commutes with $J_x$ and $J_y$ individually, so it also commutes with $J_+$. Therefore,

$$J^2\phi = J^2(J_+\psi) = J_+(J^2\psi) = \alpha J_+\psi = \alpha\phi$$

so $\phi$ is also an eigenfunction of $J^2$ with the same eigenvalue as $\psi$. On the other hand, $J_+$ does *not* commute with $J_z$. Instead, we have

$$[J_z, J_+] = [J_z, J_x + iJ_y]$$
$$= [J_z, J_x] + i[J_z, J_y]$$
$$= i\hbar J_y + \hbar J_x$$
$$= \hbar J_+$$

This means that

$$J_z \phi = J_z(J_+\psi)$$
$$= J_+ J_z \psi + \hbar J_+ \psi$$
$$= \beta J_+ \psi + \hbar J_+ \psi$$
$$= (\beta + \hbar)\phi$$

To summarize, $J_+$ transforms the wave function $\psi$ with eigenvalues $\alpha$ and $\beta$ (for operators $J^2$ and $J_z$) into a new wave function with eigenvalues $\alpha$ and $\beta + \hbar$. So the eigenvalue of $J^2$ is unchanged, but the eigenvalue of $J_z$ is increased by $\hbar$. If we apply $J_+$ to our new function $\phi$, we will get yet another eigenfunction of $J^2$ and $J_z$ with the same eigenvalue of $J^2$ but a new eigenvalue for $J_z$, namely $\beta + 2\hbar$. We can continue this process, increasing the eigenvalue of $J_z$ at each step; hence the name ladder operator. (As one might have guessed, $J_-$ has the opposite effect: it keeps the eigenvalue of $J^2$ fixed, but *lowers* the eigenvalue of $J_z$ by $\hbar$.)

It would appear that we could continue this process indefinitely, obtaining an infinite number of different eigenfunctions for $J_z$. However, this is not the case; it turns out that there is an upper limit on the possible eigenvalues of $J_z$. To see this, assume once more that $\psi$ is a normalized eigenfunction of both $J^2$ and $J_z$ with eigenvalues $\alpha$ and $\beta$, respectively. Since $J^2 = J_x^2 + J_y^2 + J_z^2$, we can write

$$(\psi, J^2\psi) - (\psi, J_z^2\psi) = (\psi, J_x^2\psi) + (\psi, J_y^2\psi)$$

Now the left-hand side is $\alpha(\psi, \psi) - \beta^2(\psi, \psi) = \alpha - \beta^2$, while the right-hand side is (since $J_x$ and $J_y$ are both Hermitian) just $(J_x\psi, J_x\psi) + (J_y\psi, J_y\psi)$. This latter quantity is nonnegative from Equation (5.14). Hence, we have

$$\alpha - \beta^2 \geq 0$$

so

$$\beta^2 \leq \alpha$$

Note that there is a classical analogy to this result: the $z$ component of a vector cannot be larger than the vector itself! Therefore, classically, $J_z^2 \leq J^2$.

How do we reconcile this upper bound on $\beta$ with the fact that we can repeatedly apply $J_+$ to $\psi$ and increase the value of $\beta$ each time? This is possible only if there is an eigenfunction of $J^2$ and $J_z$, which we will call $\psi_{max}$, for which

$$J_+\psi_{max} = 0$$

Thus, as we apply $J_+$ repeatedly, we get larger and larger values for $\beta$, but the process terminates when we reach $\psi_{max}$. A similar argument shows that there must be an eigenfunction $\psi_{min}$ for which

$$J_-\psi_{min} = 0$$

We can use these two eigenfunctions to derive the possible eigenvalues of $J^2$ and $J_z$. From the definitions of $J_+$ and $J_-$, it is tedious but straightforward to show that

$$J^2 = J_-J_+ + J_z^2 + \hbar J_z \tag{6.15}$$

$$J^2 = J_+J_- + J_z^2 - \hbar J_z \tag{6.16}$$

Using Equation (6.15), we see that

$$J^2\psi_{max} = J_-J_+\psi_{max} + J_z^2\psi_{max} + \hbar J_z\psi_{max}$$

$$\alpha\psi_{max} = 0 + \beta_{max}^2\psi + \hbar\beta_{max}\psi$$

where $\beta_{max}$ is the eigenvalue of $J_z$ for the eigenstate $\psi_{max}$. Thus,

$$\alpha = \beta_{max}^2 + \hbar\beta_{max} \tag{6.17}$$

Similarly, applying $J^2$ to the state $\psi_{min}$, and using Equation (6.16), we get

$$\alpha = b_{min}^2 - \hbar b_{min} \tag{6.18}$$

Subtracting Equation (6.18) from (6.17) gives

$$b_{max}^2 - b_{min}^2 + \hbar(b_{max} + b_{min}) = 0$$

which has the solution

$$b_{min} = -b_{max} \tag{6.19}$$

But recall that we can apply $J_+$ repeatedly to $\psi$ and with each application, increase the eigenvalue of $J_z$ by $\hbar$. Hence, all of the eigenvalues of $J_z$ must differ from each other by an integer multiple of $\hbar$, so

$$b_{max} = b_{min} + n\hbar \tag{6.20}$$

where $n$ is an integer. Combining Equations (6.19) and (6.20) gives

$$b_{max} - (-b_{max}) = n\hbar$$

so that

$$b_{max} = n\frac{\hbar}{2} \tag{6.21}$$

Define a new number $j$ given by $j = n/2$, so that $j$ is an integer or a half-integer:

$$j = 0, \frac{1}{2}, 1, \frac{3}{2}, 2, \ldots \tag{6.22}$$

Then from Equations (6.17) and (6.21), we have

$$\alpha = j^2\hbar^2 + j\hbar^2 = \hbar^2 j(j+1)$$

with the possible values of $j$ given by Equation (6.22).

We also know that $\beta$ must go in integer steps of $\hbar$ from $-j\hbar$ to $+j\hbar$, so

$$\beta = -j\hbar, -j\hbar + \hbar, \ldots, (j-1)\hbar, j\hbar$$

A convenient way to write this is

$$\beta = m_j\hbar, \quad m_j = -j, -j+1, \ldots, j$$

To summarize, then, if $\psi$ is an eigenfunction of both $J^2$ and $J_z$, then its possible eigenvalues for $J^2$ are $\hbar^2 j(j+1)$, i.e.,

$$J^2\psi = \hbar^2 j(j+1)\psi \tag{6.23}$$

where $j$ is an integer or a half-integer,

$$j = 0, \frac{1}{2}, 1, \frac{3}{2}, \ldots \tag{6.24}$$

and the eigenvalues of $J_z$ depend on the value of $j$, namely,

$$J_z\psi = m_j\hbar\psi \tag{6.25}$$

where

$$m_j = -j, -j+1, \ldots, j-1, j \tag{6.26}$$

Here is another example of quantization: the total squared angular momentum and the $z$ component cannot have arbitrary values. Instead, they are restricted to the discrete values given in Equations (6.23)–(6.26). In fact, the quantization of angular momentum differs from energy quantization in an important respect: although a given potential will have a lowest-energy state, there is no absolute lower bound on the energy that can exist in nature; the potential can always be altered to produce a lower-energy ground state. With angular momentum, on the other hand, there is an absolute lowest angular momentum state. For example, the $z$ component of angular momentum, if it is nonzero, cannot be smaller than $\hbar/2$; this represents the smallest "unit" of angular momentum found in nature.

---

### Example 6.2. Solving the One-dimensional Simple Harmonic Oscillator Using Ladder Operators

The simple harmonic oscillator in one dimension can also be solved by the method of ladder operators. This solution is simpler and more elegant than the one in Chapter 4.

For a particle of mass $m$ in a one-dimensional simple harmonic oscillator potential $V(x) = \frac{1}{2}Kx^2$, the Hamiltonian operator is

$$H = -\frac{\hbar^2}{2m}\frac{d^2}{dx^2} + \frac{1}{2}Kx^2$$

Define the ladder operators $a_-$ and $a_+$ to be given by

$$a_+ = \sqrt{\frac{K}{2}}x - \frac{\hbar}{\sqrt{2m}}\frac{d}{dx}$$

and

$$a_- = \sqrt{\frac{K}{2}}x + \frac{\hbar}{\sqrt{2m}}\frac{d}{dx}$$

We can now show that

$$[H, a_+] = \hbar\omega a_+$$

and

$$[H, a_-] = -\hbar\omega a_-$$

where $\omega = \sqrt{K/m}$. For $a_+$, we have

$$[H, a_+] = \left[-\frac{\hbar^2}{2m}\frac{d^2}{dx^2} + \frac{1}{2}Kx^2, \sqrt{\frac{K}{2}}x - \frac{\hbar}{\sqrt{2m}}\frac{d}{dx}\right]$$

$$= \left[-\frac{\hbar^2}{2m}\frac{d^2}{dx^2}, \sqrt{\frac{K}{2}}x\right] + \left[\frac{1}{2}Kx^2, \sqrt{\frac{K}{2}}x\right] + \left[-\frac{\hbar^2}{2m}\frac{d^2}{dx^2}, -\frac{\hbar}{\sqrt{2m}}\frac{d}{dx}\right]$$

$$+ \left[\frac{1}{2}Kx^2, -\frac{\hbar}{\sqrt{2m}}\frac{d}{dx}\right]$$

The second and third terms are zero, and the remaining two terms simplify to

$$[H, a_+] = -\frac{\hbar^2}{2m}\sqrt{\frac{K}{2}}\left[\frac{d^2}{dx^2}, x\right] - \frac{K}{2}\frac{\hbar}{\sqrt{2m}}\left[x^2, \frac{d}{dx}\right]$$

Using the commutator identities, this expands out to

$$[H, a_+] = -\frac{\hbar^2}{2m}\sqrt{\frac{K}{2}}\left(\frac{d}{dx}\left[\frac{d}{dx}, x\right] + \left[\frac{d}{dx}, x\right]\frac{d}{dx}\right)$$
$$-\frac{K}{2}\frac{\hbar}{\sqrt{2m}}\left(x\left[x, \frac{d}{dx}\right] + \left[x, \frac{d}{dx}\right]x\right)$$

Now note that

$$\left[x, \frac{d}{dx}\right] = -1$$

(This is trivially derived from $[X, P_x] = i\hbar$.) Then we get

$$[H, a_+] = -\frac{\hbar^2}{2m}\sqrt{\frac{K}{2}}\frac{d}{dx}(+2) - \frac{K}{2}\frac{\hbar}{\sqrt{2m}}x(-2)$$
$$= \hbar\sqrt{\frac{K}{m}}\left(\sqrt{\frac{K}{2}}x - \frac{\hbar}{\sqrt{2m}}\frac{d}{dx}\right)$$
$$= \hbar\omega a_+$$

Similarly, for $a_-$, we get

$$[H, a_-] = \left[-\frac{\hbar^2}{2m}\frac{d^2}{dx^2} + \frac{1}{2}Kx^2, \sqrt{\frac{K}{2}}x + \frac{\hbar}{\sqrt{2m}}\frac{d}{dx}\right]$$
$$= \left[-\frac{\hbar^2}{2m}\frac{d^2}{dx^2}, \sqrt{\frac{K}{2}}x\right] + \left[\frac{1}{2}Kx^2, \sqrt{\frac{K}{2}}x\right] + \left[-\frac{\hbar^2}{2m}\frac{d^2}{dx^2}, \frac{\hbar}{\sqrt{2m}}\frac{d}{dx}\right]$$
$$+ \left[\frac{1}{2}Kx^2, \frac{\hbar}{\sqrt{2m}}\frac{d}{dx}\right]$$
$$= -\frac{\hbar^2}{2m}\sqrt{\frac{K}{2}}\left[\frac{d^2}{dx^2}, x\right] + \frac{K}{2}\frac{\hbar}{\sqrt{2m}}\left[x^2, \frac{d}{dx}\right]$$
$$= -\frac{\hbar^2}{2m}\sqrt{\frac{K}{2}}\left(\frac{d}{dx}\left[\frac{d}{dx}, x\right] + \left[\frac{d}{dx}, x\right]\frac{d}{dx}\right)$$
$$+ \frac{K}{2}\frac{\hbar}{\sqrt{2m}}\left(x\left[x, \frac{d}{dx}\right] + \left[x, \frac{d}{dx}\right]x\right)$$
$$= -\frac{\hbar^2}{2m}\sqrt{\frac{K}{2}}\frac{d}{dx}(+2) + \frac{K}{2}\frac{\hbar}{\sqrt{2m}}x(-2)$$
$$= -\hbar\sqrt{\frac{K}{m}}\left(\sqrt{\frac{K}{2}}x + \frac{\hbar}{\sqrt{2m}}\frac{d}{dx}\right)$$
$$= \hbar\omega a_-$$

We can now show that if $\psi(x)$ is a solution of the time-independent Schrödinger equation for the harmonic oscillator with energy $E$, then $a_+\psi(x)$ is also a solution of the time-independent Schrödinger equation for the harmonic oscillator with energy $E + \hbar\omega$, and $a_-\psi(x)$ is a solution with energy $E - \hbar\omega$.

Applying $[H, a_+] = \hbar\omega a_+$ to $\psi$ gives

$$Ha_+\psi - a_+H\psi = \hbar\omega a_+\psi$$

But $\psi$ is a solution to the time-independent Schrödinger equation with energy $E$, so $H\psi = E\psi$. Substituting this into the equation above gives

$$Ha_+\psi - a_+E\psi = \hbar\omega a_+\psi$$

which can be written as

$$H(a_+\psi) = (E + \hbar\omega)(a_+\psi)$$

so $a_+\psi$ is a solution of the Schrödinger equation with energy $E + \hbar\omega$. Similarly, applying $[H, a_-] = -\hbar\omega a_+$ to $\psi$ gives

$$Ha_-\psi - a_-H\psi = -\hbar\omega a_-\psi$$

and $H\psi = E\psi$, so

$$Ha_-\psi - a_-E\psi = -\hbar\omega a_-\psi$$

so

$$H(a_-\psi) = (E - \hbar\omega)a_-\psi$$

so $a_-\psi$ is a solution of the Schrödinger equation with energy $E - \hbar\omega$.

Further, we have

$$a_+a_- = \left(\sqrt{\frac{K}{2}}x - \frac{\hbar}{\sqrt{2m}}\frac{d}{dx}\right)\left(\sqrt{\frac{K}{2}}x + \frac{\hbar}{\sqrt{2m}}\frac{d}{dx}\right)$$

$$= -\frac{\hbar^2}{2m}\frac{d^2}{dx^2} + \frac{K}{2}x^2 + \frac{\hbar}{\sqrt{2m}}\sqrt{\frac{K}{2}}\left(x\frac{d}{dx} - \frac{d}{dx}x\right)$$

$$= H + \frac{\hbar}{\sqrt{2m}}\sqrt{\frac{K}{2}}(-1)$$

$$= H - \hbar\omega/2$$

There is no upper bound on the possible values for $E$, but there is a lower bound; the energy cannot be negative. This means that if $\psi_0(x)$ is the ground state wave function, then $a_-\psi_0(x) = 0$. Suppose that $\psi_0$ is the ground state wave function. Then

$$a_+a_-\psi_0 = (H - \hbar\omega/2)\psi_0$$

But $a_-\psi_0 = 0$, so $(H - \hbar\omega/2)\psi_0 = 0$, so

$$H\psi_0 = (\hbar\omega/2)\psi_0$$

But this is just the Schrödinger equation with energy $E = \hbar\omega/2$, which is therefore the energy of the ground state wave function.

The expression $a_-\psi_0 = 0$ produces the differential equation:

$$\left( \sqrt{\frac{K}{2}}x + \frac{\hbar}{\sqrt{2m}}\frac{d}{dx} \right)\psi_0 = 0$$

which can be written as

$$\frac{d\psi_0}{\psi_0} = -\sqrt{\frac{K}{2}}\frac{\sqrt{2m}}{\hbar}x$$

with solution

$$\ln\psi_0 = -\frac{1}{2}\sqrt{\frac{K}{2}}\frac{\sqrt{2m}}{\hbar}x^2 + C$$

or

$$\psi_0 = Ae^{-(1/2)(\sqrt{Km}/\hbar)x^2}$$

where $A$ is the normalization constant. This is the desired ground-state wave function. Then the excited state wave functions derived in Chapter 4 are given by applying $a_+$ repeatedly to this wave function.

---

## 6.3    The Schrödinger Equation in Spherical Coordinates

In quantum mechanics, the case of spherical symmetry is often encountered. The most obvious example of this is the potential experienced by the electron in a hydrogen atom, given by

$$V(r) = -\frac{1}{4\pi\epsilon_0}\frac{e^2}{r}$$

In this case the potential depends only on the radial distance $r$; such potentials are called *central potentials*. For central potentials, the spherical coordinate system is the most convenient.

The spherical coordinate system is described by the coordinates $r$, $\theta$, and $\phi$, where $r$ is the distance from the origin, $\theta$ is the angle relative to the

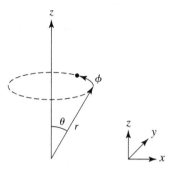

Fig. 6.3    The spherical coordinate system expresses all positions as a function of $r$, $\theta$, and $\phi$.

$z$-axis, and $\phi$ gives the azimuthal angle relative to the $x$-axis (Figure 6.3). Spherical coordinates are related to the familiar rectangular coordinates by

$$x = r \sin \theta \cos \phi \qquad (6.27)$$
$$y = r \sin \theta \sin \phi \qquad (6.28)$$
$$z = r \cos \theta \qquad (6.29)$$

Using these relations, any operator in rectangular coordinates can be expressed in terms of spherical coordinates, and vice versa.

---

**Example 6.3. Expressing the Operator $\partial/\partial\phi$ in Rectangular Coordinates**

To transform from spherical coordinates to rectangular coordinates, we use

$$\frac{\partial}{\partial \phi} = \frac{\partial x}{\partial \phi}\frac{\partial}{\partial x} + \frac{\partial y}{\partial \phi}\frac{\partial}{\partial y} + \frac{\partial z}{\partial \phi}\frac{\partial}{\partial z}$$

Substituting Equations (6.27)–(6.29) for the derivatives gives

$$\frac{\partial}{\partial \phi} = -r \sin \theta \sin \phi \frac{\partial}{\partial x} + r \sin \theta \cos \phi \frac{\partial}{\partial y} + 0$$

$$= -y \frac{\partial}{\partial x} + x \frac{\partial}{\partial y}$$

---

Now we can express the angular momentum operators in spherical coordinates. The result of Example 6.3 can be used to find $L_z$:

$$-i\hbar \frac{\partial}{\partial \phi} = -Y P_x + X P_y$$

which gives the desired expression for $L_z$:

$$L_z = -i\hbar \frac{\partial}{\partial \phi} \qquad (6.30)$$

Note the resemblance to the one-dimensional linear momentum operator, $P_x = -i\hbar(\partial/\partial x)$. In both cases the operator is based on the derivative in the direction of the classical motion. Using similar methods (but much more algebra!) the operator $L^2$ can also be derived in spherical coordinates:

$$L^2 = -\hbar^2 \left[ \frac{1}{\sin\theta} \frac{\partial}{\partial\theta} \left( \sin\theta \frac{\partial}{\partial\theta} \right) + \frac{1}{\sin^2\theta} \frac{\partial}{\partial\phi^2} \right] \qquad (6.31)$$

These operators can be used to calculate explicitly the eigenfunctions and eigenvalues of $L^2$ and $L_z$.

Consider first the eigenvalues of $L_z$. From the previous section, we know that such eigenvalues must be of the form $m\hbar$ with $m$ given by an integer or half-integer. However, this represents the set of all possible eigenvalues for a general angular momentum operator; we are dealing with a special case (orbital angular momentum) characterized by a specific angular momentum operator, so it is possible that some of these eigenvalues are excluded.

Assume that $\psi(r, \theta, \phi)$ is an eigenfunction of $L_z$ with eigenvalue $m_l\hbar$:

$$L_z\psi = m_l\hbar\psi$$

where the $l$ subscript on $m$ indicates that we are dealing with orbital angular momentum. Once again, we use separation of variables and assume that the solution is of the form

$$\psi(r, \theta, \phi) = R(r)F(\theta)G(\phi)$$

Using this form for $\psi$, along with the expression for $L_z$ from Equation (6.30), our eigenvalue equation becomes

$$-i\hbar R(r)F(\theta)\frac{dG}{d\phi} = m_l\hbar R(r)F(\theta)G(\phi)$$

$$-i\hbar\frac{dG}{d\phi} = m_l\hbar G$$

which has the solution

$$G(\phi) = e^{im_l\phi}$$

so that

$$\psi(r, \theta, \phi) = R(r)F(\theta)e^{im_l\phi} \qquad (6.32)$$

Equation (6.32) gives the general $\phi$-dependence for any wave function that is an eigenfunction of $L_z$.

Now note that we must impose an additional condition on $\psi$. In spherical coordinates, increasing $\phi$ by $2\pi$ just amounts to a 360 degree rotation, so it brings any system back to the same position in physical space. Hence, it must be true that

$$\psi(r, \theta, \phi + 2\pi) = \psi(r, \theta, \phi)$$

so Equation (6.32) gives

$$e^{im_l(\phi + 2\pi)} = e^{im_l\phi}$$

which implies

$$e^{2\pi im_l} = 1$$

This equation is satisfied if and only if $m_l$ is a positive or negative integer:

$$m_l = 0, \pm 1, \pm 2, \pm 3, \ldots$$

This is obviously more restrictive than the general condition on angular momentum quantum numbers, for which $m$ could be an integer or a half-integer. Since $m_l$ ranges from $-l$ to $+l$ in integer steps, the only way to insure that $m_l$ is an integer is for $l$ to be an integer as well:

$$l = 0, 1, 2, \ldots$$

This, then, is a distinguishing property of orbital angular momentum: while the eigenvalues of $L^2$ are $\hbar^2 l(l+1)$ and the eigenvalues of $L_z$ are $\hbar m_l$, just as in the case of general angular momentum, $l$ must be an *integer*, which forces $m_l$ to be a positive or negative integer:

$$m_l = -l, -l+1, \ldots, -1, 0, 1, \ldots, l$$

Half-integer values for $l$ and $m_l$ are excluded.

Now consider the Schrödinger equation in spherical coordinates for some arbitrary potential $V(r, \theta, \phi)$. The Hamiltonian still has the familiar form

$$H = -\frac{\hbar^2}{2m}\nabla^2 + V(r, \theta, \phi)$$

but now $\nabla^2$ must be expressed in terms of the spherical coordinates $r$, $\theta$, and $\phi$. This transformation is tedious but straightforward; it yields

$$H = -\frac{\hbar^2}{2m}\left[\frac{1}{r^2\sin\theta}\frac{\partial}{\partial\theta}\sin\theta\frac{\partial}{\partial\theta} + \frac{1}{r^2}\frac{\partial}{\partial r}r^2\frac{\partial}{\partial r} + \frac{1}{r^2\sin^2\theta}\frac{\partial^2}{\partial\phi^2}\right] + V(r, \theta, \phi)$$

But now compare the first and third terms in this expression with the expression for $L^2$ in spherical coordinates (Equation (6.31)). It is clear that $H$ can be rewritten in the form

$$H = -\frac{\hbar^2}{2m}\frac{1}{r^2}\frac{\partial}{\partial r}r^2\frac{\partial}{\partial r} + \frac{L^2}{2mr^2} + V(r, \theta, \phi) \tag{6.33}$$

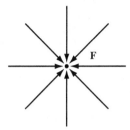

Fig. 6.4   The potential $V(r)$ corresponds to a classical central force. The direction of the force is always radial (toward or away from the origin), and it produces no torque.

Note the similarity between this expression and the classical expression for the energy of a body moving in a central potential $V(r)$:

$$E = \frac{1}{2}m\left(\frac{dr}{dt}\right)^2 + \frac{l^2}{2mr^2} + V(r)$$

In the classical central force problem, the term $l^2/2mr^2$ leads to a fictitious "centrifugal force", producing a minimum in the effective potential for some choices of $V(r)$.

Now we would like to find solutions of the Schrödinger equation which are also eigenfunctions of $L^2$ and $L_z$, i.e., they are states of definite total angular momentum and of the $z$ component of angular momentum. In order for this to be possible, the Hamiltonian in Equation (6.33) must commute with $L^2$ and $L_z$. Since $L^2$ and $L_z$ are functions only of $\theta$, $\phi$, and the derivatives with respect to $\theta$ and $\phi$, they commute with the first term in Equation (6.33), which is a function only of $r$ and derivatives with respect to $r$. Further, $L^2$ and $L_z$ commute with the second term, since they commute with both $L^2$ and $1/r^2$. Thus, the question of whether or not $L^2$ and $L_z$ commute with $H$ boils down to whether or not they commute with $V(r, \theta, \phi)$. A very simple way to insure that $L^2$ and $L_z$ do, in fact, commute with $V$ is to take $V$ to be a function only of $r$ and to be independent of $\theta$ and $\phi$, since, as we have noted, $L$ and $L_z$ are functions only of $\theta$, $\phi$, and the derivatives with respect to $\theta$ and $\phi$. Classically, a potential of the form $V(r)$ corresponds to a central force, i.e., one that is directed radially inward or radially outward (Figure 6.4). Such a force produces no torque, so that angular momentum is conserved. We will assume for the remainder of this chapter that we are dealing with a potential of the form $V = V(r)$. This will be the case, for example, for an electron in a hydrogen atom.

The full Schrödinger equation, $H\psi = E\psi$, then becomes

$$-\frac{\hbar^2}{2m}\frac{1}{r^2}\frac{\partial}{\partial r}r^2\frac{\partial\psi}{\partial r} + \frac{L^2}{2mr^2}\psi + V(r)\psi = E\psi$$

Now assume that $\psi$ is a eigenfunction of $L^2$ with eigenvalue $\hbar^2 l(l+1)$, so that this equation becomes

$$-\frac{\hbar^2}{2mr}\frac{\partial}{\partial r^2}(r\psi) + \frac{\hbar^2 l(l+1)}{2mr^2}\psi + V(r)\psi = E\psi \qquad (6.34)$$

where we have also used the fact that

$$\frac{1}{r}\frac{\partial}{\partial r^2}(r\psi) = \frac{2}{r}\frac{\partial\psi}{\partial r} + \frac{\partial^2\psi}{\partial r^2} = \frac{1}{r^2}\frac{\partial}{\partial r}r^2\frac{\partial\psi}{\partial r}$$

To solve this equation, we once again use separation of variables, assuming a solution of the form

$$\psi(r,\theta,\phi) = R(r)Y(\theta,\phi)$$

Substituting this solution into Equation (6.34) gives

$$-\frac{\hbar^2}{2mr}\frac{\partial}{\partial r^2}(rR(r))Y(\theta,\phi) + \frac{\hbar^2 l(l+1)}{2mr^2}R(r)Y(\theta,\phi) + V(r)R(r)Y(\theta,\phi)$$
$$= ER(r)Y(\theta,\phi)$$

The function $Y(\theta,\phi)$ can be divided out, yielding an equation for $R(r)$:

$$-\frac{\hbar^2}{2mr}\frac{\partial}{\partial r^2}(rR(r)) + \frac{\hbar^2 l(l+1)}{2mr^2}R(r) + V(r)R(r) = ER(r) \qquad (6.35)$$

Equation (6.35) is called the *radial Schrödinger equation*. It gives the energy and the radial part of the wave function $R(r)$ for an arbitrary central potential $V(r)$. Of course, both the energy and $R(r)$ will depend on the particular form of $V(r)$. Note also that the solutions for $R(r)$ and $E$ will, in general, depend on $l$, but they are completely independent of $m_l$, since $m_l$ does not appear in Equation (6.35). The physical reason for this is that for a fixed $l$, a change in the eigenvalue of $L_z$ can be produced simply by rotating the coordinate axes, and the energy of the system should not depend on the choice of the coordinate system.

It appears that the function $Y(\theta,\phi)$ has vanished from the problem completely. In fact, the functional form for $Y(\theta,\phi)$ can be derived from the eigenvalue equations,

$$L^2\psi = \hbar^2 l(l+1)\psi$$

$$L_z\psi = \hbar m_l\psi$$

Substituting the explicit forms for $L^2$ and $L_z$ from Equations (6.31) and (6.30) gives

$$-\hbar^2\left[\frac{1}{\sin\theta}\frac{\partial}{\partial\theta}\left(\sin\theta\frac{\partial}{\partial\theta}\right)+\frac{1}{\sin^2\theta}\frac{\partial}{\partial\phi^2}\right]R(r)Y(\theta,\phi)=\hbar^2 l(l+1)R(r)Y(\theta,\phi)$$

$$-i\hbar\frac{\partial}{\partial\phi}R(r)Y(\theta,\phi)=\hbar m_l R(r)Y(\theta,\phi)$$

Dividing out the factor $R(r)$ gives two equations for $Y(\theta,\phi)$:

$$-\hbar^2\left[\frac{1}{\sin\theta}\frac{\partial}{\partial\theta}\left(\sin\theta\frac{\partial}{\partial\theta}\right)+\frac{1}{\sin^2\theta}\frac{\partial}{\partial\phi^2}\right]Y(\theta,\phi)=\hbar^2 l(l+1)Y(\theta,\phi) \quad (6.36)$$

$$-i\hbar\frac{\partial}{\partial\phi}Y(\theta,\phi)=\hbar m_l Y(\theta,\phi) \quad (6.37)$$

An important point here is that $Y(\theta,\phi)$ is determined entirely by the eigenvalues $l$ and $m_l$ and is completely independent of $V(r)$. Therefore, it is customary to write these functions as $Y_l^m(\theta,\phi)$. (For clarity, we write $Y_l^m$ rather than $Y_l^{m_l}$; it is understood that the $m$ appearing in $Y_l^m$ always refers to orbital angular momentum.) These functions, which we will now calculate, provide a universal description of the angular part of the wave function for *all* central potentials. Once again we assume separation of variables, so that $Y_l^m(\theta,\phi)=F(\theta)G(\phi)$, and substitute this form for $Y_l^m$ into Equations (6.36) and (6.37). Note that we have already solved Equation (6.37) above with the result that

$$G(\phi)=e^{im_l\phi}$$

Now we need to determine the $\theta$-dependence of $Y_l^m(\theta,\phi)$. Recall that the largest possible value of $m_l$ for a given $l$ is $m_l=l$. Thus, if an expression for $Y_l^l(\theta,\phi)$ is known, it is possible to derive all of the other values of $Y_l^m(\theta,\phi)$ by repeated application of the lowering operator $L_-$. In terms of spherical coordinates, the raising and lowering operators are given by

$$L_+=\hbar e^{i\phi}\left(\frac{\partial}{\partial\theta}+i\cot\theta\frac{\partial}{\partial\phi}\right) \quad (6.38)$$

$$L_-=\hbar e^{-i\phi}\left(-\frac{\partial}{\partial\theta}+i\cot\theta\frac{\partial}{\partial\phi}\right) \quad (6.39)$$

In order to derive $Y_l^l(\theta,\phi)$, recall that the raising operator applied to the highest-$m_l$ state gives 0:

$$L_+Y_l^l(\theta,\phi)=0$$

Table 6.1　The first few normalized spherical harmonics, $Y_l^m(\theta, \phi)$.

| $l$ | $m_l$ | $Y_l^m(\theta, \phi)$ |
|---|---|---|
| 0 | 0 | $Y_l^m = \left(\dfrac{1}{4\pi}\right)^{1/2}$ |
| 1 | $-1$ | $Y_l^m = \left(\dfrac{3}{8\pi}\right)^{1/2} \sin\theta\, e^{-i\phi}$ |
| 1 | 0 | $Y_l^m = \left(\dfrac{3}{4\pi}\right)^{1/2} \cos\theta$ |
| 1 | $+1$ | $Y_l^m = -\left(\dfrac{3}{8\pi}\right)^{1/2} \sin\theta\, e^{i\phi}$ |
| 2 | $-2$ | $Y_l^m = \left(\dfrac{15}{32\pi}\right)^{1/2} \sin^2\theta\, e^{-2i\phi}$ |
| 2 | $-1$ | $Y_l^m = \left(\dfrac{15}{8\pi}\right)^{1/2} \sin\theta\cos\theta\, e^{-i\phi}$ |
| 2 | 0 | $Y_l^m = \left(\dfrac{5}{16\pi}\right)^{1/2} (3\cos^2\theta - 1)$ |
| 2 | $+1$ | $Y_l^m = -\left(\dfrac{15}{8\pi}\right)^{1/2} \sin\theta\cos\theta\, e^{i\phi}$ |
| 2 | $+2$ | $Y_l^m = \left(\dfrac{15}{32\pi}\right)^{1/2} \sin^2\theta\, e^{2i\phi}$ |

In terms of $L_+$ in spherical coordinates (Equation (6.38)), this equation becomes

$$\hbar e^{i\phi} \left[ \frac{\partial Y_l^l}{\partial \theta} + i\cot\theta \frac{\partial Y_l^l}{\partial \phi} \right] = 0$$

which has the solution

$$Y_l^l(\theta, \phi) = (\sin\theta)^l e^{il\phi}$$

Now, by repeatedly applying the operator $L_-$ from Equation (6.39), one can obtain all of the other functions $Y_l^m(\theta, \phi)$.

These $Y_l^m$ functions arise in a variety of different areas of physics; it is difficult to overemphasize their importance. They are called *spherical harmonics*, or more colloquially, "Y"-"l"-"m"'s. A list of the functions $Y_l^m(\theta, \phi)$ for the first few values of $l$ is given in Table 6.1.

The constants appearing in the $Y_l^m$ functions in Table 6.1 are chosen so as to normalize these functions, i.e.,

$$\int_{\theta=0}^{\pi} \int_{\phi=0}^{2\pi} |Y_l^m(\theta,\phi)|^2 \sin\theta \, d\theta \, d\phi = 1$$

This still leaves an arbitrary complex phase factor; the convention is to choose this phase so that $Y_l^{-m}(\theta,0) = (-1)^m Y_l^m(\theta,0)$. This accounts for the minus signs appearing in some of the $Y_l^m$ functions in Table 6.1.

As derived earlier, all of the spherical harmonics in Table 6.1 are of the form $Y_l^m(\theta,\phi) = f(\theta)e^{im\phi}$. This implies that $Y_l^0$ is a function only of $\theta$ and is independent of $\phi$. These $m = 0$ functions, $Y_l^0(\theta)$, are also encountered in the theory of electromagnetic fields, where they arise as the solution of Laplace's equation, which gives the electric potential in a vacuum. In this context they are written in terms of Legendre polynomials $P_l(\cos\theta)$, where $P_l(\cos\theta) \propto Y_l^0(\theta)$. Note further that the $l = 0$, $m = 0$ spherical harmonic is spherically symmetric, i.e., completely independent of both $\theta$ and $\phi$, and it is the only spherical harmonic with this property.

The spherical harmonics are difficult to visualize, since they are complex functions of spherical angular coordinates. However, note that $|Y_l^m|^2$, which gives the angular dependence of the probability density, is real and is independent of $\phi$ (since $|e^{im\phi}|^2 = 1$). In Figure 6.5, we show $|Y_l^m|^2$ as a function of $\theta$ for $l = 0, 1, 2$. These are polar graphs in which $\theta$ is the angle relative to the $z$-axis, and $|Y_l^m|^2$ is plotted as the radial distance from the origin.

We have now achieved a remarkably general result for central potentials. For *any* wave function that represents an eigenstate of $L^2$ and $L_z$ in a central potential $V(r)$, with angular momentum quantum numbers $l$ and $m_l$, the angular part of the wave function will be given by the appropriate $Y_l^m(\theta,\phi)$, completely independent of both $V(r)$ and the energy $E$. The potential *does* determine $E$ and the radial part of the wave function $R(r)$, so these will depend on the particular choice of $V(r)$.

## 6.4   The Hydrogen Atom

We are now in a position to determine the wave functions and energy levels of the electron in the hydrogen atom, one of the most important results in all of quantum mechanics. Before inserting the Coulomb potential experienced by the electron, we first simplify the general form for the radial Schrödinger equation (Equation (6.35)). Multiplying both sides of Equation (6.35) by

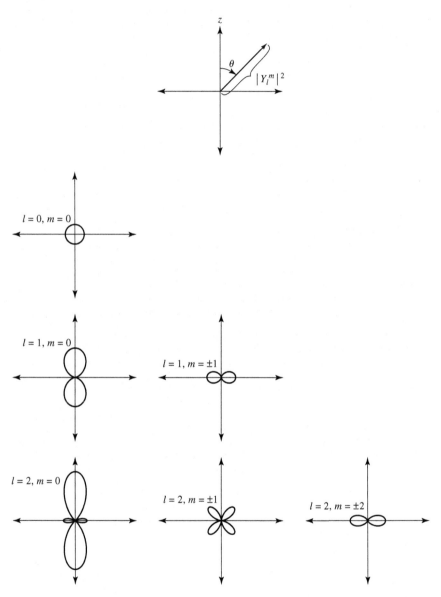

Fig. 6.5　Polar graphs showing $|Y_l^m|^2$ as a function of $\theta$ for the indicated values of $l$ and $m$.

$r$ gives

$$-\frac{\hbar^2}{2m}\frac{\partial}{\partial r^2}(rR(r)) + \frac{\hbar^2 l(l+1)}{2mr^2}rR(r) + V(r)rR(r) = ErR(r) \qquad (6.40)$$

This form of the equation suggests the substitution $u(r) = rR(r)$, simplifying the equation to the form

$$-\frac{\hbar^2}{2m}\frac{\partial}{\partial r^2}u(r) + \frac{\hbar^2 l(l+1)}{2mr^2}u(r) + V(r)u(r) = Eu(r)$$

Note again that both the wave function and the energies should depend on the particular value of $l$, since it appears in this equation. Two boundary conditions can be imposed on the wave function; as usual, as $r \to \infty$, we require $u \to 0$. However, an additional boundary condition must be imposed at the origin; since $\psi \propto u/r$, we must have $u \to 0$ as $r \to 0$ in order to keep $\psi$ finite.

Now consider the actual hydrogen atom. It consists of an electron with mass

$$m_e = 9.109 \times 10^{-31} \text{ kg}$$

orbiting a proton with mass

$$m_p = 1.672 \times 10^{-27} \text{ kg}$$

In a classical system of this sort, it would be inaccurate to take the electron to orbit a fixed proton; rather, the electron and proton orbit their common center of mass. Consider, more generally, a classical system consisting of a particle with mass $m_1$ at position $\mathbf{r}_1$, and a particle with mass $m_2$ at position $\mathbf{r}_2$, so the vector separation between the particles is

$$\mathbf{r} = \mathbf{r}_1 - \mathbf{r}_2 \tag{6.41}$$

Assume that the potential energy $V$ is a function only of the distance between the particles $r$, where $r = |\mathbf{r}|$. The energy of this system is

$$E = \frac{1}{2}m_1\left(\frac{d\mathbf{r}_1}{dt}\right)^2 + \frac{1}{2}m_2\left(\frac{d\mathbf{r}_2}{dt}\right)^2 + V(r) \tag{6.42}$$

We are free to choose any origin for the coordinate system, so we take it to lie at the center of mass of the particles, which gives

$$m_1\mathbf{r}_1 + m_2\mathbf{r}_2 = 0$$

This equation and Equation (6.41) allow us to express $\mathbf{r}_1$ and $\mathbf{r}_2$ as functions of $\mathbf{r}$:

$$\mathbf{r}_1 = \frac{m_2}{m_1 + m_2}\mathbf{r}$$

$$\mathbf{r}_2 = -\frac{m_1}{m_1 + m_2}\mathbf{r}$$

Substituting these expressions for $\mathbf{r}_1$ and $\mathbf{r}_2$ in Equation (6.42) gives an expression for $E$ as a function only of the separation between the particles:

$$E = \frac{1}{2}\mu \left(\frac{d\mathbf{r}}{dt}\right)^2 + V(r)$$

where $\mu$ is given by

$$\mu = \frac{m_1 m_2}{m_1 + m_2}$$

The quantity $\mu$ is called the *reduced mass*.

The quantum analog to this result is achieved by replacing $m$ in the Hamiltonian with $\mu$, so that

$$H = -\frac{\hbar^2}{2\mu}\nabla^2 + V$$

where, for the hydrogen atom, $\mu$ is given by

$$\mu = \frac{m_e m_p}{m_e + m_p} \tag{6.43}$$

Note that $m_e \ll m_p$, so that $\mu$ for the hydrogen atom will be very close to $m_e$. Plugging the actual numbers into Equation (6.43) gives

$$\mu = 0.9995 m_e \tag{6.44}$$

Hence, using $\mu$ instead of $m_e$ produces only a small correction in the case of the hydrogen atom. However, we will include this correction in our calculations.

The Coulomb potential experienced by the electron in the electric field of the proton is

$$V(r) = -\frac{1}{4\pi\epsilon_0}\frac{e^2}{r}$$

so the radial Schrödinger equation becomes

$$-\frac{\hbar^2}{2\mu}\frac{\partial^2 u}{\partial r^2} + \frac{\hbar^2 l(l+1)}{2\mu r^2}u - \frac{1}{4\pi\epsilon_0}\frac{e^2}{r}u = Eu \tag{6.45}$$

We will now solve this equation. Note that the potential is always negative, approaching zero as $r \to \infty$. Therefore, a bound state (which are the states we are interested in) must have $E < 0$ (Figure 6.6). To simplify the notation, we take $\varepsilon = -E$, and rewrite Equation (6.45) as

$$\frac{d^2 u}{dr^2} + \frac{2\mu e^2}{4\pi\epsilon_0 \hbar^2}\frac{u}{r} - \frac{l(l+1)u}{r^2} = \frac{2\mu}{\hbar^2}\varepsilon u \tag{6.46}$$

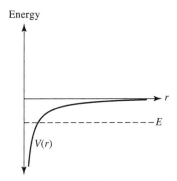

Fig. 6.6    The potential experienced by the electron in a hydrogen atom satisfies $V(r) < 0$. Hence, any bound state must have $E < 0$.

where $\varepsilon > 0$. To solve this equation, first consider its asymptotic behavior in the limit as $r \to \infty$. In this limit the second and third terms on the left-hand side go to zero, and we are left with

$$\frac{d^2 u}{dr^2} = \frac{2\mu}{\hbar^2}\varepsilon u$$

which has the solution

$$u = e^{-(\sqrt{2\mu\varepsilon}/\hbar)r} \tag{6.47}$$

Of course, this is not the exact solution to the full Equation (6.46), but it suggests that we investigate solutions of the form

$$u(r) = v(r)e^{-(\sqrt{2\mu\varepsilon}/\hbar)r} \tag{6.48}$$

where $v(r)$ is now the unknown function to be determined. Substituting this trial solution into Equation (6.46) gives an equation for $v(r)$:

$$\frac{d^2 v}{dr^2} - \frac{2\sqrt{2\mu\varepsilon}}{\hbar}\frac{dv}{dr} + \frac{2\mu e^2}{4\pi\epsilon_0\hbar^2}\frac{v}{r} - \frac{l(l+1)v}{r^2} = 0 \tag{6.49}$$

To solve for $v(r)$, we now try a series solution of the form

$$v(r) = \sum_{p=1}^{\infty} A_p r^p \tag{6.50}$$

(Note that the constant term, $A_0$, must be zero because of our previously-derived boundary condition that $u(0) = 0$.) Substituting this power-series expansion into Equation (6.49) gives an equation in powers of $r$ on the left-hand side, and in order for this to be equal to zero, the coefficient multiplying each power of $r$ must vanish. Enforcing this condition gives the following relation between the $A_p$ coefficients:

$$[p(p+1) - l(l+1)]A_{p+1} = \left[\frac{2p\sqrt{2\mu\varepsilon}}{\hbar} - \frac{2\mu e^2}{\hbar^2 4\pi\epsilon_0}\right]A_p \qquad (6.51)$$

Because $l$ is the angular momentum quantum number, it has some fixed integer value for any particular solution. Note that for $p = l$, the left-hand side of the equation vanishes, so that $A_p$ on the right-hand side must be zero. But taking $A_p = 0$ on the left-hand side of Equation (6.51) gives $A_{p-1} = 0$ on the right-hand side. This argument can be repeated to give $A_{p-2} = 0$, $A_{p-3} = 0$, and so on all of the way down. Thus, the only nonzero coefficients have $p > l$, namely, $A_{l+1}$, $A_{l+2}$, .... There is, however, one additional constraint on the polynomial series in Equation (6.50). In order for $u(r)$ to have the correct asymptotic behavior shown in Equation (6.47), we do not want the $v(r)$ factor in Equation (6.48) to give the dominant behavior at large $r$. This can be achieved by requiring the polynomial to have a finite number of terms. (A finite polynomial is always dominated by an exponential at large $r$; this need not be true for a polynomial with an infinite number of terms.) In order for the polynomial to terminate after a finite number of terms, the right-hand side of Equation (6.51) must vanish for some value of $p$, which we will call $n$. Note that $n$ must be greater than $l$, since only terms with $p \geq l+1$ are nonzero. In order for the term on the right-hand side to vanish at $p = n$, we must have

$$\frac{2n\sqrt{2\mu\varepsilon}}{\hbar} - \frac{2\mu e^2}{\hbar^2 4\pi\epsilon_0} = 0$$

Solving for $E = -\varepsilon$ gives

$$E = -\frac{\mu e^4}{(4\pi\epsilon_0)^2 2\hbar^2}\frac{1}{n^2}$$

$$= -13.6 \text{ eV } \frac{1}{n^2} \qquad (6.52)$$

This is exactly the same result derived earlier for the Bohr model of the atom (Chapter 1)! This equation also gives the physical meaning of the parameter $n$ introduced above: it determines the energy levels of the hydrogen atom. Furthermore, we have derived a constraint on $l$: since $n > l$,

$$l \leq n - 1$$

Now, however, we can move beyond the Bohr model to derive the wave functions as well. The radial wave function is

$$R(r) = \frac{u(r)}{r}$$

$$= \frac{v(r)}{r}e^{-(\sqrt{2m\varepsilon}/\hbar)r}$$

where $v(r)$ is a polynomial in powers of $r$ ranging from $r^{l+1}$ up to $r^n$. Since $R$ depends on $n$ and $l$, we write it as $R_{nl}(r)$. Solving for the power-series coefficients using Equation (6.51) gives explicit expressions for $R_{nl}(r)$. To simplify these expressions, we introduce a physical quantity $a_0$ with units of length, given by

$$a_0 = \frac{4\pi\epsilon_0\hbar^2}{\mu e^2} = 5.3 \times 10^{-11} \text{ m}$$

This length $a_0$ is called the *Bohr radius*. Then the first few normalized radial wave functions are

$$R_{10} = \left(\frac{1}{a_0}\right)^{3/2} 2e^{-r/a_0}$$

$$R_{20} = \left(\frac{1}{2a_0}\right)^{3/2} 2\left(1 - \frac{1}{2}\frac{r}{a_0}\right) e^{-r/2a_0}$$

$$R_{21} = \left(\frac{1}{2a_0}\right)^{3/2} \frac{1}{\sqrt{3}} \left(\frac{r}{a_0}\right) e^{-r/2a_0}$$

Each radial wave function can be combined with the corresponding $Y_l^m$ to obtain the full wave function. Since $R_{nl}(r)$ is a function of $n$ and $l$, and $Y_l^m(\theta, \phi)$ is a function of $l$ and $m_l$, the full wave function will be determined by three quantum numbers: $n$, $l$, and $m_l$:

$$\psi_{nlm_l}(r, \theta, \phi) = R_{nl}(r)Y_l^m(\theta, \phi)$$

Each of these quantum numbers has a physical significance: $n$, called the *principal quantum number*, determines the energy of the electron, $l$ determines the total squared angular momentum, and $m_l$ gives the $z$ component of the angular momentum. We have also derived the allowed values for all of these quantum numbers:

$$n = 1, 2, 3, \ldots$$

$$l \leq n - 1$$

$$m_l = -l, -l + 1, \ldots, 0, \ldots, l - 1, l$$

The normalized wave functions for $n = 1, 2, 3$ are given in Table 6.2.

Note that the energy $E_n$ is entirely determined by $n$ and is independent of $l$, despite the fact that the radial Schrödinger equation (Equation (6.35)), which determines the energy levels, contains $l$. This is an "accidental" degeneracy, in the sense that it occurs only for the Coulomb potential ($V(r) \propto 1/r$). For the case of a general radial potential, $E$ will depend on $l$.

Table 6.2    The normalized hydrogen wave functions for $n = 1, 2, 3$.

| $n$ | $l$ | $m_l$ | $\psi_{nlm_l}(r, \theta, \phi)$ |
|-----|-----|-------|-------------------------------------|
| 1 | 0 | 0 | $\psi_{100} = \dfrac{1}{\sqrt{\pi}}\left(\dfrac{1}{a_0}\right)^{3/2} e^{-r/a_0}$ |
| 2 | 0 | 0 | $\psi_{200} = \dfrac{1}{4\sqrt{2\pi}}\left(\dfrac{1}{a_0}\right)^{3/2}\left(2 - \dfrac{r}{a_0}\right) e^{-r/2a_0}$ |
| 2 | 1 | 0 | $\psi_{210} = \dfrac{1}{4\sqrt{2\pi}}\left(\dfrac{1}{a_0}\right)^{3/2}\left(\dfrac{r}{a_0}\right) e^{-r/2a_0}\cos\theta$ |
| 2 | 1 | $\pm 1$ | $\psi_{21\pm 1} = \dfrac{1}{8\sqrt{\pi}}\left(\dfrac{1}{a_0}\right)^{3/2}\left(\dfrac{r}{a_0}\right) e^{-r/2a_0}\sin\theta e^{\pm i\phi}$ |
| 3 | 0 | 0 | $\psi_{300} = \dfrac{1}{81\sqrt{3\pi}}\left(\dfrac{1}{a_0}\right)^{3/2}\left(27 - 18\dfrac{r}{a_0} + 2\dfrac{r^2}{a_0^2}\right) e^{-r/3a_0}$ |
| 3 | 1 | 0 | $\psi_{310} = \dfrac{\sqrt{2}}{81\sqrt{\pi}}\left(\dfrac{1}{a_0}\right)^{3/2}\left(6 - \dfrac{r}{a_0}\right)\left(\dfrac{r}{a_0}\right) e^{-r/3a_0}\cos\theta$ |
| 3 | 1 | $\pm 1$ | $\psi_{31\pm 1} = \dfrac{1}{81\sqrt{\pi}}\left(\dfrac{1}{a_0}\right)^{3/2}\left(6 - \dfrac{r}{a_0}\right)\left(\dfrac{r}{a_0}\right) e^{-r/3a_0}\sin\theta e^{\pm i\phi}$ |
| 3 | 2 | 0 | $\psi_{320} = \dfrac{1}{81\sqrt{6\pi}}\left(\dfrac{1}{a_0}\right)^{3/2}\left(\dfrac{r^2}{a_0^2}\right) e^{-r/3a_0}(3\cos^2\theta - 1)$ |
| 3 | 2 | $\pm 1$ | $\psi_{32\pm 1} = \dfrac{1}{81\sqrt{\pi}}\left(\dfrac{1}{a_0}\right)^{3/2}\left(\dfrac{r^2}{a_0^2}\right) e^{-r/3a_0}\sin\theta\cos\theta e^{\pm i\phi}$ |
| 3 | 2 | $\pm 2$ | $\psi_{32\pm 2} = \dfrac{1}{162\sqrt{\pi}}\left(\dfrac{1}{a_0}\right)^{3/2}\left(\dfrac{r^2}{a_0^2}\right) e^{-r/3a_0}\sin^2\theta e^{\pm 2i\phi}$ |

The total number of states with a given energy $E_n$ is straightforward to compute. A given value of $n$ corresponds to $n - 1$ possible states for $l$, and each $l$ state has $2l + 1$ possible values for $m_l$. Thus, the total degeneracy for energy $E_n$ is $1 + 3 + 5 + \cdots + 2n - 1 = n^2$. We will see in Chapter 8 that each electron also has two possible spin states corresponding to two possible values of the $z$ component of the spin angular momentum, so the true degeneracy for a given $n$ is actually $2n^2$.

Although this model for the hydrogen atom agrees with the Bohr model as far as the predictions of the energy levels are concerned, it presents a very different physical picture of the behavior of the electrons. In the Bohr model, electrons lie in distinct "orbits" around the proton, with the length of each orbit fixed by the need for it to comprise an integer number of

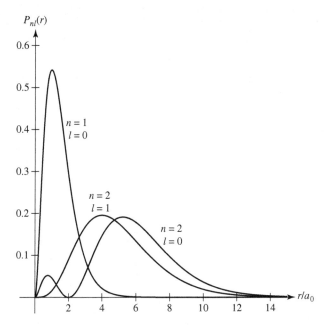

Fig. 6.7   The radial probability density, $P_{nl}(r)$, as a function of $r$ for the $n = 1$ and $n = 2$ states of the hydrogen atom.

de Broglie wavelengths. In the picture we have just derived (which is the best current picture of the atom), the behavior of the electrons is defined entirely by the wave functions. The idea of an electron "orbiting" the central proton is essentially abandoned. The electron does not follow a well-defined trajectory; instead, one can only talk about the probability of finding the electron inside a given volume $V$, which is given by

$$P = \int_V |\psi_{nlm_l}(r, \theta, \phi)|^2 r^2 \, dr \sin\theta \, d\theta \, d\phi$$

We can also define radial probability densities, $P_{nl}(r)$, by integrating the full probability density over $\theta$ and $\phi$ but retaining the $r$-dependence:

$$P_{nl}(r) \, dr = \int_{\theta=0}^{\pi} \int_{\phi=0}^{2\pi} |\psi_{nlm_l}(r, \theta, \phi)|^2 r^2 \, dr \sin\theta \, d\theta \, d\phi$$

Then $P_{nl}(r) \, dr$ gives the probability of finding the electron in a small interval $dr$ at a radius $r$ from the proton. A graph of $P_{nl}(r)$ as a function of $r$ is given in Figure 6.7. Clearly, the larger $n$ states, corresponding to larger energies, have larger mean radii for the electron. Although it is meaningless to talk about a single fixed "radius" of the atom in this picture, the probability of finding the electron at a given $r$ is largest where $P_{nl}(r)$ is peaked.

It is also common to use $\langle r \rangle$ as a reasonable definition of the atomic radius for any given $n, l$ state.

For instance, for the ground state of hydrogen, $\langle r \rangle$ is given by

$$\langle r \rangle = \int_{r=0}^{\infty} \int_{\theta=0}^{\pi} \int_{\phi=0}^{2\pi} r^2 \, dr \sin\theta \, d\theta \, d\phi |\psi_{100}|^2 r$$

with the ground state wave function

$$\psi_{100} = \frac{1}{\sqrt{\pi}} \left(\frac{1}{a_0}\right)^{3/2} e^{-r/a_0}$$

This wave function is independent of $\theta$ and $\phi$, so

$$\int_{\theta=0}^{\pi} \int_{\phi=0}^{2\pi} \sin\theta \, d\theta \, d\phi = 4\pi$$

and

$$\begin{aligned}
\langle r \rangle &= \int_{r=0}^{\infty} 4\pi r^2 \, dr \frac{1}{\pi} \left(\frac{1}{a_0}\right)^3 e^{-2r/a_0} r \\
&= \frac{4}{a_0^3} \int_{r=0}^{\infty} e^{-2r/a_0} r^3 \, dr \\
&= \frac{4}{a_0^3} \frac{6}{(2/a_0)^4} \\
&= \frac{3}{2} a_0
\end{aligned}$$

Thus, $\langle r \rangle = 3a_0/2 = 8 \times 10^{-11}$ m. For this reason, the "radius of the hydrogen atom" is often taken to be about $10^{-10}$ m.

This picture of the hydrogen atom has explanatory power far beyond what is possible with the Bohr model. In particular, a more detailed analysis of the hydrogen spectrum reveals that the energy levels are not exactly given by Equation (6.52), but are slightly perturbed from the predicted values. It is possible to explain these perturbations within the context of the model we have just derived; this will be done in Chapter 9.

## PROBLEMS

**6.1** A particle with mass $m$ is confined inside of a rectangular box given by $0 \leq x \leq a$, $0 \leq y \leq b$, and $0 \leq z \leq c$. The solution to the Schrödinger equation is

$$\psi(x, y, z) = A \sin\left(\frac{n_x \pi x}{a}\right) \sin\left(\frac{n_y \pi y}{b}\right) \sin\left(\frac{n_z \pi z}{c}\right)$$

where $A$ is a constant. The energy levels are given by
$E = (\hbar^2 \pi^2/2m)(n_x^2/a^2 + n_y^2/b^2 + n_z^2/c^2)$.
(a) Normalize the wave functions. You should obtain $A = \sqrt{8/V}$,
where $V = abc$ is the volume of the box.
(b) Suppose the particle is in the ground state. Calculate the probability of finding the particle in the lower fourth of the box, i.e., in the region $z \leq c/4$.

**6.2** A particle with mass $m$ is confined inside a cubic box with edges of length $a$. Show that there are six different wave functions that have $E = 14(\hbar^2 \pi^2/2ma^2)$. (This is called sixfold degeneracy.)

**6.3** (a) A particle with mass $m$ is confined inside a rectangular box with sides of length $a$, $a$, and $2a$. What is the energy of the first excited state? Is this state degenerate? If so, determine how many different wave functions have this energy.
(b) Now assume the rectangular box has sides of length $a$, $2a$, and $2a$. What is the energy of the first excited state? Is this state degenerate? If so, determine how many different wave functions have this energy.

**6.4** (a) A particle with mass $m$ and energy $E$ is inside a square tube with infinite potential barriers at $x = 0$, $x = a$, $y = 0$, and $y = a$. The tube is infinitely long in the $z$ direction. Inside the tube, $V = 0$. The particle is moving in the $+z$ direction. Solve the Schrödinger equation to derive the allowed wave functions for this particle. Do not try to normalize the wave functions, but make sure they correspond to motion in the $+z$ direction.
(b) Energy should not be quantized in this case because the particle is not in a bound state. Use the answer from part (a) to show that this is indeed the case.

**6.5** Show that $L_x$ and $L_y$ are Hermitian.

**6.6** Verify that $[X, Y] = 0$ and $[P_x, P_y] = 0$.

**6.7** Calculate these commutators:

$$[L_z, P_x], \ [L_z, X]$$
$$[L_z, P_y], \ [L_z, Y]$$
$$[L_z, P_z], \ [L_z, Z]$$

**6.8** Show that $\hbar$ has units of angular momentum.

**6.9** The operator $Q$ obeys the commutation relation $[Q, H] = E_0 Q$, where $E_0$ is a constant with units of energy. Show that if $\psi(x)$ is a solution of the time-independent Schrödinger equation with energy $E$, then $Q\psi(x)$ is also a solution of the time-independent Schrödinger equation, and determine the energy corresponding to $Q\psi(x)$.

**6.10** A particle is confined in a cubic box with edge of length $a$, with $V = 0$ inside the box. The particle is in its ground state; determine whether or not the particle is in an eigenstate of $L_z$.

**6.11** Consider a three-dimensional system with wave function $\psi$. If $\psi$ is in the $l = 0$ state, we already know that $L_z\psi = 0$. Show that $L_x\psi = 0$ and $L_y\psi = 0$ as well. (Note this is the only exception to the rule that a wave function cannot be simultaneously an eigenfunction of $L_x$, $L_y$, and $L_z$.)

**6.12** A particle is in an eigenstate of $L^2$ and $L_z$, with quantum numbers $l$ and $m_l$. By symmetry, we must have $\langle L_x^2 \rangle = \langle L_y^2 \rangle$. Show that $\hbar^2 l/2 \le \langle L_x^2 \rangle \le \hbar^2 l(l + 1)/2$.

**6.13** The "radius of the hydrogen atom" is often taken to be on the order of about $10^{-10}$ m. If a measurement is made to determine the location of the electron for hydrogen in its ground state, what is the probability of finding the electron within $10^{-10}$ m of the nucleus?

**6.14** (a) The electron in a hydrogen atom is in the $l = 1$ state having the lowest possible energy and the highest possible value for $m_l$. What are the $n$, $l$, and $m_l$ quantum numbers?
(b) A particle is moving in an unknown central potential. The wave function of the particle is spherically symmetric. What are the values of $l$ and $m_l$?

**6.15** The deuteron is a nucleus of "heavy hydrogen" consisting of one proton and one neutron. As a simple model for this nucleus, consider a single particle of mass $m$ moving in a fixed spherically-symmetric potential $V(r)$, defined by $V(r) = -V_0$ for $r < r_0$ and

$V(r) = 0$ for $r > r_0$. This is called a spherical square-well potential. Assume that the particle is in a bound state with $l = 0$.

(a) Find the general solutions $R(r)$ to the radial Schrödinger equation for $r < r_0$ and $r > r_0$. Use the fact that the wave function must be finite at 0 and $\infty$ to simplify the solution as much as possible. (You do not have to normalize the solutions.)

(b) The deuteron is only just bound; i.e., $E$ is nearly equal to 0. Take $m$ to be the proton mass, $m = 1.67 \times 10^{-27}$ kg, and take $r_0$ to be a typical nuclear radius, $r_0 = 1 \times 10^{-15}$ m. Find the value of $V_0$ (the depth of the potential well) in MeV (1 MeV = $1.6 \times 10^{-13}$ J). (Hint: The continuity conditions at $r_0$ must be used. The radial wave function $R(r)$ and its derivative $R'(r)$ must both be continuous at $r_0$; this is equivalent to requiring that $u(r)$ and $u'(r)$ must both be continuous at $r_0$, where $u(r) = rR(r)$. The resulting equations cannot be solved exactly but can be used to derive the value for $V_0$.)

**6.16** Determine all potentials $V(r, \theta, \phi)$ for which it is possible to find solutions of the time-independent Schrödinger equation which are also eigenfunctions of the operator $L_z$.

**6.17** A particle with mass $m$ is confined inside of a spherical cavity of radius $r_0$. The potential is spherically symmetric and can be written in the form: $V(r) = 0$ for $r < r_0$, and $V(r) = \infty$ for $r = r_0$; in other words, there is an infinite potential barrier at $r = r_0$. The particle is in the $l = 0$ state.

(a) Solve the radial Schrödinger equation and use the appropriate boundary conditions to find the ground state radial wave function $R(r)$ and the ground state energy. (You do not have to normalize the solution).

(b) What is the pressure exerted by the particle (in the $l = 0$ ground state) on the surface of the sphere?

**6.18** A particle of mass $m$ is in a three-dimensional, spherically-symmetric harmonic oscillator potential given by $V(r) = (1/2)Kr^2$. The particle is in the $l = 0$ state. Find the ground-state radial wave function $R(r)$ and the ground-state energy. (You do not have to normalize the solution).

**6.19** Deuterium is an isotope of hydrogen with a nucleus consisting of one proton and one neutron. Let $\lambda(D)_{2\to1}$ be the wavelength of the photon emitted when the electron in a deuterium atom drops from the $n = 2$ state to the $n = 1$ state, and let $\lambda(H)_{2\to1}$ be the corresponding wavelength for ordinary hydrogen. Calculate $\lambda(D)_{2\to1} - \lambda(H)_{2\to1}$.

Chapter 7

# Math interlude C: Matrices, Dirac notation, and the Dirac delta function

In Chapter 5, we examined some of the general properties of vector spaces. We have, so far, treated functions as general vectors in an abstract vector space, with the inner product represented as an integral over pairs of the functions, and linear operators transforming one function into another. Here we examine two other ways of treating vector spaces. In the next section, we examine finite-dimensional vectors, for which linear operators are represented by matrices. In the following section, we introduce Dirac notation, which is simply a general means to represent arbitrary vectors in an abstract vector space. Finally, we discuss briefly an unrelated topic, but one which comes up frequently in quantum mechanics as well as in other areas of physics: the Dirac delta function.

## 7.1   The Matrix Formulation of Linear Operators

As we have seen in Chapter 5, the set of wave functions $\psi(x)$ can be treated as an infinite-dimensional vector space with the appropriate definitions for linear operators and inner products. However, we will often be interested in finite-dimensional vector spaces. These vector spaces occur, for instance, in the study of angular momentum. For example, suppose that a particle has total angular momentum $j = 1/2$, so that the possible $m_j$ states are $-1/2$ and $+1/2$. We can represent these two states as column vectors; for instance, $\begin{pmatrix} 1 \\ 0 \end{pmatrix}$ can represent the $m_j = +1/2$ state, and $\begin{pmatrix} 0 \\ 1 \end{pmatrix}$ can represent the $m_j = -1/2$ state. In general, we will use column vectors of the form $\begin{pmatrix} x_1 \\ x_2 \\ x_3 \end{pmatrix}$ to represent general finite-dimensional vectors; the dimension of the vector space is then just the number of entries in the column.

$$\begin{matrix} C & A & B \\ \begin{pmatrix} x \end{pmatrix} & = \begin{pmatrix} \longrightarrow \end{pmatrix} \begin{pmatrix} \downarrow \end{pmatrix} \end{matrix}$$

Fig. 7.1  If $C = AB$, then to obtain a given element in matrix $C$, scan across the corresponding row of matrix $A$ and the corresponding column of matrix $B$, multiply each pair of numbers and add the result.

How is a linear operator represented using this formulation? If we are dealing with an $n$-dimensional vector space, so that the vectors are column vectors with $n$ entries, then a general linear operator is an $n \times n$ matrix, and the act of operating on the vector is simply given by multiplying the matrix by the column vector.

Recall how matrix multiplication works. Suppose that $A$ is an $l \times m$ matrix, and $B$ is an $m \times n$ matrix. Then the elements of the product, $C = AB$, are given by

$$C_{jk} = \sum_{i=1}^{m} A_{ji} B_{ik} \qquad (7.1)$$

where $A_{ji}$, for example, is the element in the $j$th row and the $i$th column of $A$. Thus, multiplication of an $l \times m$ matrix by an $m \times n$ matrix yields an $l \times n$ matrix, and such multiplication is possible only if the number of columns in $A$ is equal to the number of rows in $B$.

More graphically, we can think of a single element in matrix $C$ being generated by scanning across the corresponding row of matrix $A$ and the corresponding column of matrix $B$, multiplying each pair of numbers and adding the result (Figure 7.1).

---

### Example 7.1. Matrix Multiplication Is not Commutative

Consider the two matrices $A = \begin{pmatrix} 1 & 2 \\ 3 & 4 \end{pmatrix}$ and $B = \begin{pmatrix} 5 & 6 \\ 7 & 8 \end{pmatrix}$. The rules for matrix multiplication give

$$AB = \begin{pmatrix} 1 & 2 \\ 3 & 4 \end{pmatrix} \begin{pmatrix} 5 & 6 \\ 7 & 8 \end{pmatrix} = \begin{pmatrix} 19 & 22 \\ 43 & 50 \end{pmatrix}$$

while

$$BA = \begin{pmatrix} 5 & 6 \\ 7 & 8 \end{pmatrix} \begin{pmatrix} 1 & 2 \\ 3 & 4 \end{pmatrix} = \begin{pmatrix} 23 & 34 \\ 31 & 46 \end{pmatrix}$$

Clearly $AB \neq BA$ in this case.

---

A column vector with $n$ entries can be treated as an $n \times 1$ matrix. Hence, multiplying an $n \times n$ matrix by an $n \times 1$ column vector yields another $n \times 1$ column vector. Thus, for a vector space consisting of $n$-dimensional column vectors, the operators are $n \times n$ matrices:

$$\begin{pmatrix} A_{11} & A_{12} & A_{13} \\ A_{21} & A_{22} & A_{23} \\ A_{31} & A_{32} & A_{33} \end{pmatrix} \begin{pmatrix} x_1 \\ x_2 \\ x_3 \end{pmatrix} = \begin{pmatrix} y_1 \\ y_2 \\ y_3 \end{pmatrix}$$

The multiplication of two operators represented by the matrices $A$ and $B$ corresponds to multiplication of the two matrices. Since matrix multiplication is not commutative (Example 7.1), it will general not be true that any two operators will commute, as we have already noted for many quantum mechanical operators.

We can also find the eigenvalues and eigenvectors of an operator represented as a matrix. Suppose we have a matrix $A$ multiplying a column vector $\mathbf{x}$, and assume that one of the eigenvalues is $c$. Then the eigenvalue equation is

$$A\mathbf{x} = c\mathbf{x} \tag{7.2}$$

Because we are dealing with matrices, this can be written as

$$A\mathbf{x} = cI\mathbf{x} \tag{7.3}$$

where $I$ is the *identity matrix* with 1's along the main diagonal and 0's everywhere else. The identity matrix has the property that $I\mathbf{x} = \mathbf{x}$ for any column vector $\mathbf{x}$. For example, in three dimensions,

$$\begin{pmatrix} 1 & 0 & 0 \\ 0 & 1 & 0 \\ 0 & 0 & 1 \end{pmatrix} \begin{pmatrix} x_1 \\ x_2 \\ x_3 \end{pmatrix} = \begin{pmatrix} x_1 \\ x_2 \\ x_3 \end{pmatrix}$$

Then Equation (7.3) can be rewritten as

$$(A - cI)\mathbf{x} = 0 \tag{7.4}$$

In three dimensions, for example, Equation (7.4) corresponds to

$$\begin{pmatrix} A_{11} - c & A_{12} & A_{13} \\ A_{21} & A_{22} - c & A_{23} \\ A_{31} & A_{32} & A_{33} - c \end{pmatrix} \begin{pmatrix} x_1 \\ x_2 \\ x_3 \end{pmatrix} = \begin{pmatrix} 0 \\ 0 \\ 0 \end{pmatrix}$$

A matrix equation of this form always has a trivial solution of the form

$$\begin{pmatrix} x_1 \\ x_2 \\ x_3 \end{pmatrix} = \begin{pmatrix} 0 \\ 0 \\ 0 \end{pmatrix}$$

i.e., a vector for which all of the entries are zero. This is not a very interesting solution. In order for Equation (7.4) to have a nonzero solution, the determinant of $A - cI$ must be zero. For example, in three dimensions, the condition for a nonzero solution is

$$\begin{vmatrix} A_{11} - c & A_{12} & A_{13} \\ A_{21} & A_{22} - c & A_{23} \\ A_{31} & A_{32} & A_{33} - c \end{vmatrix} = 0$$

An $n \times n$ determinant of this kind, in which each term on the main diagonal is of the form $A_{ii} - c$, corresponds to a polynomial of degree $n$ in the variable $c$. Thus, there will always be $n$ complex values of $c$ for which the determinant is zero, corresponding to $n$ complex eigenvalues (not necessarily all distinct). The first step is to find these eigenvalues; they can then each be inserted, in turn, into Equation (7.2) to find the corresponding eigenvectors. Note from Equation (7.2) that an eigenvector multiplied by an arbitrary constant will remain an eigenvector with the same eigenvalue, a result that is already familiar from our earlier work with eigenvectors.

---

**Example 7.2. The Eigenvalues and Eigenvectors of the Matrix** $A = \begin{pmatrix} 0 & 1 \\ 1 & 0 \end{pmatrix}$

Assume an eigenvalue $c$. Then the determinant equation is

$$\begin{vmatrix} -c & 1 \\ 1 & -c \end{vmatrix} = 0$$

Expanding out the determinant gives

$$c^2 - 1 = 0$$

so

$$c = \pm 1$$

Thus, the eigenvalues are $c = -1$ and $c = 1$. These can now be inserted into the eigenvalue equation. First, for $c = 1$,

$$\begin{pmatrix} 0 & 1 \\ 1 & 0 \end{pmatrix} \begin{pmatrix} x_1 \\ x_2 \end{pmatrix} = 1 \begin{pmatrix} x_1 \\ x_2 \end{pmatrix}$$

which yields the two equations:

$$x_2 = x_1$$
$$x_1 = x_2$$

Note that these two equations are not independent. This will *always* be the case, since there must always be one extra degree of freedom, which allows multiplication of the final solution by an arbitrary constant. If a set of completely independent equations has been obtained (so that every component of the vector $\mathbf{x}$ can be calculated exactly), then the eigenvalues have been calculated incorrectly. Now we fix $x_1$ to any value, and $x_2$ will be determined. So, for example, we can choose $x_1 = 1$, which gives $x_2 = 1$, yielding the eigenvector $\begin{pmatrix} 1 \\ 1 \end{pmatrix}$, with eigenvalue $c = 1$.

Now consider $c = -1$. For this case, we get

$$\begin{pmatrix} 0 & 1 \\ 1 & 0 \end{pmatrix} \begin{pmatrix} x_1 \\ x_2 \end{pmatrix} = -1 \begin{pmatrix} x_1 \\ x_2 \end{pmatrix}$$

so

$$x_2 = -x_1$$
$$x_1 = -x_2$$

Again, these are not independent equations. We can take, for example, $x_1 = 1$, which gives $x_2 = -1$, yielding the eigenvector $\begin{pmatrix} 1 \\ -1 \end{pmatrix}$. However, we could just as easily have taken $x_1 = -1$, giving $x_2 = 1$ and producing the eigenvector $\begin{pmatrix} -1 \\ 1 \end{pmatrix}$. It would appear that we have found two different eigenvectors. However, this is not the case: $\begin{pmatrix} 1 \\ -1 \end{pmatrix}$ and $\begin{pmatrix} -1 \\ 1 \end{pmatrix}$ are two different representations of the *same* vector; they differ by an overall multiplicative factor of $-1$.

---

The inner product of two vectors $\mathbf{x}$ and $\mathbf{y}$ is given by

$$(\mathbf{x}|\mathbf{y}) = \begin{pmatrix} x_1^* & x_2^* & \cdots & x_n^* \end{pmatrix} \begin{pmatrix} y_1 \\ y_2 \\ \vdots \\ y_n \end{pmatrix} = x_1^* y_1 + x_2^* y_2 + \cdots + x_n^* y_n \qquad (7.5)$$

This is identical to the familiar dot product of two vectors, except that all of the elements of the first vector are complex-conjugated (which will matter only when they are not real numbers). This definition of the inner product can be used to normalize any $n$-dimensional column vector. In order that $\mathbf{x}$ be normalized, we require $(\mathbf{x}|\mathbf{x}) = 1$, which corresponds to

$$\begin{pmatrix} x_1^* & x_2^* & \cdots & x_n^* \end{pmatrix} \begin{pmatrix} x_1 \\ x_2 \\ \vdots \\ x_n \end{pmatrix} = |x_1|^2 + |x_2|^2 + \cdots + |x_n|^2 = 1 \qquad (7.6)$$

To normalize $\mathbf{x}$, we multiply $\mathbf{x}$ by a constant $c$ so that Equation (7.6) is satisfied.

---

**Example 7.3. Normalizing the Vector** $\begin{pmatrix} 1 \\ 3 \\ 2i \end{pmatrix}$

We multiply this vector by the constant $c$, and then require that Equation (7.6) be satisfied:

$$c^* \begin{pmatrix} 1 & 3 & -2i \end{pmatrix} c \begin{pmatrix} 1 \\ 3 \\ 2i \end{pmatrix} = 1$$

Multiplying out the matrices gives

$$|c|^2 (1 + 9 + 4) = 1$$

for which we choose the positive real solution, $c = 1/\sqrt{14}$. As in our previous derivation of normalization constants, there is an arbitrary phase factor, i.e., $c$ can be multiplied by $e^{i\phi}$, where $\phi$ is any real number, and the vector will still be normalized. However, it is usually easiest to take $c$ to be real and positive, which we will do here. Hence, the normalized vector is

$$\mathbf{x} = \frac{1}{\sqrt{14}} \begin{pmatrix} 1 \\ 3 \\ 2i \end{pmatrix}$$

---

If we insert an operator into the inner product, we end up with the product of three matrices:

$$(\mathbf{x}|A\mathbf{y}) = \begin{pmatrix} x_1^* & x_2^* & \cdots & x_n^* \end{pmatrix} \begin{pmatrix} A_{11} & A_{12} & \cdots \\ A_{21} & & \\ \vdots & & A_{nn} \end{pmatrix} \begin{pmatrix} y_1 \\ y_2 \\ \vdots \\ y_n \end{pmatrix}$$

Using these definitions, it is possible to determine the adjoint of a matrix operator. Recall that the adjoint $A^\dagger$ of the operator $A$ has the property that

$$(\mathbf{x}|A\mathbf{y}) = (A^\dagger \mathbf{x}|\mathbf{y}) \tag{7.7}$$

for all $\mathbf{x}$ and $\mathbf{y}$. Using Equation (7.5) to write Equation (7.7) in matrix form, and expanding the matrix products out in terms of their components as in Equation (7.1), gives

$$\sum_{i,j=1}^{n} x_i^* A_{ij} y_j = \sum_{i,j=1}^{n} (A_{ij}^\dagger x_j)^* y_i$$

Note that matrix *elements* are just numbers, so they do commute, and we can rewrite the right-hand side of the equation:

$$\sum_{i,j=1}^{n} x_i^* A_{ij} y_j = \sum_{i,j=1}^{n} x_j^* A_{ij}^{\dagger *} y_i$$

$$= \sum_{j,i=1}^{n} x_i^* A_{ji}^{\dagger *} y_j \tag{7.8}$$

where, in the last line, we have simply interchanged the summation labels $i$ and $j$. Equation (7.8) implies that

$$A_{ij}^\dagger = A_{ji}^*$$

so that the adjoint of a matrix operator corresponds to the conjugate transpose of the matrix, i.e., the interchange of rows and columns and the complex-conjugating of all of the entries.

---

**Example 7.4. Determining the Adjoint of a Matrix**

Consider the matrix operator

$$L = \begin{pmatrix} 1 & i \\ -3 & e^{i\theta} \end{pmatrix}$$

where $\theta$ is a real number. What is the matrix corresponding to the adjoint operator $L^\dagger$?

First transpose the matrix to get $\begin{pmatrix} 1 & -3 \\ i & e^{i\theta} \end{pmatrix}$. Then take the complex-conjugate of all of the entries to obtain the adjoint operator $L^\dagger$. This gives us:

$$L^\dagger = \begin{pmatrix} 1 & -3 \\ -i & e^{-i\theta} \end{pmatrix}$$

---

The matrix corresponding to a Hermitian operator must be self-adjoint. Thus, such a matrix must be equal to its conjugate transpose. For a real matrix, this means that the matrix is symmetric, e.g., $\begin{pmatrix} 1 & 2 \\ 2 & 7 \end{pmatrix}$. But Hermitian complex matrices need not be symmetric. For example, the matrix $\begin{pmatrix} 0 & -i \\ i & 0 \end{pmatrix}$ corresponds to a Hermitian operator.

## 7.2   Dirac Notation

In Chapter 5 we introduced the properties of an abstract vector space, as well as operators and inner products acting on that space. We then made use of a special case in which the elements of the vector space were functions, and the inner product was simply an integral:

$$(f|g) = \int f(x)^* g(x) \, dx$$

However, as noted in the previous section, another perfectly acceptable example of a vector space is the set of column vectors for which the inner product is, instead,

$$(\mathbf{x}|\mathbf{y}) = \begin{pmatrix} x_1^* & x_2^* & \cdots & x_n^* \end{pmatrix} \begin{pmatrix} y_1 \\ y_2 \\ \vdots \\ y_n \end{pmatrix}$$

Often, however, we will want to deal with general vector spaces and general inner products, without reference to whether we are talking about functions, column vectors, or some other special class of vectors. In fact, we have already developed the abstract notation to deal with general vector spaces, operators, and inner products in Chapter 5. This notation is commonly used in the world of mathematics. In quantum mechanics, however, most physicists use a slightly different notation called *Dirac notation*, which we now discuss.

In Dirac notation, a general vector is written like this: $|\psi\rangle$. These vectors satisfy all of the properties of a vector space, as defined in Chapter 5, e.g., the sum of two vectors is a vector:

$$|\psi_1\rangle + |\psi_2\rangle = |\psi_3\rangle$$

and the product of a vector and a complex number is a vector:

$$c|\phi_1\rangle = |\phi_2\rangle$$

To represent the inner product of two vectors, $|\psi\rangle$ and $|\phi\rangle$, we simply write $\langle\phi|\psi\rangle$. Now suppose that the vectors $|\psi_1\rangle$, $|\psi_2\rangle$, ..., $|\psi_n\rangle$ represent an orthonormal basis set for our vector space, i.e.,

$$\langle\psi_i|\psi_j\rangle = \delta_{ij}$$

We saw in Chapter 5 that an arbitrary three-dimensional vector $\mathbf{r}$ could be expressed as a sum of orthonormal basis set vectors as

$$\mathbf{r} = (\mathbf{r}\cdot\hat{x})\hat{x} + (\mathbf{r}\cdot\hat{y})\hat{y} + (\mathbf{r}\cdot\hat{z})\hat{z}$$

The generalization of this to arbitrary vectors in Dirac notation is

$$|\phi\rangle = |\psi_1\rangle\langle\psi_1|\phi\rangle + |\psi_2\rangle\langle\psi_2|\phi\rangle + \cdots + |\psi_n\rangle\langle\psi_n|\phi\rangle \qquad (7.9)$$

Note that in this equation, the constants multiplying the basis vectors are the inner products $\langle\psi_1|\phi\rangle$, $\langle\psi_2|\phi\rangle$, ..., $\langle\psi_n|\phi\rangle$; we have written these constants to the *right* of the basis vectors themselves, $|\psi_1\rangle$, $|\psi_2\rangle$, ..., $|\psi_n\rangle$. This is completely equivalent to the notation developed in Chapter 5, where we used $(\phi|\psi)$ to represent an inner product. We have not introduced any new mathematics so far; we have only rewritten the ideas developed in Chapter 5 in this new notation.

Now we introduce a new property of vector spaces. Suppose we have an inner product of two Dirac vectors, given by

$$\langle\phi|\psi\rangle$$

The right-hand part of this inner product is just the vector $|\psi\rangle$. But suppose we detach the left-hand side, and write it as the quantity $\langle\phi|$. Does this have any meaning at all? We can combine $\langle\phi|$ with any vector $|\psi\rangle$ to get the complex number $\langle\phi|\psi\rangle$. Thus, the quantity $\langle\phi|$ is a function which maps the vector space into the set of complex numbers. It is clumsier to express this concept in terms of the inner product notation developed in Chapter 5; in that notation we would need to write $\langle\phi|$ as $(\phi| \quad )$, where any vector could be inserted in the blank space to produce a complex number. It turns out that this set of all mappings of the vector space into the set of complex numbers has all of the properties of a vector space itself, so it is given a special name: the *dual space*. Given a set of vectors $|\psi_1\rangle$, $|\psi_2\rangle$, ..., $|\psi_n\rangle$, the dual space consists of $\langle\psi_1|$, $\langle\psi_2|$, ..., $\langle\psi_n|$, with all of the properties of a vector space, namely, the sum of two elements of the dual space is a member of the dual space:

$$\langle\psi_1| + \langle\psi_2| = \langle\psi_3|$$

as is the product of a complex number with a member of the dual space:

$$c\langle\phi_1| = \langle\phi_2|$$

The inner product in Dirac notation $\langle\phi|\psi\rangle$ is sometimes called a *bracket*, so that $\langle\psi_1|, \langle\psi_2|, \ldots, \langle\psi_n|$ are called *bra* vectors, and $|\psi_1\rangle, |\psi_2\rangle, \ldots, |\psi_n\rangle$ are called *ket* vectors.

With this new idea of dual vectors, Equation (7.9) can be written in a particularly elegant fashion:

$$\sum_n |\psi_n\rangle\langle\psi_n| = 1$$

In Dirac notation, linear operators behave in the standard way, e.g., if $P$ is a linear operator, we have

$$P(c|\psi\rangle) = cP|\psi\rangle$$

and

$$P(|\psi_1\rangle + |\psi_2\rangle) = P|\psi_1\rangle + P|\psi_2\rangle$$

Inner products with an operator inserted are written as $\langle\phi|P|\psi\rangle$; this is equivalent to $(\phi|P\psi)$ in the notation of Chapter 5. Finally, the dual of the vector $P|\psi\rangle$ is simply $\langle\psi|P^\dagger$.

It is important to remember that vectors and operators in Dirac notation are completely general; they will sometimes be used to represent finite-dimensional vector spaces, for which the operators can be represented as matrices and the vectors as column vectors, but they can also represent infinite-dimensional vector spaces as well. For example, the time-independent Schrödinger equation is simply

$$H|\psi\rangle = E|\psi\rangle$$

Sometimes this will represent the familiar Schrödinger equation with derivatives of wave functions, but sometimes it will represent matrix quantities, such as in the next chapter.

## 7.3   The Dirac Delta Function

In this section we will introduce a function that comes up frequently in physics and has some rather remarkable properties: the Dirac delta function. The delta function, $\delta(x)$, has the following properties:

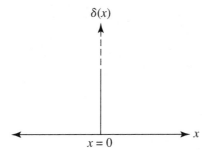

Fig. 7.2   The delta function is zero everywhere except at $x = 0$, where it is infinitely peaked.

$$\delta(x) = 0, \quad \text{for } x \neq 0 \tag{7.10}$$

$$\int_{-\infty}^{\infty} \delta(x)\, dx = 1 \tag{7.11}$$

$$\int_{-\infty}^{\infty} \delta(x) f(x)\, dx = f(0) \tag{7.12}$$

Although the range of these integrals has been taken from $-\infty$ to $\infty$, these equations are valid as long as the range of integration includes the point $x = 0$.

A function which can satisfy these equations must be very strange indeed. It is zero everywhere except at $x = 0$, but in order for it to integrate to 1 (Equation (7.11)), it must be infinitely sharply peaked at the origin (Figure 7.2). In fact, from a mathematician's point of view, the delta function is not really a function at all, but it can be defined in a rigorous way as the limit of a set of functions of constant (unit) area, but with an increasingly narrower width (Figure 7.3). The property for which the delta

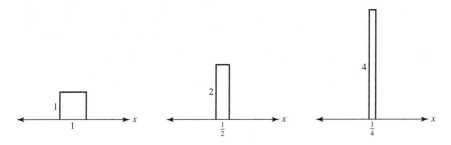

Fig. 7.3   The delta function can be treated as the limit of a set of functions with unit area but decreasing width.

function is primarily used is given by Equation (7.12): the integral of the product of a delta function with any other function $f(x)$ "picks out" the value of $f(x)$ at the origin. If we wish to pick out the value of $f(x)$ at some other value of $x$, a change of variables gives

$$\int_{-\infty}^{\infty} \delta(x-a)f(x) = f(a) \tag{7.13}$$

The Kronecker delta, which we have already encountered, is a discrete version of the delta function; the expression corresponding to Equation (7.13) for the Kronecker delta is

$$\sum_{i=1}^{\infty} \delta_{in} c_i = c_n$$

We can also define a three-dimensional delta function, $\delta^3(\mathbf{r})$, given by

$$\delta^3(\mathbf{r}) = \delta(x)\delta(y)\delta(z)$$

Then the equation corresponding to Equation (7.13) in three dimensions is

$$\int \delta^3(\mathbf{r}-\mathbf{r}_0)f(\mathbf{r})\,d^3\mathbf{r} = f(\mathbf{r}_0)$$

where the range of integration must include the point $\mathbf{r}_0$. Thus, the integral over the three-dimensional delta function picks out the value of the function at a single point in three-dimensional space.

The delta function is useful in representing the kinds of idealized point distributions that are frequently encountered in physics. For example, consider the charge density, $\rho(\mathbf{r})$, produced by a point charge $e$ at the position $\mathbf{r}_0$. This charge density can be written in terms of a delta function:

$$\rho(\mathbf{r}) = e\delta^3(\mathbf{r}-\mathbf{r}_0)$$

This charge density is infinite at the point $\mathbf{r}_0$ and zero everywhere else, but the total charge $Q$ is well-defined:

$$Q = \int \rho(\mathbf{r})\,d^3\mathbf{r} = \int e\delta^3(\mathbf{r}-\mathbf{r}_0)\,d^3\mathbf{r} = e$$

## PROBLEMS

**7.1** The operator $Q$ is given by the matrix:

$$Q = \begin{pmatrix} 1 & i \\ -i & -1 \end{pmatrix}$$

(a) Determine the matrix corresponding to $Q^\dagger$.
(b) Is $Q$ Hermitian?
(c) Find the eigenvalues of $Q$.
(d) For each eigenvalue in part (c), determine the corresponding eigenvector.

**7.2** The operator $A$ is given by the matrix:

$$A = \begin{pmatrix} 1 & 0 & 1 \\ 0 & 0 & 0 \\ 1 & 0 & 1 \end{pmatrix}$$

(a) Is $A$ Hermitian?
(b) Find the eigenvalues and corresponding eigenvectors.
(c) What is unusual about the eigenvectors corresponding to the eigenvalue $c = 0$?

**7.3** Suppose that an $n \times n$ matrix $A$ is diagonal so that $A_{ij} = 0$ for $i \neq j$, but the diagonal elements $A_{11}, A_{22}, \ldots$ need not be zero. Assume $A_{11} \neq A_{22} \neq \ldots A_{nn}$. Find the eigenvalues and eigenvectors of this matrix.

**7.4** The *trace* of a matrix $A$, written $tr(A)$, is defined to be the sum of its diagonal elements:

$$tr(A) = \sum_{i=1}^{n} A_{ii}$$

(a) Show that for any two square matrices, $tr(AB) = tr(BA)$.
(b) Show that for any matrix $A$, the trace is equal to the sum of its eigenvalues (where multiple eigenvalues must be included in the sum multiple times).

**7.5** Normalize these vectors: $\begin{pmatrix} 1 \\ -1 \end{pmatrix}$, $\begin{pmatrix} 1 \\ i \\ -1 \end{pmatrix}$, $\begin{pmatrix} -1 \\ 2 \\ -3 \\ 4 \end{pmatrix}$.

**7.6** A particle is in the state $|\phi\rangle$, and let $|\psi_1\rangle$, $|\psi_2\rangle, \ldots, |\psi_n\rangle$ be an orthonormal basis for the vector space which contains $|\phi\rangle$. $Q$ is a Hermitian operator. Show that

$$\langle Q^2 \rangle = \sum_{j=1}^{n} |\langle \phi | Q | \psi_j \rangle|^2$$

**7.7** Suppose that $|\psi_1\rangle$, $|\psi_2\rangle, \ldots, |\psi_n\rangle$ is an orthonormal basis set, and all of the basis vectors are eigenvectors of the operator $Q$ with $Q|\psi_j\rangle = q_j|\psi_j\rangle$ for all $j$. A particle is in the state $|\phi\rangle$. Show that for this particle, the expectation value of $Q$ is

$$\langle Q \rangle = \sum_{j=1}^{n} q_j |\langle \phi | \psi_j \rangle|^2$$

**7.8** If the operator $U$ has the property that $U^\dagger U = I$ (where $I$ is the identity operator), then $U$ is called a *unitary* operator. Show that if $|\psi_1\rangle$, $|\psi_2\rangle, \ldots, |\psi_n\rangle$ are a set of orthonormal vectors, then $U|\psi_1\rangle$, $U|\psi_2\rangle, \ldots, U|\psi_n\rangle$ are also a set of orthonormal vectors.

**7.9** Suppose that $U$ is a unitary operator, as defined in Exercise 7.8, and $U$ is represented by a matrix. Show that the columns of $U$ form a set of orthonormal column vectors.

**7.10** A particle is in the state $|\phi\rangle$. Let $|\psi_1\rangle$, $|\psi_2\rangle, \ldots, |\psi_n\rangle$ be an orthonormal basis for the vector space which contains $|\phi\rangle$, and assume that all of these basis vectors are eigenvectors of $H$ with $H|\psi_j\rangle = E_j|\psi_j\rangle$ for all $j$. Suppose that the operator $Q$ satisfies

$$\langle Q \rangle = \sum_{m} E_m |\langle \phi | Z | \psi_m \rangle|^2$$

where $Z$ is just the usual position operator in the $z$ direction. Derive an expression for $Q$ as a function of $H$ and $Z$, but not containing $E_m$.

**7.11** The delta function is sometimes represented as

$$\delta(x) = \frac{1}{2\pi} \int_{-\infty}^{\infty} e^{ikx}\, dk$$

Show that this definition satisfies the properties of the delta function given in Equations (7.10)–(7.12).

**7.12** Show that

$$\delta(ax) = \frac{\delta(x)}{|a|}$$

**7.13** Show that

$$\int_{-\infty}^{\infty} f(x)\delta(g(x)) \, dx = \frac{f(x_0)}{|dg/dx|_{x=x_0}}$$

where $x_0$ is determined by $g(x_0) = 0$.

Chapter 8

# Spin angular momentum

In Chapter 6 we noted that there are two types of angular momentum most relevant in quantum mechanics: orbital angular momentum, produced at the classical level by the physical motion of a particle, and spin angular momentum, which represents a type of angular momentum internal to the particle. Here we will examine the latter in more detail. Although it is tempting to think of "spin" as being produced by the actual internal rotation of a particle, this is misleading. It is more accurate to treat spin as an intrinsic property of the particle, like charge or mass.

## 8.1 Spin Operators

All of the formalism derived for angular momentum operators in Chapter 6 carries over to spin operators. In particular, the operators $S_x$, $S_y$, and $S_z$ give the components of spin angular momentum in the $x$, $y$, and $z$ directions, respectively, while $S^2 = S_x^2 + S_y^2 + S_z^2$ is the operator corresponding to the square of the total angular momentum. Further, these operators satisfy the standard angular momentum commutation relations derived in Chapter 6:

$$[S_x, S_y] = i\hbar S_z \tag{8.1}$$
$$[S_z, S_x] = i\hbar S_y \tag{8.2}$$
$$[S_y, S_z] = i\hbar S_x \tag{8.3}$$

As in the case of orbital angular momentum, it is impossible to measure all three components of spin simultaneously, but it is possible to measure a single component of **S** and the total squared angular momentum $S^2$. As before, we will normally make the somewhat arbitrary decision to measure the $z$ component of angular momentum. Then a particle which is in an eigenstate of the operators $S^2$ and $S_z$ has a wave function $|\psi\rangle$ satisfying

$$S^2|\psi\rangle = \hbar^2 s(s+1)|\psi\rangle \tag{8.4}$$
$$S_z|\psi\rangle = \hbar m_s|\psi\rangle$$

However, there are also significant differences between spin angular momentum and orbital angular momentum. First, recall that the eigenfunctions of orbital angular momentum could be written as spatial wave functions, namely the spherical harmonics $Y_l^m(\theta, \phi)$. There are no spatial wave functions corresponding to eigenstates of spin, since spin is a purely internal property of a particle. Second, the total spin eigenvalue $s$ has a fixed value for any given particle, and, unlike orbital angular momentum, it cannot be increased or decreased for a single particle. For example, the electron in a hydrogen atom can attain arbitrarily large values for $l$ (for arbitrarily large values of $n$). However, its spin is always $s = 1/2$. This is one reason that viewing spin as the physical rotation of a particle is misleading; an elementary particle cannot be "spun up" to obtain larger values of spin. Finally, we saw in Chapter 6 that the orbital angular momentum quantum number was restricted to integer values. This is not true for spin; the spin quantum number can take on the full range of both integer and half-integer values: $s = 0, 1/2, 1, 3/2, \ldots$.

All elementary particles in nature have an intrinsic spin, given by the $s$ quantum number of Equation (8.4). In fact, spin is such a fundamental property that particles are classified according to their spin: particles with integer spin ($s = 0, 1, 2, \ldots$) are called *bosons*, while particles with half-integer spin ($s = 1/2, 3/2, \ldots$) are called *fermions*. The properties of fermions and bosons in multi-particle collections are fundamentally different, as we will see in Chapter 13. All of the particles making up the ordinary matter in the universe, i.e., protons, neutrons, and electrons, have $s = 1/2$ and are therefore fermions. Bosons include the photon ($s = 1$), the pion ($s = 0$), and the graviton ($s = 2$), which is hypothesized to transmit the gravitational force.

## 8.2    Evidence for Spin

What is the reason for believing that spin angular momentum even exists? Today, the entire edifice of particle physics is built on the idea that particles have spin, so there are countless experiments confirming it. But there were two main pieces of experimental evidence available to the pioneers of quantum mechanics, and both are based on the same mechanism: the link between angular momentum and the production of magnetic fields.

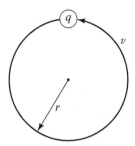

Fig. 8.1   A particle of charge $q$ moving in a circular orbit of radius $r$ at velocity of magnitude $v$ produces a magnetic moment $\mu = qvr/2$.

Consider first the case of orbital angular momentum, and recall how magnetic fields are generated by a classical charged particle moving in a circular trajectory. A classical particle with charge $q$, moving in a circular orbit of radius $r$ with velocity of magnitude $v$ (Figure 8.1) produces the current

$$I = \frac{qv}{2\pi r}$$

and the magnetic moment is

$$\mu = IA$$
$$= \frac{qv}{2\pi r}\pi r^2$$
$$= \frac{qvr}{2}$$

Further, the classical angular momentum is

$$L = mvr$$

so that the relation between the magnetic moment and the angular momentum for a classical charged particle in a circular orbit is

$$\mu = \frac{q}{2m}L$$

For an electron with charge $q = -e$, this expression gives

$$\mu = -\frac{e}{2m_e}L \tag{8.5}$$

It is conventional to write the magnetic moment in terms of the *Bohr magneton*, which is defined by

$$\mu_B = \frac{e\hbar}{2m_e} = 9.3 \times 10^{-24} \ \mathrm{A \cdot m^2}$$

Then the expression for the magnetic moment of an electron can be rewritten as

$$\mu = -\frac{\mu_B}{\hbar} L$$

Note that $L$ and $\hbar$ have the same units, and $\mu_B$ has units of magnetic moment. Now recall that both the angular momentum and the magnetic moment are vectors. Further, we insert an extra factor $g_l$ into the expression for the magnetic moment, setting $g_l = 1$. This gives

$$\boldsymbol{\mu}_l = -\frac{g_l \mu_B}{\hbar} \mathbf{L} \tag{8.6}$$

Why did we insert this extra factor $g_l$, only to set it to one? The reason is that when we generalize our result to include spin angular momentum, we will have another factor $g_s$, but it will not necessarily be equal to one. Using $g_l$ in Equation (8.6) preserves the parallel between the orbital magnetic moment and the spin magnetic moment.

Arguing from analogy to Equation (8.6), we now postulate that the spin angular momentum will also generate a magnetic field with magnetic moment of the form

$$\boldsymbol{\mu}_s = -\frac{g_s \mu_B}{\hbar} \mathbf{S} \tag{8.7}$$

It is observed experimentally that Equation (8.7) does, in fact, apply to the electron, but now $g_s \neq 1$. The value of $g_s$ for the electron is, in fact, one of the best measured quantities in nature (X. Fan, et al., *Physical Review Letters* **130**, 071801, 2023):

$$g_s = 2.00231930436118 \pm 0.00000000000026$$

This measured value for $g_s$ leads to some obvious questions: Why is it almost exactly equal to 2? And why is it not exactly equal to 2? It turns out that the answers to both of these questions are very significant. The answer to the first question can be derived from the Dirac equation, which is the basic equation of relativistic quantum mechanics (Chapter 15). The Dirac equation predicts that $g_s$ should be *exactly* equal to 2. The small deviation from 2 is a real effect, but it was not explained until the invention of *quantum field theory* by Richard Feynmann and others in the 1940's. This small deviation from 2 is called the *anomalous magnetic moment* of the electron, and its calculation is far beyond the scope of this book. For most practical purposes, it is sufficient to take $g_s = 2$.

These derivations of $\boldsymbol{\mu}_l$ and $\boldsymbol{\mu}_s$ are based, for $\mu_l$, on purely classical arguments, and, for $\mu_s$, on analogy from the $\mu_l$ derivation, so why should they be believed? The evidence, as always, is based on experiment. The first piece of evidence comes from the energy levels of the atom. Our derivation of the hydrogen energy levels in Chapter 6 is incomplete. In the spectrum of hydrogen, for example, experimental evidence shows that certain of the degenerate energy levels are not, in fact, completely degenerate but are separated by a tiny amount in energy. This can be explained by the interaction between the spin magnetic moment of the electron and its orbital magnetic moment; this interaction splits the degenerate levels apart in energy. This calculation will be performed in detail in Chapter 9 after the necessary mathematical machinery has been developed. Under the assumption that the orbital and spin magnetic moments are given by Equations (8.6) and (8.7), respectively, this splitting can be predicted correctly. Obviously, this mechanism will work only if the electron has *both* an orbital magnetic moment and a spin magnetic moment. The former will be produced by any charged particle with orbital angular momentum, but the latter makes sense only if the electron has its own intrinsic angular momentum. This explanation was first proposed by Goudsmit and Ulhenbeck in 1925, and it is now known to be correct.

A "cleaner" piece of evidence for the spin magnetic moment of the electron (and, therefore, for the spin of the electron) comes from the *Stern–Gerlach experiment*. Consider what happens to a classical bar magnet placed in an *inhomogeneous* magnetic field, like the one shown in Figure 8.2. In this figure, the field strength increases in the upward direction. The small magnet on the left will be attracted upward, since the field strength at the $S$ pole, which pulls the magnet up, is larger than the field strength at the $N$ pole, which pulls the magnet down. The magnet on the right is repelled in both directions, and the same argument shows that it will be pushed downward.

Now consider an ideal dipole $\boldsymbol{\mu}$ in an inhomogeneous field. The force exerted on the dipole is $\mathbf{F} = -\nabla V$, where $V$ is the potential energy of the dipole in the magnetic field. Assume for simplicity that $\mathbf{B}$ is in the $z$ direction, so

$$\mathbf{F} = -\frac{\partial V}{\partial z}\hat{z}$$

The potential energy of a magnetic dipole $\boldsymbol{\mu}$ in a magnetic field $\mathbf{B}$ is

$$V = -\boldsymbol{\mu}{\cdot}\mathbf{B}$$

Therefore, the force on the dipole can be written as

$$\mathbf{F} = \frac{\partial B_z}{\partial z}\mu_z\hat{z} \tag{8.8}$$

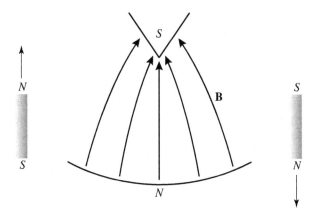

Fig. 8.2   The classical bar magnet on the left will be attracted upward in the inhomogeneous magnetic field, while the magnet on the right will be repelled downward.

Imagine that we shoot a beam of atoms into the page through the magnetic field shown in Figure 8.2 and measure their deflection on a screen behind the magnetic field. Equation (8.8) says that the force should be proportional to the $z$ component of the magnetic moment of the atom. In the absence of spin, Equation (8.6) indicates that this $z$ component should be

$$\mu_{lz} = -\frac{g_l \mu_B}{\hbar} L_z$$

So in the absence of spin, the force, and therefore the deflection of each atom, should depend on the $z$ component of the orbital angular momentum. If the atom is in an eigenstate of $L_z$ with quantum numbers $l$ and $m_l$, then

$$L_z |l\ m_l\rangle = \hbar m_l |l\ m_l\rangle$$

so the deflection should depend on $m_l$. Since $m_l$ can range from $-l$ to $+l$, we expect that the atoms would be split into $2l + 1$ discrete beams.

    Stern and Gerlach actually performed this experiment with silver atoms in 1922 and discovered that the beam formed two bands on the screen behind the magnet; in other words, the inhomogeneous magnetic field split the beam into exactly two separate beams. This experiment was repeated using hydrogen atoms by Phipps and Taylor in 1927. Here the experiment is even clearer in its predictions: hydrogen in its ground state has $l = 0$ and $m_l = 0$, so if orbital angular momentum is the only source of magnetic moments in the hydrogen atom, there should be no deflection in the inhomogeneous magnetic field at all. However, Phipps and Taylor also got the same answer: the beam of hydrogen atoms was split into two bands.

This result can be explained if the electron has spin angular momentum. Assume that the electron in the hydrogen atom has spin $s$. Then the $z$ component of the spin magnetic moment is

$$S_z |s \ m_s\rangle = \hbar m_s |s \ m_s\rangle$$

For the hydrogen atoms to be split into two beams requires that $m_s$ take on exactly two possible values, namely, $m_s = \pm 1/2$. This will be true if the electron has $s = 1/2$.

## 8.3 Adding Angular Momentum

The electrons in an atom have both orbital and spin angular momentum. Even in the simplest atom, hydrogen, both types of angular momentum will be present. It is important, therefore, to understand how to calculate the *total* angular momentum, including both the orbital and spin components. Classically, the addition of angular momentum is straightforward: it corresponds to a vector sum of the two individual components of angular momentum. In quantum mechanics, things are more complicated. We will define the total angular momentum operator $\mathbf{J}$ as the sum of the $\mathbf{L}$ and $\mathbf{S}$ operators:

$$\mathbf{J} = \mathbf{L} + \mathbf{S} \tag{8.9}$$

An eigenstate of total angular momentum and of the $z$ component of total angular momentum is written as $|j \ m_j\rangle$ with

$$J^2 |j \ m_j\rangle = \hbar^2 j(j+1)|j \ m_j\rangle$$
$$J_z |j \ m_j\rangle = \hbar m_j |j \ m_j\rangle \tag{8.10}$$

Suppose that we have a state with quantum numbers $l$, $m_l$, $s$, $m_s$, and we would like to calculate the possible values of $j$ and $m_j$. Clearly, $J^2 \neq L^2 + S^2$, so there is not a simple relationship between $j$, $l$, and $s$. However, it is true from Equation (8.9) that

$$J_z = L_z + S_z$$

Thus, if $|m_l \ m_s\rangle$ is an eigenstate of both $S_z$ and $L_z$ such that

$$L_z |m_l \ m_s\rangle = \hbar m_l |m_l \ m_s\rangle$$
$$S_z |m_l \ m_s\rangle = \hbar m_s |m_l \ m_s\rangle$$

then

$$J_z |m_l \ m_s\rangle = (L_z + S_z)|m_l \ m_s\rangle = \hbar(m_l + m_s)|m_l \ m_s\rangle$$

A comparison of this equation with Equation (8.10) shows that the $z$-component quantum numbers are additive:

$$m_j = m_l + m_s \tag{8.11}$$

This result also provides a clue as to the possible values of $j$. Since $m_l \leq l$ and $m_s \leq s$, Equation (8.11) implies

$$m_j \leq l + s \tag{8.12}$$

Since $m_j \leq j$, Equation (8.12) will automatically be satisfied as long as

$$j \leq l + s$$

This quantum mechanical relation is in agreement with the upper bound on the classical value of $|\mathbf{J}|$: classically, the largest possible value of $|\mathbf{J}|$ occurs when $\mathbf{L}$ and $\mathbf{S}$ are parallel, so that $|\mathbf{J}| = |\mathbf{L}| + |\mathbf{S}|$. The smallest possible value of $|\mathbf{J}|$ in the classical case occurs when $\mathbf{L}$ and $\mathbf{S}$ are pointing in opposite directions, so that $|\mathbf{J}| = ||\mathbf{L}| - |\mathbf{S}||$. This suggests the quantum analog

$$j \geq |l - s|$$

Note further that $j$ can vary between $|l-s|$ and $l+s$ only in integer steps, not half-integer steps. The reason for this can be derived from Equation (8.11). Suppose that $j$ and $m_j$ both have their maximum values; $j = l + s$ and $m_j = m_l + m_s$. Since $m_l$ and $m_s$ vary by integer steps between their maximum and minimum values, the next largest value of $m_j$ possible is $m_j = m_l + m_s - 1$. However, if $j$ could vary by half-integer steps, we would expect to see $m_j = m_l + m_s - 1/2$, which does not exist. To summarize, the possible values of $j$ are

$$j = |l - s|, \ |l - s| + 1, \ldots, l + s - 1, \ l + s \tag{8.13}$$

---

### Example 8.1. Adding Angular Momenta in the Hydrogen Atom

The electron in a hydrogen atom is in an $l = 1$ state. What are the possible values of $j$ and $m_j$?

The electron has $s = 1/2$ and $l = 1$, so Equation (8.13) gives $j = 1/2$ or $j = 3/2$. In the usual way, $m_j$ can vary from $-j$ to $+j$ in integer steps, so the possible values of $j$ and $m_j$ are

$$j = 1/2, \quad m_j = -1/2, +1/2$$

$$j = 3/2, \quad m_j = -3/2, -1/2, +1/2, +3/2$$

In general, if the electron in a hydrogen atom has quantum number $l$, then the possible values for $j$ are $j = l - 1/2$ and $j = l + 1/2$. The exception to this is the case $l = 0$ for which $j = 1/2$ is the only possible state.

Nothing in this derivation is peculiar to orbital and spin angular momenta; these arguments apply to the addition of *any* angular momenta. For instance, for two particles with spins $s_1$ and $s_2$, the possible values for the total spin quantum number $s$ are

$$s = |s_1 - s_2|, \ |s_1 - s_2| + 1, \ldots, s_1 + s_2 - 1, \ s_1 + s_2$$

A frequently encountered situation is that of two particles each with spin $1/2$, such as the electron and proton in a hydrogen atom. In this case, the possible total spin states are $s = 1$ and $s = 0$. The $s = 1$ state is called the *triplet state* because it has three possible values for $m_s$ (i.e., $-1$, $0$, and $+1$), while the $s = 0$ state is called the *singlet state* because it can have only $m_s = 0$.

## 8.4 The Matrix Representation of Spin

In this section we will develop a mathematical representation of spin states. Unlike orbital angular momentum, the spin states cannot be written as functions of position. Instead, they can be represented as column vectors.

Consider first a particle, such as the electron, which has $s = 1/2$. It has two possible values for $m_s$ which are $m_s = -1/2$ and $m_s = +1/2$. Since $m_s$ represents the $z$ component of spin, it is convenient to think of these states as two different orientations of the angular momentum vector, namely, spin "down" and spin "up" (Figure 8.3). These states can be represented in Dirac notation as $| \downarrow \rangle$ for spin down and $| \uparrow \rangle$ for spin up. Thus, we have

$$S_z| \downarrow \rangle = -\frac{\hbar}{2}| \downarrow \rangle$$

$$S_z| \uparrow \rangle = +\frac{\hbar}{2}| \uparrow \rangle$$

There are only a finite number of these states (i.e., two). This suggests that they can be represented in terms of a two-dimensional vector space consisting of column vectors. Thus, the spin up state is written as

$$| \uparrow \rangle \Leftrightarrow \begin{pmatrix} 1 \\ 0 \end{pmatrix}$$

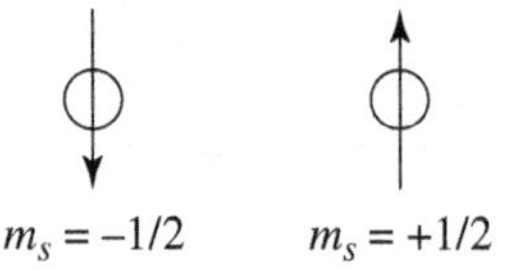

$$m_s = -1/2 \qquad m_s = +1/2$$

Fig. 8.3   For a particle with $s = 1/2$, the state $m_s = -1/2$ is spin down, and $m_s = +1/2$ is spin up.

and the spin down state as

$$|\downarrow\rangle \Leftrightarrow \begin{pmatrix} 0 \\ 1 \end{pmatrix}$$

These two vectors form an orthonormal basis set, since

$$\langle \uparrow | \uparrow \rangle = \begin{pmatrix} 1 & 0 \end{pmatrix} \begin{pmatrix} 1 \\ 0 \end{pmatrix} = 1$$

$$\langle \downarrow | \downarrow \rangle = \begin{pmatrix} 0 & 1 \end{pmatrix} \begin{pmatrix} 0 \\ 1 \end{pmatrix} = 1$$

$$\langle \uparrow | \downarrow \rangle = \begin{pmatrix} 1 & 0 \end{pmatrix} \begin{pmatrix} 0 \\ 1 \end{pmatrix} = 0$$

$$\langle \downarrow | \uparrow \rangle = \begin{pmatrix} 0 & 1 \end{pmatrix} \begin{pmatrix} 1 \\ 0 \end{pmatrix} = 0$$

Although the spin $1/2$ case will be investigated in the most detail, this matrix representation can be extended to other spin states. For example, for $s = 1$ there are three spin vectors, namely,

$$|s = 1, \, m_s = 1\rangle \Leftrightarrow \begin{pmatrix} 1 \\ 0 \\ 0 \end{pmatrix}$$

$$|s = 1, \, m_s = 0\rangle \Leftrightarrow \begin{pmatrix} 0 \\ 1 \\ 0 \end{pmatrix}$$

$$|s = 1, \, m_s = -1\rangle \Leftrightarrow \begin{pmatrix} 0 \\ 0 \\ 1 \end{pmatrix}$$

Now consider the spin operators $S_x$, $S_y$, and $S_z$. From the discussion in Chapter 7, these should be $2 \times 2$ matrices, which we will now calculate. Consider an arbitrary $2\times2$ matrix $A$. It is possible to pick out the individual elements of the matrix by the following sorts of matrix multiplications:

$$(1\ 0) \begin{pmatrix} A_{11} & A_{12} \\ A_{21} & A_{22} \end{pmatrix} \begin{pmatrix} 1 \\ 0 \end{pmatrix} = A_{11}$$

$$(0\ 1) \begin{pmatrix} A_{11} & A_{12} \\ A_{21} & A_{22} \end{pmatrix} \begin{pmatrix} 1 \\ 0 \end{pmatrix} = A_{21}$$

and so forth. In general, for a matrix of arbitrary size, the element $A_{ij}$ can be picked out by multiplying on the left side with a row vector having 1 in the $i$th entry and 0's everywhere else, and on the right by a column vector having 1 in the $j$th entry and 0's everywhere else:

$$\begin{pmatrix} 0 & 0 & \cdots & 1 & \cdots & 0 \end{pmatrix} \begin{pmatrix} & & A & & \end{pmatrix} \begin{pmatrix} 0 \\ 0 \\ \vdots \\ 1 \\ \vdots \\ 0 \end{pmatrix} = A_{ij}$$

Since the basis vectors $|\downarrow\rangle$ and $|\uparrow\rangle$ are eigenvectors of the operator $S_z$, it is straightforward to use this method to calculate the matrix elements of $S_z$:

$$S_{z11} = \langle\uparrow|S_z|\uparrow\rangle = \langle\uparrow|\frac{\hbar}{2}|\uparrow\rangle = \frac{\hbar}{2}\langle\uparrow|\uparrow\rangle = \frac{\hbar}{2}$$

$$S_{z12} = \langle\uparrow|S_z|\downarrow\rangle = \langle\uparrow|-\frac{\hbar}{2}|\downarrow\rangle = -\frac{\hbar}{2}\langle\uparrow|\downarrow\rangle = 0$$

$$S_{z21} = \langle\downarrow|S_z|\uparrow\rangle = \langle\downarrow|\frac{\hbar}{2}|\uparrow\rangle = \frac{\hbar}{2}\langle\downarrow|\uparrow\rangle = 0$$

$$S_{z22} = \langle\downarrow|S_z|\downarrow\rangle = \langle\downarrow|-\frac{\hbar}{2}|\downarrow\rangle = -\frac{\hbar}{2}\langle\downarrow|\downarrow\rangle = -\frac{\hbar}{2}$$

Thus, $S_z$ is given by the matrix

$$S_z = \begin{pmatrix} \hbar/2 & 0 \\ 0 & -\hbar/2 \end{pmatrix}$$

---

## Example 8.2. Using the Matrix Representation of $S_z$

Here we verify that $S_z$ gives the correct result when operating on $|\downarrow\rangle$ and $|\uparrow\rangle$. The matrix representations for $S_z$, $|\downarrow\rangle$, and $|\uparrow\rangle$ give

$$S_z|\uparrow\rangle = \begin{pmatrix} \hbar/2 & 0 \\ 0 & -\hbar/2 \end{pmatrix} \begin{pmatrix} 1 \\ 0 \end{pmatrix} = \frac{\hbar}{2}\begin{pmatrix} 1 \\ 0 \end{pmatrix} = \frac{\hbar}{2}|\uparrow\rangle$$

and

$$S_z |\downarrow\rangle = \begin{pmatrix} \hbar/2 & 0 \\ 0 & -\hbar/2 \end{pmatrix} \begin{pmatrix} 0 \\ 1 \end{pmatrix} = -\frac{\hbar}{2} \begin{pmatrix} 0 \\ 1 \end{pmatrix} = -\frac{\hbar}{2} |\downarrow\rangle$$

so $S_z$ does indeed give the correct result when applied to $|\downarrow\rangle$ and $|\uparrow\rangle$.

---

Obtaining matrix representations for $S_x$ and $S_y$ is a bit more complicated. To do this, we make use of the spin ladder operators, which are the analogs of the operators defined in Chapter 6 for orbital angular momentum:

$$S_+ = S_x + iS_y \tag{8.14}$$

$$S_- = S_x - iS_y \tag{8.15}$$

As in Chapter 6, these operators raise and lower the values of $m_s$, and they give 0 when attempting to raise $m_s$ above its highest allowed value or attempting to lower $m_s$ below its lowest allowed value. Thus,

$$S_- |s\, m_s\rangle \propto |s\, m_s - 1\rangle \tag{8.16}$$

$$S_+ |s\, m_s\rangle \propto |s\, m_s + 1\rangle \tag{8.17}$$

The constants of proportionality in Equations (8.16) and (8.17) need to be determined. Note that $S_-^\dagger = S_+$ and $S_+^\dagger = S_-$, so if

$$S_- |s\, m_s\rangle = c|s\, m_s - 1\rangle \tag{8.18}$$

where $c$ is a constant to be determined, then the dual of this equation is

$$\langle s\, m_s|S_+ = \langle s\, m_s - 1|c^* \tag{8.19}$$

and taking the inner product of the quantities in Equations (8.18) and (8.19) gives

$$\langle s\, m_s|S_+S_-|s\, m_s\rangle = c^*c\langle s\, m_s - 1|s\, m_s - 1\rangle = |c|^2 \tag{8.20}$$

In order to determine $c$, we must express $S_+S_-$ in terms of operators whose behavior, when applied to $|s\, m_s\rangle$, is known. Note that

$$S_+S_- = (S_x + iS_y)(S_x - iS_y)$$

$$= S_x^2 + S_y^2 - i[S_x, S_y]$$

$$= S_x^2 + S_y^2 - i(i\hbar S_z)$$

$$= S^2 - S_z^2 + \hbar S_z$$

and we know how $S^2$ and $S_z$ operate on $|s\ m_s\rangle$. Thus Equation (8.20) can be written as

$$\langle s\ m_s|S^2 - S_z^2 + \hbar S_z|s\ m_s\rangle = |c|^2$$

which gives

$$|c|^2 = \hbar^2[s(s+1) - m_s^2 + m_s]$$

Hence, Equation (8.18) can be written as

$$S_-|s\ m_s\rangle = \hbar\sqrt{s(s+1) - m_s(m_s - 1)}|s\ m_s - 1\rangle \qquad (8.21)$$

A similar argument gives

$$S_+|s\ m_s\rangle = \hbar\sqrt{s(s+1) - m_s(m_s + 1)}|s\ m_s + 1\rangle$$

For the special case of $s = 1/2$, we get $S_-|\uparrow\rangle = \hbar|\downarrow\rangle$ and $S_+|\downarrow\rangle = \hbar|\uparrow\rangle$.

These expressions for $S_+$ and $S_-$ can now be used to determine the matrix elements for $S_x$ and $S_y$, because we have explicit expressions for the way in which $S_+$ and $S_-$ operate on $|\downarrow\rangle$ and $|\uparrow\rangle$, and $S_x$ and $S_y$ can be written as

$$S_x = \frac{1}{2}(S_+ + S_-) \qquad (8.22)$$

$$S_y = \frac{1}{2i}(S_+ - S_-) \qquad (8.23)$$

The matrix elements of $S_+$ are

$$S_{+11} = \langle\uparrow|S_+|\uparrow\rangle = 0 \quad (\text{because } S_+|\uparrow\rangle = 0)$$
$$S_{+12} = \langle\uparrow|S_+|\downarrow\rangle = \langle\uparrow|\hbar|\uparrow\rangle = \hbar\langle\uparrow|\uparrow\rangle = \hbar$$
$$S_{+21} = \langle\downarrow|S_+|\uparrow\rangle = 0 \quad (\text{because } S_+|\uparrow\rangle = 0)$$
$$S_{+22} = \langle\downarrow|S_+|\downarrow\rangle = \langle\downarrow|\hbar|\uparrow\rangle = 0 \quad (\text{because } \langle\downarrow|\uparrow\rangle = 0)$$

Similarly, for the matrix representing $S_-$,

$$S_{-21} = \hbar$$

and all of the other matrix elements are 0. Then $S_+$ and $S_-$ are

$$S_+ = \begin{pmatrix} 0 & \hbar \\ 0 & 0 \end{pmatrix}$$

$$S_- = \begin{pmatrix} 0 & 0 \\ \hbar & 0 \end{pmatrix}$$

Then Equations (8.22) and (8.23) give the matrices for $S_x$ and $S_y$:

$$S_x = \begin{pmatrix} 0 & \hbar/2 \\ \hbar/2 & 0 \end{pmatrix}$$

$$S_y = \begin{pmatrix} 0 & -i\hbar/2 \\ i\hbar/2 & 0 \end{pmatrix}$$

It is convenient to factor out $\hbar/2$ from $S_x$, $S_y$, and $S_z$, and to write all of these matrices in terms of the *Pauli spin matrices*, $\sigma_x$, $\sigma_y$, and $\sigma_z$:

$$S_x = \frac{\hbar}{2}\sigma_x$$

$$S_y = \frac{\hbar}{2}\sigma_y$$

$$S_z = \frac{\hbar}{2}\sigma_z$$

where $\sigma_x$, $\sigma_y$, and $\sigma_z$ are

$$\sigma_x = \begin{pmatrix} 0 & 1 \\ 1 & 0 \end{pmatrix}$$

$$\sigma_y = \begin{pmatrix} 0 & -i \\ i & 0 \end{pmatrix}$$

$$\sigma_z = \begin{pmatrix} 1 & 0 \\ 0 & -1 \end{pmatrix}$$

The next step is to determine the eigenvectors and eigenvalues of the spin operators. Note that we have already done this for $S_z$; the eigenvectors and eigenvalues for $S_z$ are

$$|\uparrow\rangle = \begin{pmatrix} 1 \\ 0 \end{pmatrix}, \quad \text{eigenvalue} = +\frac{\hbar}{2}$$

and

$$|\downarrow\rangle = \begin{pmatrix} 0 \\ 1 \end{pmatrix}, \quad \text{eigenvalue} = -\frac{\hbar}{2}$$

Now consider the spin operator in the $x$ direction $S_x$, and assume an eigenvector $\begin{pmatrix} \psi_1 \\ \psi_2 \end{pmatrix}$ with eigenvalue $c$. The eigenvalue equation is

$$\frac{\hbar}{2}\begin{pmatrix} 0 & 1 \\ 1 & 0 \end{pmatrix}\begin{pmatrix} \psi_1 \\ \psi_2 \end{pmatrix} = c\begin{pmatrix} \psi_1 \\ \psi_2 \end{pmatrix}$$

This leads to the determinant equation:

$$\begin{vmatrix} -c & \hbar/2 \\ \hbar/2 & -c \end{vmatrix} = 0$$

which gives $c^2 - (\hbar/2)^2 = 0$, so that $c = \pm\hbar/2$. Thus, the eigenvalues of $S_x$ are identical to the eigenvalues of $S_z$. However, this is exactly what we would expect. There is no preferred direction for the particle, and the coordinate axes can always be rotated so that the $x$-axis points in the $z$ direction, so any quantities measured along the $z$-axis should have the same set of possible values if they are measured along the $x$-axis (or any other axis, for that matter).

The eigenvector of $S_x$ for $c = +\hbar/2$ is given by

$$\frac{\hbar}{2}\begin{pmatrix} 0 & 1 \\ 1 & 0 \end{pmatrix}\begin{pmatrix} \psi_1 \\ \psi_2 \end{pmatrix} = \frac{\hbar}{2}\begin{pmatrix} \psi_1 \\ \psi_2 \end{pmatrix}$$

which yields the two equations

$$(\hbar/2)\psi_2 = (\hbar/2)\psi_1$$
$$(\hbar/2)\psi_1 = (\hbar/2)\psi_2$$

As expected, these two equations are not independent of each other; both are satisfied as long as $\psi_1 = \psi_2$. Thus, the eigenvector, which we will designate $|\rightarrow\rangle$, can be any multiple of $\begin{pmatrix} 1 \\ 1 \end{pmatrix}$. However, the eigenvector must be normalized. Writing $|\rightarrow\rangle = c\begin{pmatrix} 1 \\ 1 \end{pmatrix}$, the normalization requirement is

$$\langle\rightarrow\,|\rightarrow\rangle = |c|^2\begin{pmatrix} 1 & 1 \end{pmatrix}\begin{pmatrix} 1 \\ 1 \end{pmatrix} = 2|c|^2 = 1$$

so that $c = 1/\sqrt{2}$. Hence, the normalized eigenvector of $S_x$ with spin in the $+x$ direction is

$$|\rightarrow\rangle = \frac{1}{\sqrt{2}}\begin{pmatrix} 1 \\ 1 \end{pmatrix} \tag{8.24}$$

A similar calculation for the eigenvector corresponding to spin in the $-x$ direction (eigenvalue $= -\hbar/2$) yields

$$|\leftarrow\rangle = \frac{1}{\sqrt{2}}\begin{pmatrix} 1 \\ -1 \end{pmatrix} \tag{8.25}$$

For $S_y$ the eigenvectors are (Problem 8.4):

$$| \nearrow \rangle = \frac{1}{\sqrt{2}} \begin{pmatrix} 1 \\ i \end{pmatrix}, \quad (+y \text{ direction}) \tag{8.26}$$

$$| \swarrow \rangle = \frac{1}{\sqrt{2}} \begin{pmatrix} 1 \\ -i \end{pmatrix}, \quad (-y \text{ direction}) \tag{8.27}$$

Note that the set of eigenvectors of $S_z$, namely, $| \uparrow \rangle = \begin{pmatrix} 1 \\ 0 \end{pmatrix}$ and $| \downarrow \rangle = \begin{pmatrix} 0 \\ 1 \end{pmatrix}$, form an orthonormal basis set, so any other spin state can be represented as a sum of $| \uparrow \rangle$ and $| \downarrow \rangle$. From Equations (8.24) and (8.25), $| \rightarrow \rangle$ and $| \leftarrow \rangle$ can be expressed in terms of $| \uparrow \rangle$ and $| \downarrow \rangle$:

$$| \rightarrow \rangle = \frac{1}{\sqrt{2}}(| \uparrow \rangle + | \downarrow \rangle)$$

$$| \leftarrow \rangle = \frac{1}{\sqrt{2}}(| \uparrow \rangle - | \downarrow \rangle)$$

Thus, a particle with a spin in the $+x$ or $-x$ direction is a linear combination of states with spins in the $+z$ and $-z$ directions. This is one of the counterintuitive predictions of quantum mechanics with no classical analog. As we will see in the next section, it leads to some rather strange consequences.

## 8.5　The Stern–Gerlach Experiment

As noted in Section 8.2, the Stern–Gerlach experiment provided some of the first evidence for the existence of spin angular momentum. In this section, we will use an idealized Stern–Gerlach experiment to examine some of the more bizarre consequences of quantum mechanics.

　　Imagine a Stern–Gerlach apparatus with an inhomogeneous magnetic field oriented in the $z$ direction, so that the apparatus separates particles based on the $z$ component of their magnetic moment. A beam of individual electrons is passed through this apparatus, and the beam splits into two parts with $m_s = +1/2$ and $m_s = -1/2$, respectively (Figure 8.4). This allows us to separate the electrons into states of pure $| \uparrow \rangle$ and $| \downarrow \rangle$.

　　Now suppose that the beam containing the electrons in the $| \uparrow \rangle$ state is run through a second Stern–Gerlach apparatus with the field now aligned in the $x$ direction (Figure 8.5).

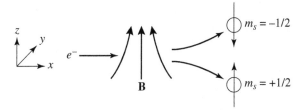

Fig. 8.4  A Stern–Gerlach apparatus with an inhomogeneous magnetic field in the $z$ direction will separate a beam of electrons into two states with $m_s = +1/2$ and $m_s = -1/2$.

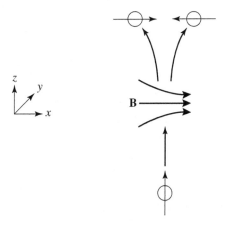

Fig. 8.5  A beam of electrons in the $|\uparrow\,\rangle$ state is run through a Stern–Gerlach apparatus with an inhomogeneous magnetic field in the $x$ direction.

This beam now splits into two separate beams having spins of $+1/2$ and $-1/2$ in the $x$ direction. This result can be explained from a quantum mechanical point of view. Since $S_x$ and $S_z$ do not commute, the electrons cannot simultaneously be in a state of definite $S_z$ and definite $S_x$. The first Stern–Gerlach apparatus forces the electrons into an eigenstate of $S_z$, namely, $|\uparrow\,\rangle$. However, this state is a linear combination of the eigenstates of $S_x$. Therefore, the second Stern–Gerlach apparatus "sees" a mixture of $|\rightarrow\rangle$ and $|\leftarrow\rangle$ and separates the beam into these two states.

Given a set of electrons in the state $|\uparrow\,\rangle$, it is possible to calculate the probability that a second measurement, such as the one just described, will yield a spin in the $+x$ or $-x$ direction. In Dirac notation, if a particle is in a particular state $|\psi\rangle$, and we make a measurement to determine whether or not it is in the state $|\phi\rangle$, the probability that it will be found to be in

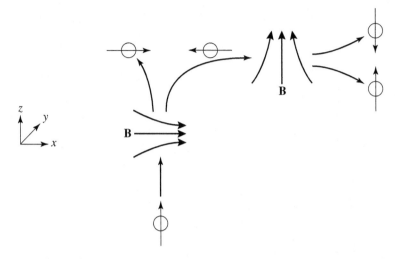

Fig. 8.6   A beam of electrons in the $|\uparrow\rangle$ state is run through a Stern–Gerlach apparatus with an inhomogeneous magnetic field in the $x$ direction and then another Stern–Gerlach apparatus with an inhomogeneous magnetic field in the $z$ direction.

the state $|\phi\rangle$ is

$$P = |\langle\phi|\psi\rangle|^2$$

---

### Example 8.3. Calculating Spin State Probabilities

A particle is initially in the $|\uparrow\rangle$ state, and the spin is measured in the $x$ direction. What is the probability that the spin is found to be in the $+x$ direction?

The particle is initially in the state $|\uparrow\rangle = \begin{pmatrix} 1 \\ 0 \end{pmatrix}$. We wish to know, when the spin is measured in the $x$ direction, whether the particle will be found to be in the state $|\rightarrow\rangle = \frac{1}{\sqrt{2}} \begin{pmatrix} 1 \\ 1 \end{pmatrix}$. This probability is

$$P = |\langle\rightarrow|\uparrow\rangle|^2$$

$$= \left| \frac{1}{\sqrt{2}} \begin{pmatrix} 1 & 1 \end{pmatrix} \begin{pmatrix} 1 \\ 0 \end{pmatrix} \right|^2$$

$$= \frac{1}{2}$$

---

Now here is where quantum mechanics gives a truly strange result. Suppose we set up a triple Stern–Gerlach apparatus (Figure 8.6). The first apparatus picks out the electrons in the $|\uparrow\rangle$ state, the second apparatus splits this beam into the $|\rightarrow\rangle$ and $|\leftarrow\rangle$ states, and then we run the $|\rightarrow\rangle$ electrons back through a third Stern–Gerlach apparatus that is identical to the first one: it separates the electrons on the basis of the $z$ component of their spins. In a classical system, since we selected out only electrons with spins in the $+z$ direction, the final Stern–Gerlach apparatus would produce only a single beam of electrons with spins in the $+z$ direction. However, this is not what happens at all; instead, the beam splits again into a beam with spin in the $+z$ direction and a beam with spin in the $-z$ direction!

To see what is happening, note that the first Stern–Gerlach apparatus picks out electrons in the state $|\uparrow\rangle$. The second Stern–Gerlach apparatus separates this beam into the states $|\rightarrow\rangle$ and $|\leftarrow\rangle$, and we keep only the $|\rightarrow\rangle$ electrons. But $|\rightarrow\rangle$ is a mixture of $|\uparrow\rangle$ and $|\downarrow\rangle$ states, namely, $|\rightarrow\rangle = (1/\sqrt{2})(|\uparrow\rangle + |\downarrow\rangle)$. So in making this measurement, we have added back in the $|\downarrow\rangle$ component, which shows up in the third apparatus. The very act of measuring the spin changes the state of the particle. This is one of the characteristics of quantum mechanics not present in classical physics: in an experiment of this kind, there is no way to avoid changing the state of the system through the act of measurement. This idea is examined in more detail in the discussion of measurement theory in Section 8.8.

## 8.6  Spin Precession

In this section we will see an example of how spin can be incorporated into the Schrödinger equation. Imagine that we have a classical magnetic dipole $\mu$ in a magnetic field $\mathbf{B}$ (Figure 8.7). The magnetic field exerts a torque $\mu\times\mathbf{B}$ on the dipole, which will cause it to line up parallel with the field. But now suppose that in addition, the dipole has angular momentum (Figure 8.8). A rotating body to which a torque is applied will *precess* in a direction perpendicular to the angular momentum vector.

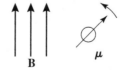

Fig. 8.7  A classical magnetic dipole $\mu$ in a magnetic field $\mathbf{B}$ experiences a torque $\mu\times\mathbf{B}$, which tends to align the dipole with the field.

Of course, there is no way to know if these classical analogies will carry over into the quantum realm until we solve the Schrödinger equation. Consider an electron with magnetic moment $\mu$ at rest in an external magnetic field **B**. Since we are interested in how the particle spin evolves in time, we will use the time-dependent Schrödinger equation,

$$H|\psi\rangle = i\hbar\frac{\partial}{\partial t}|\psi\rangle \qquad (8.28)$$

where $|\psi\rangle$ is the Dirac wave function for the particle. This leads to two obvious questions: What is the form for $H$, and what is the form for $|\psi\rangle$? Since the electron is at rest, the kinetic energy part of $H$ will be zero, and $H$ will be given purely by the potential energy which is

$$V = -\boldsymbol{\mu}\cdot\mathbf{B}$$

Since we are interested in the evolution of the orientation of the spin of the electron, $V$ must be expressed in terms of the spin rather than the magnetic moment, using the relation

$$\boldsymbol{\mu} = -\frac{2\mu_B}{\hbar}\mathbf{S}$$

where we have taken $g_s = 2$ for the electron. Then the potential is

$$V = \frac{2\mu_B}{\hbar}\mathbf{B}\cdot\mathbf{S}$$

We choose a coordinate system so that **B** is pointing in the $z$ direction, so $\mathbf{B}\cdot\mathbf{S} = BS_z$, and we express $S_z$ as

$$S_z = \frac{\hbar}{2}\sigma_z$$

so that the potential becomes

$$V = \mu_B B\sigma_z$$

Then the Schrödinger equation (Equation (8.28)) becomes

$$\mu_B B\sigma_z|\psi\rangle = i\hbar\frac{\partial}{\partial t}|\psi\rangle \qquad (8.29)$$

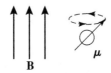

Fig. 8.8    A classical magnetic dipole with angular momentum will precess in a magnetic field.

This equation indicates the form for $|\psi\rangle$: since $\sigma_z$ is a $2\times2$ Pauli spin matrix, $|\psi\rangle$ is just the spin state of the electron written as a two-component column vector

$$|\psi\rangle = \begin{pmatrix} \psi_+ \\ \psi_- \end{pmatrix}$$

where $\psi_+$ and $\psi_-$ will be functions of time. Then Equation (8.29) takes the form

$$\mu_B B \begin{pmatrix} 1 & 0 \\ 0 & -1 \end{pmatrix} \begin{pmatrix} \psi_+ \\ \psi_- \end{pmatrix} = i\hbar \frac{\partial}{\partial t} \begin{pmatrix} \psi_+ \\ \psi_- \end{pmatrix}$$

Carrying out the matrix multiplication yields two ordinary differential equations:

$$\mu_B B \psi_+ = i\hbar \frac{d\psi_+}{dt}$$

$$\mu_B B(-\psi_-) = i\hbar \frac{d\psi_-}{dt}$$

The general solutions of these two equations give the time evolution of $\psi_+$ and $\psi_-$

$$\psi_+ = A_+ e^{-i(\mu_B B/\hbar)t}$$

$$\psi_- = A_- e^{i(\mu_B B/\hbar)t}$$

where $A_+$ and $A_-$ are constants to be determined. In matrix form, the solution is then

$$|\psi\rangle = \begin{pmatrix} A_+ e^{-i(\mu_B B/\hbar)t} \\ A_- e^{i(\mu_B B/\hbar)t} \end{pmatrix} \tag{8.30}$$

The constants $A_+$ and $A_-$ are determined from the initial conditions. In particular, if we take $t = 0$ to be the initial time, then

$$|\psi(t = 0)\rangle = \begin{pmatrix} A_+ \\ A_- \end{pmatrix}$$

For example, suppose that the electron has spin up (i.e., in the $+z$ direction) at $t = 0$. This corresponds to the state

$$|\psi(t = 0)\rangle = \begin{pmatrix} 1 \\ 0 \end{pmatrix}$$

so that $A_+ = 1$ and $A_- = 0$. Then Equation (8.30) gives the wave function at any later time $t$:

$$|\psi\rangle = \begin{pmatrix} e^{-i(\mu_B B/\hbar)t} \\ 0 \end{pmatrix}$$

Although this wave function is a function of time, it represents a state of constant spin. To see this, note that $P = |\langle\uparrow|\psi(t)\rangle|^2$ gives the probability of finding the particle in the spin up state. This probability is

$$P = \left| (1\ 0) \begin{pmatrix} e^{-(i\mu_B B/\hbar)t} \\ 0 \end{pmatrix} \right|^2 = 1$$

Thus, the electron starts out in the $|\uparrow\rangle$ state, and it stays there forever. This is consistent with the classical analog; a classical dipole pointing parallel to a magnetic field experiences no torque and does not rotate.

Now consider the more interesting case of an electron with spin initially in the $+x$ direction. In this case the initial spin state (correctly normalized) is

$$|\psi(t = 0)\rangle = \frac{1}{\sqrt{2}} \begin{pmatrix} 1 \\ 1 \end{pmatrix}$$

Then $A_+ = 1/\sqrt{2}$ and $A_- = 1/\sqrt{2}$, so the full time-dependent wave function from Equation (8.30) is

$$|\psi\rangle = \begin{pmatrix} \dfrac{1}{\sqrt{2}} e^{-i(\mu_B B/\hbar)t} \\ \dfrac{1}{\sqrt{2}} e^{i(\mu_B B/\hbar)t} \end{pmatrix}$$

To simplify the equation, define a new quantity $\omega$ given by

$$\omega = 2\mu_B B/\hbar$$

where $\omega$ has units of 1/time or, equivalently, frequency. Then the wave function is

$$|\psi\rangle = \frac{1}{\sqrt{2}} \begin{pmatrix} e^{-i(\omega/2)t} \\ e^{i(\omega/2)t} \end{pmatrix}$$

To understand the physical meaning of this wave function, we can evaluate it at a variety of times. In particular, we have

$$|\psi(t = 0)\rangle = \frac{1}{\sqrt{2}} \begin{pmatrix} 1 \\ 1 \end{pmatrix}$$

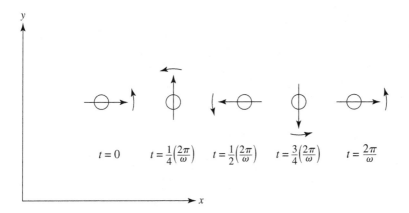

Fig. 8.9   An electron precessing in the $x$-$y$ plane with frequency $\omega = 2\mu_B B/\hbar$.

which is just our initial condition: the spin is in the $+x$ direction at $t = 0$. Further,

$$|\psi(t = 2\pi/\omega)\rangle = \frac{1}{\sqrt{2}} \begin{pmatrix} e^{-i\pi} \\ e^{i\pi} \end{pmatrix} = -\frac{1}{\sqrt{2}} \begin{pmatrix} 1 \\ 1 \end{pmatrix}$$

This is the original wave function multiplied by a phase factor of $-1$. Thus, at $t = 2\pi/\omega$, the spin is once again pointing in the $+x$ direction. This suggests that we investigate intermediate times. At the halfway point between $t = 0$ and $t = 2\pi/\omega$, i.e., at $t = (1/2)2\pi/\omega$, we get

$$|\psi(t = (1/2)2\pi/\omega)\rangle = \frac{1}{\sqrt{2}} \begin{pmatrix} e^{-i\pi/2} \\ e^{i\pi/2} \end{pmatrix} = -\frac{i}{\sqrt{2}} \begin{pmatrix} 1 \\ -1 \end{pmatrix}$$

which is the wave function for spin in the $-x$ direction. Taking half of this interval again, we get, at $t = (1/4)(2\pi/\omega)$,

$$|\psi(t = (1/4)2\pi/\omega)\rangle = \frac{1}{\sqrt{2}} \begin{pmatrix} e^{-i\pi/4} \\ e^{i\pi/4} \end{pmatrix} = \frac{1}{2}(1 - i) \begin{pmatrix} 1 \\ i \end{pmatrix}$$

which is the spin eigenstate in the $+y$ direction. Similarly, taking $t = (3/4)(2\pi/\omega)$ gives the spin eigenstate in the $-y$ direction.

Putting all of this information together, we see that the electron is precessing, with the direction of its spin vector rotating in the counterclockwise direction (Figure 8.9). The period of precession is $2\pi/\omega$, so the angular frequency is $\omega = 2\mu_B B/\hbar$. This phenomenon is the basis of magnetic resonance imaging, which will be examined in more detail in Chapter 14.

## 8.7 Spin Systems with Two Particles

In this section we consider systems of two particles with spin. We first examine the case in which there is no spin-dependent interaction between the two particles. We then consider what happens if there is such an interaction, so that the Hamiltonian depends on the spins.

### *Noninteracting Spins*

Consider the general system of two particles, each with spin $1/2$, in which there is no spin-dependent interaction between the particles. The spin operators for particle 1 and particle 2 can be written as $\mathbf{S}_1$ and $\mathbf{S}_2$, respectively. As in the case for a single particle, we will consider states which are eigenfunctions of $S_1^2$, $S_2^2$, $S_{1z}$, and $S_{2z}$. We have been using the notation $|\uparrow\rangle$ and $|\downarrow\rangle$ to refer to eigenstates of the single particle spin operator $S_z$. For the two-particle system, we write the eigenstate as $|m_{s1}\, m_{s2}\rangle$. So, for example, $|\uparrow\downarrow\rangle$ represents the state with spin up for particle 1 and spin down for particle 2 (i.e., $m_{s1} = +1/2$ and $m_{s2} = -1/2$). Thus,

$$S_{1z}|\uparrow\downarrow\rangle = +\frac{\hbar}{2}|\uparrow\downarrow\rangle$$

$$S_{2z}|\uparrow\downarrow\rangle = -\frac{\hbar}{2}|\uparrow\downarrow\rangle$$

The eigenstates of $S_1^2$ and $S_2^2$ do not need to be specified in the wave function because these are fixed by the fact that the particles have spin $1/2$. Thus,

$$S_1^2|\uparrow\downarrow\rangle = \hbar^2\left(\frac{1}{2}\right)\left(\frac{1}{2}+1\right)|\uparrow\downarrow\rangle$$

$$S_2^2|\uparrow\downarrow\rangle = \hbar^2\left(\frac{1}{2}\right)\left(\frac{1}{2}+1\right)|\uparrow\downarrow\rangle$$

So far, this is all straightforward and rather uninteresting; all we have done is to write down a wave function which combines the spin information for two particles together.

Now, however, consider the total spin of the system. We can write a total spin operator

$$\mathbf{S} = \mathbf{S}_1 + \mathbf{S}_2$$

so

$$S_z = S_{1z} + S_{2z}$$

and

$$S^2 = S_1^2 + S_2^2 + 2\mathbf{S}_1 \cdot \mathbf{S}_2 \tag{8.31}$$

From our previous discussion about the addition of angular momentum, we know that the quantum number for the $z$ component of the total spin $m_s$ is

$$m_s = m_{1s} + m_{2s}$$

while the quantum number for the total spin $s$ can have the values $s = 0$ or $s = 1$. Thus, the two-particle state can also be expressed in terms of $s$ and $m_s$ instead of $m_{1s}$ and $m_{2s}$. The wave function corresponding to a state of definite $s$ and $m_s$ can be written as $|s\ m_s\rangle$. For example, for $s = 0$, the only possible value of $m_s$ is $m_s = 0$, and the wave function is $|0\ 0\rangle$, the singlet state. If $s = 1$, then $m_s = -1, 0$, or $1$ with corresponding wave functions $|1\ -1\rangle$, $|1\ 0\rangle$, and $|1\ 1\rangle$, the triplet state.

This leads to an obvious question: can we simultaneously measure $s$, $m_s$, $m_{1s}$, and $m_{2s}$? This would be possible only if all of the operators $S^2$, $S_z$, $S_{1z}$, and $S_{2z}$ commuted with each other. Unfortunately, this is not the case. The expression for $S^2$ (Equation (8.31)) contains the term $2\mathbf{S}_1 \cdot \mathbf{S}_2$, which expands out as $2(S_{1x}S_{2x} + S_{1y}S_{2y} + S_{1z}S_{2z})$, and, for example, $S_{1x}$ and $S_{1y}$ do not commute with $S_{1z}$, and $S_{2x}$ and $S_{2y}$ do not commute with $S_{2z}$. More explicitly,

$$[S^2, S_{1z}] = [S_1^2 + S_2^2 + 2(S_{1x}S_{2x} + S_{1y}S_{2y} + S_{1z}S_{2z}), S_{1z}]$$

$$= 2[S_{1x}, S_{1z}]S_{2x} + 2[S_{1y}, S_{1z}]S_{2y}$$

$$= -2i\hbar S_{1y}S_{2x} + 2i\hbar S_{1x}S_{2y}$$

which is nonzero. A similar argument indicates that $S^2$ does not commute with $S_{2z}$. Therefore, it is *not* possible to measure $s$ and $m_{1s}$, $m_{2s}$ simultaneously. The particles can be in a state in which the $z$ component of both spins is known exactly (written as $|m_{1s}\ m_{2s}\rangle$), or they can be in a state in which $s$ and $m_s$ are known exactly (written as $|s\ m_s\rangle$), but not both at the same time.

Suppose that we are in a state of definite $s$ and $m_s$. Although we cannot uniquely specify the values of $m_{1s}$ and $m_{2s}$, we can write the state $|s\ m_s\rangle$ as a linear combination of the four possible $|m_{1s}\ m_{2s}\rangle$ states, since the latter form a basis set. In other words, we can write

$$|s\ m_s\rangle = c_1 |\uparrow\uparrow\rangle + c_2 |\uparrow\downarrow\rangle + c_3 |\downarrow\uparrow\rangle + c_4 |\downarrow\downarrow\rangle \tag{8.32}$$

and we can calculate the constants $c_1$, $c_2$, $c_3$, and $c_4$. One reason for doing this is that if the particles are in a state of definite $s$, $m_s$, there is nothing to prevent us from subsequently measuring the individual spin states. Although we cannot predict the result in advance (since we are not in an eigenstate of $S_{1z}$ and $S_{2z}$), Equation (8.32) can be used to find the probabilities of measuring a particular value of $m_{1s}$ and $m_{2s}$.

Consider first the state $|1 \ 1\rangle$. It must always be true that $m_{1s} + m_{2s} = m_s$, and the only way to have $m_{1s} + m_{2s} = 1$ is for both $m_{1s}$ and $m_{2s}$ to be $+1/2$, i.e., for both particles to be in the spin up state. Thus,

$$|1 \ 1\rangle = |\uparrow \uparrow\rangle \tag{8.33}$$

A similar argument gives

$$|1 \ -1\rangle = |\downarrow \downarrow\rangle \tag{8.34}$$

On the other hand, both $|1 \ 0\rangle$ and $|0 \ 0\rangle$ have $m_s = 0$, so they are both linear combinations of $|\uparrow \downarrow\rangle$ and $|\downarrow \uparrow\rangle$. To find the desired linear combinations, recall that the lowering operator $S_-$ acts on $|1 \ 1\rangle$ to give a multiple of $|1 \ 0\rangle$. Since $S_-$ is just $S_x - iS_y$, it is correct to write

$$S_- = S_{1-} + S_{2-}$$

We can therefore begin with Equation (8.33), apply $S_-$ to the left-hand side, and $S_{1-} + S_{2-}$ to the right-hand side (using Equation (8.21)), and derive an expression for $|1 \ 0\rangle$ as a function of $|\uparrow \downarrow\rangle$ and $|\downarrow \uparrow\rangle$. This procedure gives

$$S_-|1 \ 1\rangle = S_{1-}|\uparrow \uparrow\rangle + S_{2-}|\uparrow \uparrow\rangle$$
$$\hbar\sqrt{1(1+1) - 1(1-1)}|1 \ 0\rangle = \hbar\sqrt{(1/2)(1/2+1) - (1/2)(1/2-1)}|\downarrow \uparrow\rangle$$
$$+ \hbar\sqrt{(1/2)(1/2+1) - (1/2)(1/2-1)}|\uparrow \downarrow\rangle$$

which simplifies to

$$|1 \ 0\rangle = \frac{1}{\sqrt{2}}|\downarrow \uparrow\rangle + \frac{1}{\sqrt{2}}|\uparrow \downarrow\rangle \tag{8.35}$$

This method cannot be applied to the state $|0 \ 0\rangle$, but now note that all of our spin states should be orthonormal. The normalized linear combination of $|\uparrow \downarrow\rangle$ and $|\downarrow \uparrow\rangle$ which is orthogonal to $\frac{1}{\sqrt{2}}|\downarrow \uparrow\rangle + \frac{1}{\sqrt{2}}|\uparrow \downarrow\rangle$ is

$$|0 \ 0\rangle = \frac{1}{\sqrt{2}}|\downarrow \uparrow\rangle - \frac{1}{\sqrt{2}}|\uparrow \downarrow\rangle \tag{8.36}$$

(Of course, there is always the freedom to multiply the right-hand side of Equation (8.36) by a factor with unit absolute value such as $-1$.)

We have now expressed all four $|s\ m_s\rangle$ states as linear combinations of the $|m_{1s}\ m_{2s}\rangle$ states [Equations (8.33), (8.34), (8.35), and (8.36)]. Although we have examined the specific case of two particles with spin $1/2$, this result can be generalized to particles with other spins. The constants which appear in Equations (8.33), (8.34), (8.35), and (8.36) (and in those more general expressions) are called *Clebsch–Gordon coefficients*.

---

### Example 8.4. Probabilities for a Two-Particle System

A system of 2 particles, each with spin $1/2$, is in the singlet state. A measurement is made of the $z$ component of the spin of the first particle. What is the probability that $m_{1s} = +1/2$?

In the singlet state, the wave function is

$$|0\ \ 0\rangle = \frac{1}{\sqrt{2}}|\downarrow\uparrow\rangle - \frac{1}{\sqrt{2}}|\uparrow\downarrow\rangle \tag{8.37}$$

The only state with $m_{1s} = +1/2$ appearing on the right-hand side of Equation (8.37) is $|\uparrow\downarrow\rangle$, so the probability that particle 1 is in the spin up state is

$$P = |\langle\uparrow\downarrow|0\ \ 0\rangle|^2$$

$$= |\langle\uparrow\downarrow|\left(\frac{1}{\sqrt{2}}|\downarrow\uparrow\rangle - \frac{1}{\sqrt{2}}|\uparrow\downarrow\rangle\right)|^2$$

$$= 1/2$$

Thus, there is a 50% chance that the first particle is found to be in the spin up state and a 50% chance it is found to be in the spin down state.

---

### Interacting Spins

Now consider two spin-$1/2$ particles with a spin-dependent interaction. One of the simplest possible spin interactions is given by the Hamiltonian

$$H = \lambda\mathbf{S}_1\cdot\mathbf{S}_2 \tag{8.38}$$

where $\lambda$ is a real constant. Here it is assumed that the particles are fixed in space, and there is no other interaction between them, so that Equation (8.38) gives the entire interaction. Since this Hamiltonian depends on all three components of $\mathbf{S}_1$ and $\mathbf{S}_2$, it will not commute with either $S_{1z}$ or $S_{2z}$. Hence, states of definite $m_{1s}$ and $m_{2s}$, such as $|\uparrow\uparrow\rangle$, $|\uparrow\downarrow\rangle$, etc., cannot be eigenfunctions of $H$.

On the other hand, the Hamiltonian given in Equation (8.38) *does* commute with $S^2$. This is because Equation (8.31) can be rewritten as

$$\mathbf{S}_1 \cdot \mathbf{S}_2 = \frac{1}{2}(S^2 - S_1^2 - S_2^2)$$

which allows the Hamiltonian in Equation (8.38) to be written as

$$H = \frac{\lambda}{2}(S^2 - S_1^2 - S_2^2)$$

When written in this form, it is clear that $H$ commutes with $S^2$ and $S_z$. Thus, states of definite $s$ and $m_s$ are also eigenstates of $H$, and we can compute their energies. We obtain

$$H|s \quad m_s\rangle = \frac{\lambda}{2}\hbar^2 \left[ s(s+1) - \frac{1}{2}\left(\frac{1}{2}+1\right) - \frac{1}{2}\left(\frac{1}{2}+1\right) \right] |s \quad m_s\rangle$$

Thus, the energy $E$ is a function entirely of $s$ and is independent of $m_s$; all three triplet states are degenerate. Inserting the actual values for $s$ gives the energy levels of the system:

$$s = 1 \quad \text{(triplet state)} \quad E = \frac{1}{4}\lambda\hbar^2$$

$$s = 0 \quad \text{(singlet state)} \quad E = -\frac{3}{4}\lambda\hbar^2$$

The energy levels of more complex spin-dependent Hamiltonians can be calculated using similar methods.

---

### Example 8.5. The Magnetic Dipole-Dipole Interaction Between Two Particles

The magnetic dipole-dipole interaction between two particles with magnetic moments $\boldsymbol{\mu}_1$ and $\boldsymbol{\mu}_2$ fixed in space at a separation $\mathbf{r}$ is given by the Hamiltonian

$$H = \frac{(\boldsymbol{\mu}_1 \cdot \boldsymbol{\mu}_2)}{r^3} - 3\frac{(\boldsymbol{\mu}_1 \cdot \mathbf{r})(\boldsymbol{\mu}_2 \cdot \mathbf{r})}{r^5}$$

Consider two neutrons, which are uncharged but have spin $1/2$ and magnetic moment $\boldsymbol{\mu} = (g_n e/2m_n)\mathbf{S}$. If they are fixed in space at a distance $a$ apart, find the energy levels of the system.

Rewriting the Hamiltonian in terms of spins gives

$$H = \left(\frac{g_n e}{2m_n}\right)^2 \left( \frac{(\mathbf{S}_1 \cdot \mathbf{S}_2)}{r^3} - 3\frac{(\mathbf{S}_1 \cdot \mathbf{r})(\mathbf{S}_2 \cdot \mathbf{r})}{r^5} \right)$$

The choice of the coordinate system is arbitrary, so we choose the vector **r** separating the two neutrons to lie along the $z$-axis. Then $\mathbf{r} = a\hat{z}$ and $r = a$, so the Hamiltonian becomes

$$H = \left(\frac{g_n e}{2m_n}\right)^2 \frac{1}{a^3} (\mathbf{S}_1 \cdot \mathbf{S}_2 - 3S_{1z}S_{2z})$$

As above, $\mathbf{S}_1 \cdot \mathbf{S}_2$ can be written as $\mathbf{S}_1 \cdot \mathbf{S}_2 = \frac{1}{2}(S^2 - S_1^2 - S_2^2)$. It is also true that $S_z^2 = (S_{1z} + S_{2z})^2 = S_{1z}^2 + S_{2z}^2 + 2S_{1z}S_{2z}$, so

$$S_{1z}S_{2z} = \frac{1}{2}(S_z^2 - S_{1z}^2 - S_{2z}^2)$$

and the Hamiltonian becomes

$$H = \left(\frac{g_n e}{2m_n}\right)^2 \frac{1}{a^3} \left(\frac{1}{2}(S^2 - S_1^2 - S_2^2) - \frac{3}{2}(S_z^2 - S_{1z}^2 - S_{2z}^2)\right) \tag{8.39}$$

If the neutrons are in a state $|s \ m_s\rangle$, this state is an eigenfunction of $H$. The only possible confusion arises from the operators $S_{1z}^2$ and $S_{2z}^2$ in Equation (8.39), since the state $|s \ m_s\rangle$ is *not* a state of definite $m_{1s}$ and $m_{2s}$. However, note that for a spin-1/2 particle, $S_z^2|\uparrow\rangle = (\hbar/2)^2|\uparrow\rangle$ and $S_z^2|\downarrow\rangle = (\hbar/2)^2|\downarrow\rangle$, so $S_z^2$ applied to *any* spin state gives an eigenvalue of $\hbar^2/4$.

Then applying $H$ to the state $|s \ m_s\rangle$ yields

$$H|s \ m_s\rangle = \left(\frac{g_n e}{2m_n}\right)^2 \frac{1}{a^3} \left(\frac{1}{2}(S^2 - S_1^2 - S_2^2) - \frac{3}{2}(S_z^2 - S_{1z}^2 - S_{2z}^2)\right)|s \ m_s\rangle$$

$$= \left(\frac{g_n^2 e^2}{4m_n^2 a^3}\right) \left(\frac{1}{2}\hbar^2 \left[s(s+1) - \frac{1}{2}\left(\frac{1}{2}+1\right) - \frac{1}{2}\left(\frac{1}{2}+1\right)\right]\right.$$

$$\left. -\frac{3}{2}\hbar^2 \left[(m_s^2 - \left(\frac{1}{2}\right)^2 - \left(\frac{1}{2}\right)^2\right]\right)|s \ m_s\rangle$$

$$= \frac{g_n^2 e^2 \hbar^2}{8m_n^2 a^3}[s(s+1) - 3m_s^2]|s \ m_s\rangle$$

so the energy levels are

$$E = \frac{g_n^2 e^2 \hbar^2}{8m_n^2 a^3}[s(s+1) - 3m_s^2]$$

The dipole-dipole interaction splits all of the $s$, $m_s$ states into distinct energy levels with the exception of the states $m_s = \pm 1$, which remain degenerate. These energy levels are

$$s = 1, \quad m_s = \pm 1, \ E = \frac{g_n^2 e^2 \hbar^2}{8 m_n^2 a^3}(-1)$$

$$s = 0, \quad m_s = 0, \quad E = 0$$

$$s = 1, \quad m_s = 0, \quad E = \frac{g_n^2 e^2 \hbar^2}{8 m_n^2 a^3}(2)$$

The dipole-dipole interaction between the proton and electron is the basis for *hyperfine* splitting in hydrogen, which is discussed in more detail in Chapter 9.

## 8.8   Measurement Theory

Consider a particle in a one-dimensional infinite square well of width $a$ centered at the origin. If the particle is in the ground state, then the time-dependent wave function can be written as

$$\Psi(x, t) = \sqrt{\frac{2}{a}} \cos\left(\frac{\pi x}{a}\right) e^{-iEt/\hbar}$$

Rewriting the spatial part of the wave function as a sum of complex exponentials, the full wave function is

$$\Psi(x, t) = \sqrt{\frac{1}{2a}} e^{i(\pi x/a - Et/\hbar)} + \sqrt{\frac{1}{2a}} e^{-i(\pi x/a + Et/\hbar)} \tag{8.40}$$

This wave function looks like the superposition of two waves moving in opposite directions. However, the particle, when observed, can obviously only be moving in a single direction. Suppose that a measurement is made of the direction of motion of the particle, and assume, for instance, that the particle is found to be moving to the right. Then the wave function becomes

$$\Psi(x, t) = e^{i(\pi x/a - Et/\hbar)} \tag{8.41}$$

Thus, the act of making a measurement alters the wave function, collapsing it from the form given by Equation (8.40) to the form given by Equation (8.41); the wave function changes from a superposition of states into a single state.

The idea that the particle is in an indeterminate superposition of states, and is only forced into a single definite state by the act of measurement is called the "Copenhagen interpretation," and it is the most widely-held view of the nature of measurement in quantum mechanics. However, this interpretation has some disturbing consequences. Consider, for example, a set of two particles, each with spin $1/2$, that is in the singlet state (e.g., as in Example 8.4). If the $z$ component of the spin of the first particle is measured and found to be $+1/2$, then the second particle must have $z$ component of spin equal to $-1/2$, since $m_{1s} + m_{2s} = m_s = 0$. Thus, the final wave function in this case is $| \uparrow \downarrow \rangle$. The act of measuring the individual spins of the particles alters the wave function from $\frac{1}{\sqrt{2}} | \downarrow \uparrow \rangle - \frac{1}{\sqrt{2}} | \uparrow \downarrow \rangle$ to $| \uparrow \downarrow \rangle$. This is called the "collapse of the wave function." Before the measurement is made, the particles have the potential to be in either the $| \uparrow \downarrow \rangle$ state or the $| \downarrow \uparrow \rangle$ state, but they are not actually in either state. The act of measuring the spin of one particle collapses the wave function into one of these two states.

Now consider the following thought experiment. Two electrons are prepared in a singlet state. Without either spin being measured, one electron is left on earth while the second is transported to Alpha Centauri. A scientist on earth measures the spin of the first electron and finds it to be in the spin up state. Immediately following this measurement, a scientist orbiting Alpha Centauri measures the spin of the second electron and, of course, finds it to be in the spin down state.

This may not seem particularly remarkable. For example, if we began with a white marble and a black marble, then we leave one on Earth and transported the second one to a nearby star, the discovery that the marble left on earth was white would immediately indicate that the other marble must be black. The quantum mechanical situation is more subtle, however. Strictly speaking, the two electron spins remain in a linear superposition of the $| \uparrow \downarrow \rangle$ and $| \downarrow \uparrow \rangle$ states until one of the spins is measured. The scientist on earth collapses the wave function by measuring the spin of the first electron, and somehow the second electron, 4 light-years away, immediately "knows" to assume the opposite spin state! This rather strange result is a version of the *Einstein–Rosen–Podolsky paradox*.

As an even more extreme case, consider the example of "Schrödinger's cat." A cat is inside a closed box (Figure 8.10). A radioactive substance is decaying; if a decay occurs in a fixed time interval, poison gas will be released into the box killing the cat. On the other hand, if no decay occurs in the given time interval, the cat will remain alive. What does the Copenhagen interpretation say about the state of the cat at any given time? Before we open up the box to check on the cat, it can be treated as the

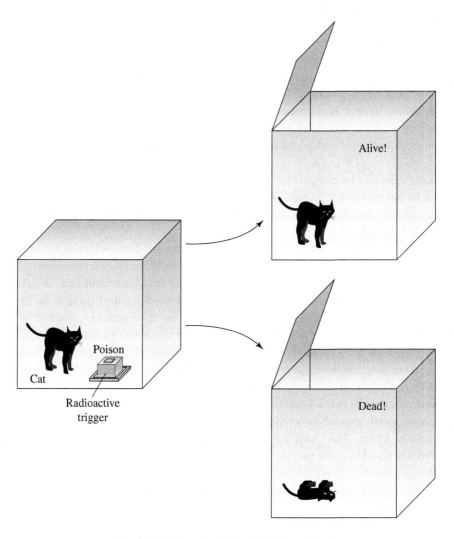

Fig. 8.10    The unfortunate Schrödinger's cat.

superposition of two states, $|\psi_{alive}\rangle$ and $|\psi_{dead}\rangle$, and we can write

$$|\psi_{cat}\rangle = \frac{1}{\sqrt{2}}|\psi_{alive}\rangle + \frac{1}{\sqrt{2}}|\psi_{dead}\rangle$$

Now we open up the box, and our act of observing collapses the wave function. If the cat is dead, for example, we find

$$|\psi_{cat}\rangle = |\psi_{dead}\rangle$$

The state of the cat prior to our observation, however, seems absurd. How can the cat be in a linear superposition of "alive" and "dead?" This seems like a very unsatisfying state of affairs. In fact, the entire Copenhagen interpretation represents a rather substantial shift in the way physicists had viewed the nature of measurement, so other interpretations have been put forward.

### Hidden Variables

An alternative way of framing quantum mechanics is to assume that it only seems like a probabilistic theory because it represents an approximation to a deeper, deterministic theory. This point of view is perhaps the most conservative, since it assumes that the more bizarre aspects of quantum mechanics are due only to our ignorance of the true underlying theory. There are certainly precedents for this idea in other areas of physics. Suppose, for instance, that several measurements are made of the speed $v$ of molecules in a sample of gas. This distribution of speeds would appear to be random, and a large enough set of measurements would uncover a probability distribution $P(v)$ for the molecular speeds, given by

$$P(v) = 4\pi \left(\frac{m}{2\pi kT}\right)^{3/2} v^2 e^{-mv^2/2kT}$$

called the *Maxwell distribution*. One might be tempted to conclude that the motion of the gas molecules is completely random, and the only thing measurable is the probability distribution.

In fact, of course, this is not the case. The molecules follow trajectories that are completely determined by classical mechanics. The motion only seems random because of the enormous number of molecules involved. This is the fundamental idea of statistical mechanics: there exist deterministic systems which are so large that an exact calculation of all of the trajectories is impractical; the best that can be done is to describe the system in terms of probabilities.

In the same way, we can postulate that quantum mechanics is actually deterministic, and only our ignorance makes it appear random. This is the *hidden variables* formulation of quantum mechanics. Unfortunately for those dissatisfied with the Copenhagen interpretation, hidden variables models face serious difficulties. In particular, J.S. Bell showed in 1964 that hidden variables models can be tested experimentally. By considering experiments of the sort discussed in the previous section, in which particles with a known total angular momentum are allowed to fly apart and the individual spins are then measured, Bell showed that the Copenhagen interpretation and the hidden variables interpretation give different results. Experiments based on Bell's idea have subsequently shown that it is the hidden variables model, not the Copenhagen interpretation, which is incompatible with the observed behavior of such particles.

There is one loophole here: Bell's theoretical work, as well as the subsequent experimental investigations, apply only to so-called *local* hidden variables models. These are models in which information propagates at a finite speed, e.g., a speed at or below the speed of light. If information can propagate instantaneously from one point to another, then hidden variables theories can still be made compatible with experiment. However, this possibility seems even more outlandish than the bizarreness of the Copenhagen interpretation, so it is normally not considered seriously.

### The Many Worlds Interpretation of Quantum Mechanics

In 1957 Hugh Everett proposed another way to avoid the problem of the collapse of the wave function: suppose that it never really collapses! Consider, for example, a particle in the one-dimensional potential well discussed in the previous section. Suppose that when you measure the direction of the particle, the wave function doesn't collapse; instead, the act of measurement splits the universe into two separate worlds: in one world the particle is observed to be moving to the left, and in the other world it is moving to the right. This is called the *many worlds interpretation* of quantum mechanics. In this interpretation, the wave function never collapses. Instead, the universe is constantly branching into multiple worlds; any possible outcome of a quantum measurement happens in one of the resulting worlds. So, for example, Schrödinger's cat is alive in one world and dead in another (Figure 8.11). As the quantum mechanics instructor of this author once said, "Strangely enough, there are grown men who believe this." (P.J.E. Peebles, Princeton University, 1979).

### PROBLEMS

**8.1** (a) A particle with spin 1 has orbital angular momentum $l = 0$. What are the possible values for the total angular momentum quantum number $j$?

(b) The same particle has $l = 3$. What are the possible values for $j$?

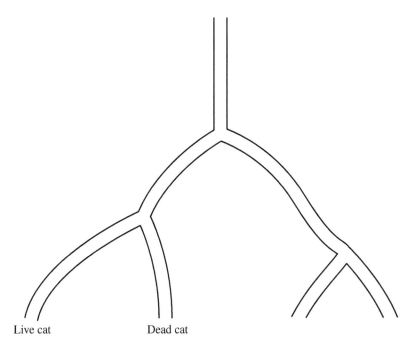

Live cat          Dead cat

Fig. 8.11   In the many-worlds interpretation, every measurement causes the universe to branch into multiple worlds; each possible outcome of the measurement occurs in one of the worlds.

**8.2** (a) A particle has spin $3/2$ and orbital angular momentum $l = 1$. What are the possible values for the total angular momentum quantum number $j$?
(b) For each value of $j$ in part (a), determine the possible values of $m_j$.

**8.3** Determine (using the matrix representation) which of the following operators are Hermitian and which are not: $S_x$, $S_y$, $S_z$, $S_+$, $S_-$.

**8.4** Derive the eigenvalues and the corresponding normalized eigenvectors of $S_y$ given in Equations (8.26) and (8.27).

**8.5** A particle has spin 1, so that $m_s = -1$, 0, or 1. Derive the matrices which correspond to $S_x$, $S_y$, and $S_z$.

**8.6** (a) A particle has $s = 3/2$. The operator $S_{++}$ is defined to be the square of the raising operator: $S_{++} = (S_+)^2$, where $S_+$ is the usual raising operator:

$$S_+|s\ m_s\rangle = \hbar\sqrt{s(s+1) - m_s(m_s+1)}|s\ m_s+1\rangle$$

Derive the matrix corresponding to the operator $S_{++}$.
(b) What is the matrix corresponding to the adjoint operator $(S_{++})^\dagger$?

**8.7** Let the operator $Q$ be given by $Q = S_+ S_-$, where $S_+$ and $S_-$ are the usual raising and lowering operators:

$$S_-|s\ m_s\rangle = \hbar\sqrt{s(s+1) - m_s(m_s-1)}|s\ m_s-1\rangle$$

$$S_+|s\ m_s\rangle = \hbar\sqrt{s(s+1) - m_s(m_s+1)}|s\ m_s+1\rangle$$

Derive the matrix corresponding to the operator $Q$ for a spin 1 particle. Determine whether or not $Q$ is Hermitian.

**8.8** Using the matrix representation of the spin operators for $s = 1/2$, verify the results for $[S_x, S_y]$, $[S_y, S_z]$, and $[S_z, S_x]$ given in Equations (8.1)–(8.3).

**8.9** A large number of spin-1/2 particles are run through a Stern–Gerlach machine. When they emerge, all of the particles have the same spin wave function $\begin{pmatrix} \psi_1 \\ \psi_2 \end{pmatrix}$ (in the usual basis in which $\begin{pmatrix} 1 \\ 0 \end{pmatrix}$ represents spin in the $+z$ direction, and $\begin{pmatrix} 0 \\ 1 \end{pmatrix}$ represents spin in the $-z$ direction). The spin of these particles is measured in the $z$ direction. On average, 9/25 of the particles have spin in the $+z$ direction, and 16/25 have spin in the $-z$ direction.

(a) Determine a possible normalized spin wave function $\begin{pmatrix} \psi_1 \\ \psi_2 \end{pmatrix}$.

(b) Is there a single unique solution to part (a), a finite number of different solutions, or an infinite number of different solutions? (Multiplying the entire wave vector by a constant does not count as a different solution.)

**8.10** A Stern–Gerlach experiment is set up with the axis of the inhomogeneous magnetic field in the $x$-$y$ plane, at an angle $\theta$ relative to the $x$-axis. Call this direction $r$:

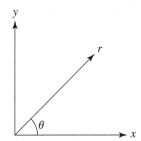

The spin operator in the $r$ direction is
$$S_r = (\cos\theta)S_x + (\sin\theta)S_y$$
(a) For a spin-1/2 particle, calculate the matrix corresponding to $S_r$. Calculate the eigenvalues and corresponding eigenvectors. Normalize the eigenvectors and express them in the form $a|\uparrow\rangle + b|\downarrow\rangle$, where $a$ and $b$ are constants.

(b) Suppose a measurement of the spin of the particle in the $r$ direction is made and it is determined that the spin is in the positive $r$ direction, i.e., $S_r|\psi\rangle = (+\hbar/2)|\psi\rangle$. Now a second measurement is made to determine $m_{sx}$ (the component of the spin in the $x$ direction). What is the probability that $m_{sx} = -1/2$? Suppose that instead of measuring $m_{sx}$, the $z$ component of the spin $m_s$ is measured. What is the probability that $m_s = +1/2$?

(c) Suppose that the particle has spin in the positive $r$ direction as in part (b). The $z$ component of the spin is measured and it is discovered that $m_s = +1/2$. Now a third measurement is made to determine $m_{sx}$. What is the probability that $m_{sx} = -1/2$?

**8.11** A spin-1/2 particle is in the state $|\psi\rangle = \sqrt{2/3}|\uparrow\rangle + i\sqrt{1/3}|\downarrow\rangle$.

(a) Verify that the wave function is correctly normalized.

(b) A measurement is made of the $x$ component of the spin. What is the probability that the spin will be in the $-x$ direction?

(c) Suppose a measurement is made of the spin in the $z$ direction and it is discovered that the particle has $m_s = -1/2$. Now a second measurement is made to determine the spin in the $x$ direction. What is the probability that the spin will be in the $+x$ direction?

**8.12** An electron is precessing in a magnetic field. The wave function for the electron is
$$|\psi\rangle = \frac{1}{\sqrt{2}}\begin{pmatrix} \cos\omega t + \sin\omega t \\ \cos\omega t - \sin\omega t \end{pmatrix}$$

(a) Describe the plane of rotation of the spin of this particle.

(b) In what direction is it rotating in this plane?

**8.13** A magnetic field pointing in the $-z$ direction produces a Hamiltonian $H = -\omega S_z$, where $\omega$ is a constant with units of frequency. A spin-1/2 particle is placed in this magnetic field. At $t = 0$, the particle is pointing in the $+y$ direction.

(a) Derive an expression for the spin vector $\begin{pmatrix} \psi_1 \\ \psi_2 \end{pmatrix}$ as a function of time.

(b) At $t = \pi/\omega$, a measurement is made of the spin in the $x$ direction. What is the probability that the spin is in the $+x$ direction?

(c) Suppose that at $t = \pi/\omega$, a measurement is made of the spin in the $x$ direction, and it is found that the spin is in the $+x$ direction. Then at the time $t = 2\pi/\omega$, another measurement is made of the spin in the $x$ direction. What is the probability that the spin is in the $+x$ direction?

**8.14** An electron is precessing in a magnetic field aligned along the $+z$-axis. At $t = 0$, the spin of the electron is in the positive $x$ direction. The wave function is

$$|\psi\rangle = \frac{1}{\sqrt{2}} \begin{pmatrix} e^{-i\omega t/2} \\ e^{i\omega t/2} \end{pmatrix}$$

For $t > 0$, calculate the probability of finding the electron in the state

(a) $m_s = +\hbar/2$

(b) $m_{sx} = +\hbar/2$, where $m_{sx}$ is the component of spin in the $x$ direction.

**8.15** A spin-1/2 particle is placed in a magnetic field pointing in the $+x$ direction which produces a Hamiltonian $H = \omega S_x$, where $\omega$ is a constant with units of frequency. At $t = 0$, the particle is pointing in the $+z$ direction. Derive an expression for the spin vector $\begin{pmatrix} \psi_1 \\ \psi_2 \end{pmatrix}$ as a function of time.

**8.16** A magnetic field pointing in the $+z$ direction produces a Hamiltonian $H = \omega S_z$, where $\omega$ is a constant with units of frequency. A spin-1 particle is placed in this magnetic field. The matrix corresponding to $S_z$ for a spin-1 particle is

$$S_z = \hbar \begin{pmatrix} 1 & 0 & 0 \\ 0 & 0 & 0 \\ 0 & 0 & -1 \end{pmatrix}$$

At $t = 0$, the particle is pointing in the $+x$ direction with normalized spin vector

$$|\psi\rangle = \begin{pmatrix} 1/2 \\ 1/\sqrt{2} \\ 1/2 \end{pmatrix}$$

Derive an expression for the spin vector $\begin{pmatrix} \psi_1 \\ \psi_2 \\ \psi_3 \end{pmatrix}$ as a function of time.

**8.17** Consider a system of two particles: particle 1 has spin 1, and particle 2 has spin 1/2. Let **S** be the total spin angular momentum operator for the two particles, where the eigenvalues of $S^2$ and $S_z$ are $\hbar^2 s(s + 1)$ and $\hbar m_s$, respectively. The particles are in the state $s = 3/2$ and $m_s = 1/2$.
(a) Calculate the wave function $|s = 3/2 \quad m_s = 1/2\rangle$ as a linear combination of the wave functions $|m_{1s} \ m_{2s}\rangle$, where $m_{1s}$ is the $z$ component of the spin of particle 1, and $m_{2s}$ is the $z$ component of the spin of particle 2.
(b) Find the probabilities that the $z$ component of the spin of particle 1 is
  (i) $m_{1s} = +1$
  (ii) $m_{1s} = 0$
  (iii) $m_{1s} = -1$

**8.18** Suppose that particle 1 (with spin 1) and particle 2 (with spin 1/2) interact via the Hamiltonian operator

$$H = \lambda \mathbf{S}_1 \cdot \mathbf{S}_2$$

where $\lambda$ is a constant. Calculate the energy of the state $|s \ m_s\rangle$.

**8.19** Two spin-1/2 particles are fixed in space with the Hamiltonian

$$H = a S_z^2 + b S^2$$

where $a$ and $b$ are positive constants, and as usual, $S^2$ is the total spin operator squared and $S_z$ is the operator which gives the $z$ component of the total spin. What are the energy levels of this system?

Chapter 9

# Time-independent perturbation theory

In theory, the Schrödinger equation allows us to solve any quantum mechanical system exactly. We simply insert the potential $V$ and solve for the wave function $\psi$ and the energy $E$. Unfortunately, there are very few potentials, such as the infinite square well or the Coulomb potential of the hydrogen atom, for which a simple exact solution exists. In order to make any further progress, we need to develop some techniques for finding approximate solutions to the Schrödinger equation. This chapter and Chapter 11 are devoted to a very important set of these techniques called *perturbation theory*.

The basic idea of perturbation theory rests on a simple general argument. Suppose we begin with a potential for which we can solve the Schrödinger equation exactly, such as the infinite square well of width $a$; recall from Chapter 4 that the ground-state energy and wave function are given by $\psi = \sqrt{2/a}\sin(\pi x/a)$ and $E = \hbar^2\pi^2/2ma^2$. Now suppose we make a tiny change in $V$ such as a small notch in the center of the potential (Figure 9.1). We cannot solve the Schrödinger equation for this new potential, but intuition suggests that a small change in $V$ ought to produce a small change in $\psi$ and in $E$. This intuition is correct. The reason is that the Schrödinger equation is a special kind of differential equation: it is *linear*, i.e., $\psi$ and its derivatives are taken only to the first power. Linear differential equations like the Schrödinger equation have the property that small changes in the parameters produce small changes in the solution. The fact that a small change in $V$ produces a small change in $\psi$ and $E$ is the fundamental idea of perturbation theory.

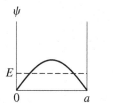
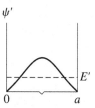

Fig. 9.1   The infinite square well on the left has the ground-state wave function and energy $\psi = \sqrt{2/a}\sin(\pi x/a)$ and $E = \hbar^2\pi^2/2ma^2$. A small change in the potential, shown on the right, produces a small change in $\psi$ and $E$.

## 9.1   Derivation of Time-Independent Perturbation Theory

Now we will calculate mathematically the change in $E$ produced by an arbitrary small change in the Hamiltonian $H$. [Although we talk generically about a change in the Hamiltonian, this usually amounts to a change in the potential.] Assume we have a Hamiltonian $H$ for which we can solve the Schrödinger equation exactly. We need to consider two possibilities for a small change in $H$: either the change in $H$ is constant in time, or it varies as a function of time. If the change in $H$ is constant in time, we are dealing with *time-independent* perturbation theory, which is the subject of this chapter. If, on the other hand, the change in $H$ varies with time, we have *time-dependent* perturbation theory, which is discussed in Chapter 11.

We will now derive what happens if we have a small, time-independent change in the Hamiltonian. Assume that we have a Hamiltonian $H_0$, for which we can find all of the eigenstates $|\psi_n\rangle$ with energies $E_n$:

$$H_0|\psi_n\rangle = E_n|\psi_n\rangle$$

Note that this is just shorthand for an infinite set of equations: $H_0|\psi_1\rangle = E_1|\psi_1\rangle$, $H_0|\psi_2\rangle = E_2|\psi_2\rangle$, and so on. For example, the $|\psi_n\rangle$ could correspond to the hydrogen wave functions, the spin eigenfunctions of an electron in a magnetic field, or any other set of wave functions that are exact solutions of the Schrödinger equation. Now add a small perturbation $\lambda H'$ to the Hamiltonian:

$$H = H_0 + \lambda H' \tag{9.1}$$

Here $\lambda$ is taken to be a dimensionless small number, $\lambda \ll 1$, so that the perturbation $\lambda H'$ is small compared to the original Hamiltonian $H_0$. We would like to solve the new Schrödinger equation:

$$H|\psi\rangle = E|\psi\rangle \tag{9.2}$$

In this equation $H$ is the new (perturbed) Hamiltonian of Equation (9.1), $|\psi\rangle$ is the new wave function after we have added the perturbation to the Hamiltonian, and $E$ is the new energy. Of course, we cannot solve this equation exactly (or we wouldn't be bothering to develop perturbation theory), but we can use some mathematical techniques to see how the change in the Hamiltonian changes the wave functions and energies.

The first step is to write the new energy and wave function in Equation (9.2) as a power series in the small number $\lambda$ that appears in Equation (9.1):

$$E = E_n + \lambda E^{[1]} + \lambda^2 E^{[2]} + \cdots \qquad (9.3)$$
$$|\psi\rangle = |\psi_n\rangle + \lambda|\phi_1\rangle + \lambda^2|\phi_2\rangle + \cdots \qquad (9.4)$$

In these equations, $|\psi_n\rangle$ and $E_n$ are the original eigenfunction and energy before we apply the perturbation; since Equation (9.2) has an infinite number of solutions (corresponding to a small perturbation applied to any of the eigenfunctions of $H_0$) we have to pick a particular eigenfunction $|\psi_n\rangle$ to perturb. The energies $E^{[1]}$, $E^{[2]}, \ldots$ and the wave functions $|\phi_1\rangle$, $|\phi_2\rangle$ are unknown; our goal is to solve for them.

Recall that we are interested in the small change in $E$ which results from our small change $\lambda H'$ in the Hamiltonian. The first term in Equation (9.3) gives us the energy before we apply the small perturbation. The rest of the terms give us the small change in the energy due to the small change in $H$. But if $\lambda$ is tiny, only the first of these terms really matters. For instance, if we take $\lambda = 10^{-6}$, then $\lambda^2 = 10^{-12}$, $\lambda^3 = 10^{-18}$, and so on. So the third term in Equation (9.3) is tiny compared to the second term, and the fourth term is tiny compared to the third term. That means the change in $E$ is essentially $\lambda E^{[1]}$, and we can ignore all of the other terms in Equation (9.3). [Exception: if, for a particular perturbation, $\lambda E^{[1]}$ is exactly zero, we will have to go further and solve for $\lambda^2 E^{[2]}$.]

Substituting the expressions for $E$ and $|\psi\rangle$ from Equations (9.3) and (9.4) into the Schrödinger equation given by Equation (9.2) gives the following rather messy result:

$$(H_0 + \lambda H')(|\psi_n\rangle + \lambda|\phi_1\rangle + \lambda^2|\phi_2\rangle + \cdots)$$
$$= (E_n + \lambda E^{[1]} + \lambda^2 E^{[2]} + \cdots)(|\psi_n\rangle + \lambda|\phi_1\rangle + \lambda^2|\phi_2\rangle + \cdots)$$
$$(9.5)$$

Although it looks like this is only making matters worse, we can now apply two ideas to simplify this equation. First, recall that if $\lambda$ is small, then $\lambda$ is much larger than $\lambda^2$, which is much larger then $\lambda^3$, and so on. This means that for very small $\lambda$, the terms in Equation (9.5) with different powers of $\lambda$ do not affect each other, so Equation (9.5) must be satisfied separately for each individual power of $\lambda$. Equating powers of $\lambda$ in this equation gives

$$\lambda^0 \qquad H_0|\psi_n\rangle = E_n|\psi_n\rangle \tag{9.6}$$

$$\lambda^1 \qquad \lambda H'|\psi_n\rangle + \lambda H_0|\phi_1\rangle = \lambda E_n|\phi_1\rangle + \lambda E^{[1]}|\psi_n\rangle \tag{9.7}$$

$$\lambda^2 \qquad \lambda^2 H_0|\phi_2\rangle + \lambda^2 H'|\phi_1\rangle = \lambda^2 E^{[1]}|\phi_1\rangle + \lambda^2 E^{[2]}|\psi_n\rangle + \lambda^2 E_n|\phi_2\rangle \tag{9.8}$$

Equation (9.6) is just the original unperturbed Schrödinger equation, which is reassuring, but we cannot make further progress with Equations (9.7) and (9.8) unless we know what happens when $H_0$ operates on $|\phi_1\rangle$ and $|\phi_2\rangle$. However, we don't even know what $|\phi_1\rangle$ and $|\phi_2\rangle$ are! Nonetheless, we can solve this problem by applying a second idea. Recall that for any Hamiltonian $H_0$, we can find a set of eigenfunctions $|\psi_m\rangle$ which form an orthonormal basis, i.e., any wave function $|\psi\rangle$ can be written as

$$|\psi\rangle = c_1|\psi_1\rangle + c_2|\psi_2\rangle + \cdots + c_m|\psi_m\rangle + \cdots$$

So even though we don't know $|\phi_1\rangle$ and $|\phi_2\rangle$ in Equations (9.7) and (9.8), we can expand them out as linear combinations of eigenfunctions of $H_0$, and we do know what happens when we apply $H_0$ to these eigenfunctions. We write

$$|\phi_1\rangle = c_1|\psi_1\rangle + c_2|\psi_2\rangle + \cdots + c_m|\psi_m\rangle + \cdots = \sum_i c_i|\psi_i\rangle \tag{9.9}$$

and

$$|\phi_2\rangle = \sum_i d_i|\psi_i\rangle \tag{9.10}$$

Then when $H_0$ is applied to the sums in Equations (9.9) and (9.10), it simply pulls out the appropriate energy in front of each term:

$$H_0|\phi_1\rangle = H_0\left(\sum_i c_i|\psi_i\rangle\right) = \sum_i E_i c_i|\psi_i\rangle \tag{9.11}$$

and

$$H_0|\phi_2\rangle = \sum_i E_i d_i|\psi_i\rangle \tag{9.12}$$

When $H_0$ is applied to $|\phi_1\rangle$ in Equation (9.7), it produces the sum given in Equation (9.11). Then we get

$$\lambda H'|\psi_n\rangle + \lambda\sum_i E_i c_i|\psi_i\rangle = \lambda E_n\sum_i c_i|\psi_i\rangle + \lambda E^{[1]}|\psi_n\rangle \tag{9.13}$$

We would like to solve this equation to find $E^{[1]}$. We take the inner product of $\langle\psi_n|$ with both sides of the equation, recalling that $\langle\psi_n|\psi_n\rangle = 1$:

$$\lambda\langle\psi_n|H'|\psi_n\rangle + \lambda\sum_i E_i c_i\langle\psi_n|\psi_i\rangle = \lambda E_n\sum_i c_i\langle\psi_n|\psi_i\rangle + \lambda E^{[1]} \qquad (9.14)$$

Note that $|\psi_n\rangle$ is the original eigenfunction which satisfies the unperturbed Schrödinger equation, while the $|\psi_i\rangle$'s are the complete set of such eigenfunctions, including $|\psi_n\rangle$ as a particular case. Hence, when we take the inner product of $\langle\psi_n|$ with $|\psi_i\rangle$, we get zero for $i \neq n$ and one for $i = n$. This selects out $i = n$ from the sum in Equation (9.14), giving

$$\lambda\langle\psi_n|H'|\psi_n\rangle + \lambda E_n c_n = \lambda E_n c_n + \lambda E^{[1]}$$

so

$$\lambda E^{[1]} = \langle\psi_n|\lambda H'|\psi_n\rangle \qquad (9.15)$$

where $\lambda E^{[1]}$ is the dominant or lowest order (i.e., lowest power of $\lambda$) change in the energy due to the small change $\lambda H'$ in the Hamiltonian.

We can now calculate the second-order change in $E$, i.e., the term in Equation (9.3) which is proportional to $\lambda^2$. Of course, it is reasonable to ask why we would ever want to know this, since $\lambda E^{[1]}$ is much larger than $\lambda^2 E^{[2]}$. Normally, the second order perturbation *is* irrelevant except for one very important case: if $\lambda E^{[1]}$ is exactly zero, then $\lambda^2 E^{[2]}$ gives the dominant change in the energy. To find $\lambda^2 E^{[2]}$, we substitute both Equations (9.9) and (9.10) into Equation (9.8). When $H_0$ operates on the sum of eigenvectors as in Equation (9.12), we get

$$\lambda^2\sum_i d_i E_i|\psi_i\rangle + \lambda^2 H'\sum_i c_i|\psi_i\rangle$$
$$= \lambda^2 E^{[1]}\sum_i c_i|\psi_i\rangle + \lambda^2 E^{[2]}|\psi_n\rangle + \lambda^2 E_n\sum_i d_i|\psi_i\rangle$$

As before, we take the inner product with $\langle\psi_n|$, and now we substitute $\langle\psi_n|H'|\psi_n\rangle$ for $E^{[1]}$. Solving for $\lambda^2 E^{[2]}$, we get

$$\lambda^2 E^{[2]} = \lambda^2\sum_i c_i\langle\psi_n|H'|\psi_i\rangle - \lambda^2 c_n\langle\psi_n|H'|\psi_n\rangle$$

The right-hand side is the sum over all eigenfunctions $|\psi_i\rangle$ minus the particular eigenfunction $|\psi_n\rangle$, so it can be written as

$$\lambda^2 E^{[2]} = \lambda^2\sum_{i\neq n} c_i\langle\psi_n|H'|\psi_i\rangle \qquad (9.16)$$

Now all we have to do is to find the coefficients $c_i$ which first appeared in Equation (9.9). We go back to Equation (9.13), but now instead of applying $\langle \psi_n |$ to both sides, we apply an arbitrary eigenfunction $\langle \psi_m |$, where $\langle \psi_m | \neq \langle \psi_n |$. This gives

$$\lambda \langle \psi_m | H' | \psi_n \rangle + \lambda E_m c_m = \lambda E_n c_m + 0$$

so that

$$c_m = \frac{\langle \psi_m | H' | \psi_n \rangle}{E_n - E_m} \tag{9.17}$$

Substituting this expression for the $c_i$'s in Equation (9.16) gives the final expression for $\lambda^2 E^{[2]}$:

$$\lambda^2 E^{[2]} = \sum_{i \neq n} \frac{\langle \psi_n | \lambda H' | \psi_i \rangle \langle \psi_i | \lambda H' | \psi_n \rangle}{E_n - E_i}$$
$$= \sum_{i \neq n} \frac{|\langle \psi_n | \lambda H' | \psi_i \rangle|^2}{E_n - E_i} \tag{9.18}$$

To summarize: if we start out with some Hamiltonian $H_0$ for which we can solve the Schrödinger equation exactly, and we begin in an eigenstate $|\psi_n\rangle$ with energy $E_n$, then after we change the Hamiltonian by the small amount $\lambda H'$, the dominant change in the energy, proportional to $\lambda$, will be given by Equation (9.15), while the next largest change, proportional to $\lambda^2$, will be given by Equation (9.18).

While we have used $\lambda$ to remember which terms are larger than others, we can now simplify our expressions by writing the change in $H$ as

$$H = H_0 + H_1$$

where

$$H_1 = \lambda H'$$

is very small compared to the unperturbed Hamiltonian $H_0$. Then we write the first and second order changes in the energy as

$$E^{(1)} = \lambda E^{[1]}$$

and

$$E^{(2)} = \lambda^2 E^{[2]}$$

These changes in the energy are given by

$$E^{(1)} = \langle \psi_n | H_1 | \psi_n \rangle \tag{9.19}$$

and

$$E^{(2)} = \sum_{i \neq n} \frac{|\langle \psi_n | H_1 | \psi_i \rangle|^2}{E_n - E_i} \qquad (9.20)$$

where the $|\psi_i\rangle$'s which appear in Equation (9.20) are all of the other eigenfunctions of $H_0$ aside from the one being perturbed. Equations (9.19) and (9.20) are the main result of this section; they give the first-order and second-order perturbations to the energy from an arbitrary perturbation to the Hamiltonian. As usual, the inner products which appear in Equations (9.19) and (9.20) can represent a variety of different mathematical possibilities. If the wave functions and the perturbation are functions of position (as in Example 9.1), then these inner products will be integrals. If the eigenstates are spin states and the perturbation is a function of spin operators (as in Example 9.2), then these inner products will be matrix products.

There is one case in which the entire argument falls apart. If the original eigenfunction $|\psi_n\rangle$ is degenerate with some other eigenfunction $|\psi_m\rangle$ of $H_0$, i.e., $E_n = E_m$, then the argument fails. This can be seen from the fact that both Equations (9.17) and (9.20) "blow up" in this case with zero in the denominator. Note that Equation (9.19) might be completely well behaved in this case, and it is a very common mistake to use Equation (9.19) in the case of this kind of degeneracy. **DON'T DO IT!** Since the second-order term in this case is infinite, the entire perturbation expansion becomes inapplicable for the case of degeneracy. Perturbation theory applied to degenerate eigenfunctions requires some further mathematical machinery (degenerate perturbation theory) which is beyond the scope of this text. Note, however, that there is one special case (which we will encounter frequently) in which we *can* use Equations (9.19) and (9.20) with degenerate wave functions: if $\langle \psi_n | H_1 | \psi_i \rangle = 0$ whenever $E_n = E_i$, then our expressions will be well behaved.

Although the change in the energy is usually the quantity that can be most easily measured directly, it is also possible to calculate the change in the wave function due to the perturbation. Returning to Equation (9.4), we see that the lowest-order change to $|\psi_n\rangle$ is given by $\lambda|\phi_1\rangle$, and $|\phi_1\rangle$ has already been expressed as a sum over the unperturbed eigenfunctions in Equation (9.9) with the $c_i$'s that appear in this equation given by Equation (9.17). Substituting these values for the $c_i$'s from Equation (9.17) into

Equation (9.9), we obtain

$$\lambda|\phi_1\rangle = \sum_{i \neq n} \frac{\langle \psi_i | \lambda H' | \psi_n \rangle}{E_n - E_i} |\psi_i\rangle$$

$$= \sum_{i \neq n} \frac{\langle \psi_i | H_1 | \psi_n \rangle}{E_n - E_i} |\psi_i\rangle$$

Note that we have dropped the $i = n$ term from the sum in Equation (9.9). We have the freedom to do this because this term is just proportional to the original unperturbed eigenfunction $|\psi_n\rangle$. Hence, in the original expansion of the wave function (Equation (9.4)), this term can be removed from the $\lambda|\phi_1\rangle$ term and absorbed into the $|\psi_n\rangle$ term. Then to first order, the new wave function in the presence of the perturbation is

$$|\psi\rangle = |\psi_n\rangle + \sum_{i \neq n} \frac{\langle \psi_i | H_1 | \psi_n \rangle}{E_n - E_i} |\psi_i\rangle$$

Thus, the effect of the perturbation is to "mix together" all of the other eigenfunctions in the new perturbed wave function.

---

### Example 9.1. The Anharmonic Oscillator
In Chapter 4, we derived the solutions for the one-dimensional harmonic oscillator potential,

$$V(x) = \frac{1}{2} K x^2$$

The energies are

$$E = \left(n + \frac{1}{2}\right) \hbar \omega, \quad n = 0, 1, \ldots$$

where $\omega = \sqrt{K/m}$ is the classical oscillation frequency, and the corresponding wave functions are

$$\psi_0(s) = \frac{1}{\pi^{1/4}} e^{-s^2/2}$$

$$\psi_1(s) = \frac{\sqrt{2}}{\pi^{1/4}} s e^{-s^2/2}$$

$$\vdots$$

with $s = [(Km)^{1/4}/\hbar^{1/2}]x$.

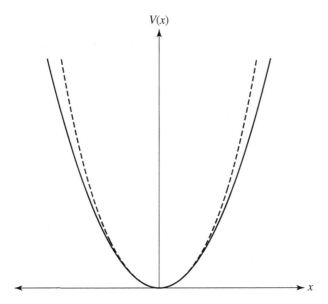

**Fig. 9.2** Solid curve: the unperturbed harmonic oscillator potential $V = (1/2)Kx^2$. Dashed curve: the perturbed potential $V = (1/2)Kx^2 + \beta x^4$.

Suppose we begin in the first excited state $\psi_1$ with energy $E = (3/2)\hbar\omega$. We will calculate what happens to the energy of this state if we add a small anharmonic term to the potential (Figure 9.2):

$$V(x) = \frac{1}{2}Kx^2 + \beta x^4$$

From Equation (9.19), the first-order perturbation is

$$E^{(1)} = \langle n = 1|\beta x^4|n = 1\rangle$$

Expressing the perturbation in terms of $s$, we get

$$E^{(1)} = \int_{s=-\infty}^{\infty} \left[\frac{\sqrt{2}}{\pi^{1/4}}se^{-s^2/2}\right]^* \left[\beta\frac{\hbar^2}{Km}s^4\right] \left[\frac{\sqrt{2}}{\pi^{1/4}}se^{-s^2/2}\right] ds$$

$$= \frac{2}{\sqrt{\pi}}\beta\frac{\hbar^2}{Km}\int_{s=-\infty}^{\infty} s^6 e^{-s^2} ds$$

$$= \frac{15}{4}\beta\frac{\hbar^2}{Km}$$

With $\omega = \sqrt{K/m}$, the total energy including the perturbation is

$$E = \hbar\omega\left(\frac{3}{2} + \frac{15}{4}\beta\frac{\hbar\omega}{K^2}\right)$$

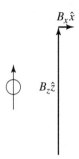

Fig. 9.3　We begin with an electron in the spin up direction in the magnetic field $B_z\hat{z}$, and we add the small perturbation $B_x\hat{x}$.

Example 9.1 shows how to apply perturbation theory when the wave function is a function of position. However, perturbation theory can also be applied to the matrix representation of spin states.

---

### Example 9.2. Spins in a Magnetic Field

Suppose we begin with an electron having spin magnetic moment $\boldsymbol{\mu}_s$ in a strong magnetic field $B_z$ in the $z$ direction (Figure 9.3). Recall from Chapter 8 that the potential for an electron with spin magnetic moment $\boldsymbol{\mu}_s$ in a magnetic field $\mathbf{B}$ is

$$V = -\boldsymbol{\mu}_s \cdot \mathbf{B}$$

where

$$\boldsymbol{\mu}_s = -\frac{g_s \mu_B}{\hbar} \mathbf{S}$$

Hence, the Hamiltonian is

$$H_0 = \frac{g_s \mu_B}{\hbar} \mathbf{B} \cdot \mathbf{S}$$

$$= \frac{g_s}{2} \mu_B \mathbf{B} \cdot \boldsymbol{\sigma}$$

and $g_s = 2$ for the electron. The eigenstates of $H_0$ are just the spin up and spin down states, $| \uparrow \rangle$ and $| \downarrow \rangle$, with energies $E_+ = +\mu_B B_z$ and $E_- = -\mu_B B_z$.

Suppose we are in the spin up state, $| \uparrow \rangle$, and we add the small magnetic field $B_x\hat{x}$ with $B_x \ll B_z$ (Figure 9.3). What is the change in the energy of the electron due to this new magnetic field? Our perturbing potential is

$$V_1 = \frac{g_s}{2} \mu_B B_x \sigma_x$$

with $g_s = 2$, the first-order change in the energy is

$$E^{(1)} = \langle \psi_n | H_1 | \psi_n \rangle$$
$$= \langle \uparrow | \mu_B B_x \sigma_x | \uparrow \rangle$$
$$= \mu_B B_x \begin{pmatrix} 1 & 0 \end{pmatrix} \begin{pmatrix} 0 & 1 \\ 1 & 0 \end{pmatrix} \begin{pmatrix} 1 \\ 0 \end{pmatrix}$$
$$= \mu_B B_x \begin{pmatrix} 1 & 0 \end{pmatrix} \begin{pmatrix} 0 \\ 1 \end{pmatrix} = 0$$

So the first-order perturbation, proportional to $B_x$, is equal to zero. The second-order perturbation is

$$E^{(2)} = \sum_{i \neq n} \frac{|\langle \psi_n | H_1 | \psi_i \rangle|^2}{E_n - E_i}$$
$$= \frac{|\langle \uparrow | \mu_B B_x \sigma_x | \downarrow \rangle|^2}{E_+ - E_-}$$
$$= \frac{[\mu_B B_x]^2}{2\mu_B B_z} \left| \begin{pmatrix} 1 \\ 0 \end{pmatrix} \begin{pmatrix} 0 & 1 \\ 1 & 0 \end{pmatrix} \begin{pmatrix} 0 \\ 1 \end{pmatrix} \right|^2$$
$$= \frac{\mu_B B_x^2}{2 B_z}$$

As expected, $E^{(2)}$ is proportional to $B_x^2$. In fact, this problem can be solved exactly, and it is instructive to see how the exact solution compares with the perturbation theory result (see Problem 9.1).

## 9.2 Perturbations to the Atomic Energy Levels

In Chapter 6 we developed an elegant model for the hydrogen atom. The energy levels were determined by the principal quantum number $n$, while the other quantum numbers $l$, $m_l$, and $m_s$ had no effect on the energy. As is often the case in physics, the elegant theory is extremely accurate, but it is not exact. There are small corrections to the theory due to internal interactions in the hydrogen atom. With perturbation theory, we now have the tools to derive these corrections.

### *Fine Structure*

Recall that for hydrogen, the energy is given by

$$E_n = (-13.6 \text{ eV}) \frac{1}{n^2}$$

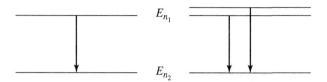

Fig. 9.4 A given spectral line corresponds to a single transition between two different energy levels. If two supposedly degenerate levels are slightly separated in energy, a double line will be produced.

In hot hydrogen gas, a series of spectral lines are observed, each one corresponding to a particular transition with $h\nu = E_{n_1} - E_{n_2}$. However, upon examining the spectrum closely, it is observed that some spectral lines are not really single lines but are closely spaced double lines. This feature is called *fine structure*. There is an obvious way to get such closely spaced spectral lines: they will be observed if the degenerate energy levels are not truly degenerate but separated in energy by a very small amount (Figure 9.4). Recall that a given energy level $E_n$ corresponds to $2n^2$ different states. Apparently, some sort of interaction, which we have not yet accounted for, splits some of these states apart in energy.

Our simple model for the hydrogen atom considered only the Coulomb interaction between the proton and the electron. What we have neglected are the various magnetic fields produced in the atom. The orbital motion of the electron sets up a magnetic field with magnetic moment given by

$$\boldsymbol{\mu}_l = -\frac{g_l e}{2m_e}\mathbf{L}, \quad g_l = 1 \tag{9.21}$$

while the spin of the electron produces a spin magnetic moment equal to

$$\boldsymbol{\mu}_s = -\frac{g_s e}{2m_e}\mathbf{S}, \quad g_s \approx 2$$

So the orbital motion of the electron produces a magnetic field, and the electron itself acts like a small magnet embedded in this magnetic field (Figure 9.5). The electron will prefer to line up with its spin magnetic field in the opposite direction to the orbital magnetic field. Hence, the "spin up" and "spin down" states of the electron will have different energies. This is the basis of fine structure: it arises from the spin-orbit coupling of the electron.

More precisely, the interaction energy between the spin magnetic moment $\boldsymbol{\mu}_s$ and the external magnetic field $\mathbf{B}$ (produced by the orbital motion of the electron) is

$$H_1 = -\boldsymbol{\mu}_s \cdot \mathbf{B} \tag{9.22}$$

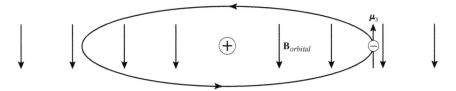

Fig. 9.5 The fine-structure splitting is produced by the interaction between the spin magnetic field of the electron and the magnetic field produced by its orbital motion.

To find **B** imagine that we are sitting in the rest frame of the electron watching the proton orbit around us. In the rest frame of the electron, the electric field of the proton,

$$\mathbf{E} = \frac{1}{4\pi\epsilon_0} \frac{e}{r^3} \mathbf{r}$$

transforms into a magnetic field, $\mathbf{B} = -\mathbf{v} \times \mathbf{E}/c^2$. Using $\mathbf{L} = \mathbf{r} \times \mathbf{p}$, we obtain, for the magnetic field induced by the orbital motion of the electron,

$$\mathbf{B} = \left(\frac{1}{4\pi\epsilon_0}\right) \frac{e}{m_e c^2 r^3} \mathbf{L}$$

Substituting our expressions for $\boldsymbol{\mu}_s$ and **B** into Equation (9.22), we obtain, for the energy of the spin-orbit interaction,

$$H_1 = \frac{e^2}{4\pi\epsilon_0 m_e^2 c^2 r^3} \mathbf{S} \cdot \mathbf{L}$$

Unfortunately, this expression is *wrong*. The problem arises because we did the calculation in the rest frame of the electron, which is an accelerating frame of reference, so we cannot simply transform back into the rest frame of the proton and expect to get the right answer. When this error is corrected, an additional factor of $1/2$ is obtained (this correction is called *Thomas precession* after L.H. Thomas, who explained this effect in 1926). The corrected expression is

$$H_1 = \frac{e^2}{8\pi\epsilon_0 m_e^2 c^2 r^3} \mathbf{S} \cdot \mathbf{L} \tag{9.23}$$

From a classical point of view, this expression makes sense, since we expect the interaction energy of the two magnetic fields to depend on the alignment of the **S** and **L** vectors (Figure 9.6).

We can get a crude order of magnitude estimate of the size of this interaction energy by taking $S \sim \hbar$, $L \sim \hbar$, and $r \sim 10^{-10}$ m. Then Equation (9.23) gives a perturbation energy of the order of $10^{-4}$ eV compared to

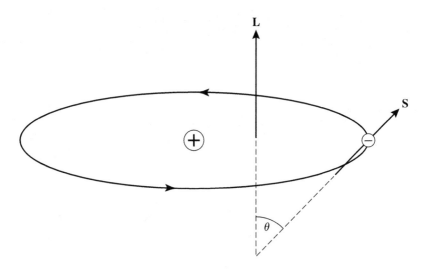

Fig. 9.6   The spin-orbit interaction Hamiltonian is proportional to **S·L**.   Classically, **S·L** = $SL\cos\theta$.

a hydrogen binding energy of 13.6 eV. It is therefore a good approximation to treat the spin-orbit interaction as a small perturbation.

In our expression for $H_1$, the interaction is proportional to $\mathbf{S \cdot L} = S_x L_x + S_y L_y + S_z L_z$. However, this expansion is essentially useless, since the electron is never in a state which is a simultaneous eigenstate of all three components of angular momentum. Instead, we use the standard procedure from Chapter 8 for dealing with dot products of operators. We define the total angular momentum operator **J** to be

$$\mathbf{J} = \mathbf{L} + \mathbf{S}$$

Then

$$J^2 = L^2 + S^2 + 2\mathbf{L \cdot S}$$

which implies

$$\mathbf{L \cdot S} = \frac{1}{2}(J^2 - L^2 - S^2)$$

and the spin-orbit perturbation becomes

$$H_1 = \frac{e^2}{16\pi\epsilon_0 m_e^2 c^2 r^3}(J^2 - L^2 - S^2)$$

In Chapter 6 we wrote the hydrogen wave functions in terms of the quantum numbers $n$, $l$, and $m_l$, and in Chapter 8 we added the spin quantum number $m_s$, but now we want to express the hydrogen wave functions as eigenfunctions of $\mathbf{J}$ and $J_z$, i.e., in the form $|n\ l\ j\ m_j\rangle$, where

$$J^2|n\ l\ j\ m_j\rangle = \hbar^2 j(j+1)|n\ l\ j\ m_j\rangle$$

$$J_z|n\ l\ j\ m_j\rangle = \hbar m_j|n\ l\ j\ m_j\rangle$$

Then the perturbation to the energy levels due to spin-orbit coupling is

$$E^{(1)}_{spin\text{-}orbit} = \langle n\ l\ j\ m_j|H_1|n\ l\ j\ m_j\rangle$$

$$= \langle n\ l\ j\ m_j|\frac{e^2}{16\pi\epsilon_0 m_e^2 c^2 r^3}(J^2 - L^2 - S^2)|n\ l\ j\ m_j\rangle$$

$$= \hbar^2[j(j+1) - l(l+1) - s(s+1)]\langle n\ l\ j\ m_j|\frac{e^2}{16\pi\epsilon_0 m_e^2 c^2 r^3}|n\ l\ j\ m_j\rangle$$

What is the value of $\langle n\ l\ j\ m_j|(e^2/16\pi\epsilon_0 m_e^2 c^2 r^3)|n\ l\ j\ m_j\rangle$? Note that this is just an integral over the radial wave function, which is a function only of $n$ and $l$; hence, we can write

$$\langle n\ l\ j\ m_j|(e^2/16\pi\epsilon_0 m_e^2 c^2 r^3)|n\ l\ j\ m_j\rangle = f_{nl}$$

where $f_{nl}$ is a function only of $n$ and $l$. Then our expression for the change in $E$ due to the spin-orbit interaction (taking $s = 1/2$ for the electron) becomes

$$E^{(1)}_{spin\text{-}orbit} = \hbar^2 f_{nl}\left[j(j+1) - l(l+1) - \frac{3}{4}\right] \tag{9.24}$$

From Chapter 9 recall that for $l \neq 0$, $j$ has two possible values, either $l - 1/2$ or $l + 1/2$, while for $l = 0$ we have only $j = 1/2$. From Equation (9.24), it is clear that the spin-orbit coupling splits each state with $l \neq 0$ into two different states with $j = l + 1/2$ having the higher energy and $j = l - 1/2$ having the lower energy.

The expression for $f_{nl}$ can be evaluated to yield a final expression for the change in $E$ from spin-orbit coupling:

$$E^{(1)}_{spin\text{-}orbit} = |E_n|\alpha^2 \frac{1}{2n}\frac{[j(j+1) - l(l+1) - 3/4]}{l(l+1/2)(l+1)} \tag{9.25}$$

where

$$\alpha = \frac{e^2}{4\pi\epsilon_0 \hbar c}$$

Note that $\alpha$ is a dimensionless number with a value of roughly $1/137$. Because of its origin in this calculation, $\alpha$ is called the *fine-structure constant*, although it crops up in many other areas of physics. Recall that the hydrogen energy levels $E_n$ are all negative, so we take the absolute value of $E_n$

in Equation (9.25) and in Equations (9.28)–(9.29) to avoid any confusion over the sign of the change in energy.

However, this is not the full story of the fine-structure splitting. We have applied our standard nonrelativistic treatment to the electron, and this is an excellent approximation, since the electron in the hydrogen atom is not highly relativistic (its kinetic energy is much smaller than its rest energy; classically, this corresponds to an electron velocity $v$ much smaller than the speed of light). However, now that we are working in the realm of tiny changes in the energy levels, we have to take into account small corrections due to relativistic effects.

In relativistic classical mechanics, the total energy of a particle with rest mass $m$ and momentum $p$ is

$$E = \sqrt{p^2 c^2 + m^2 c^4} \tag{9.26}$$

For now we are interested only in the case where the particle is only slightly relativistic so that $p \ll mc$. (We will relax this restriction in Chapter 15 when we discuss relativistic quantum mechanics in more detail.) In this limit we can expand the square root in Equation (9.26) to obtain

$$E \approx mc^2 + \frac{p^2}{2m} - \frac{p^4}{8m^3 c^2} + \cdots \tag{9.27}$$

In the limit of small $p$, each term in this expression is small compared to the preceding one. In relativity, the first term in Equation (9.27) is interpreted as the rest energy of the particle, while the remainder of the expression corresponds to the energy of motion. But what is the correct energy to use in the Schrödinger equation? In the standard nonrelativistic Schrödinger equation, the Hamiltonian operator corresponds to the kinetic energy plus the potential energy, and the rest energy plays no role. Hence, in writing the Hamiltonian, we use the second term in Equation (9.27) to give us the unperturbed Hamilton, while the third term gives the lowest-order perturbation due to relativistic effects.

Then we have, for our perturbation,

$$H_1 = -\frac{p^4}{8m_e^3 c^2}$$

and the lowest-order change in the hydrogen energy levels is

$$E^{(1)}_{relativistic} = -\frac{1}{8m_e^3 c^2} \langle n \ l \ j \ m_j | p^4 | n \ l \ j \ m_j \rangle$$

This expression can be evaluated for the hydrogen wave functions, yielding a final result of

$$E^{(1)}_{relativistic} = -|E_n| \alpha^2 \frac{1}{n^2} \left[ \frac{2n}{2l+1} - \frac{3}{4} \right] \tag{9.28}$$

Since $E^{(1)}_{relativistic}$ is a function of $n$ and $l$ but not a function of $j$, the relativistic correction does not contribute anything to the splitting of the energy levels, which is determined entirely by the spin-orbit interaction, but it does change the overall dependence of the energy levels on $n$ and $l$. Since $l \leq n - 1$, the term in brackets in Equation (9.28) is always positive, so the relativistic correction always *decreases* the energy levels.

Note that $E^{(1)}_{spin\text{-}orbit}$ and $E^{(1)}_{relativistic}$ are roughly equal in magnitude; both of them are approximately $\alpha^2 E_n$. Hence, neither contribution to the fine structure can be neglected. We therefore add Equations (9.25) and (9.28) to get the total change in energy due to both the spin-orbit coupling and relativistic effects:

$$E^{(1)}_{fine\ structure} = E^{(1)}_{spin\text{-}orbit} + E^{(1)}_{relativistic}$$

$$= |E_n|\alpha^2 \frac{1}{n^2}\left(\frac{3}{4} - \frac{n}{j + 1/2}\right) \tag{9.29}$$

The dependence on $l$ has cancelled out, so that the total change in energy is a function only of $j$ and $n$. The net effect of the fine structure is to split the $l-1/2$ and $l+1/2$ states (due to the spin-orbit coupling) and to *decrease* the energies of both states relative to the unperturbed hydrogen energy levels (see Problem 9.10). The fine-structure perturbation (both the decrease in energy relative to the unperturbed energy levels and the splitting between the $j = l + 1/2$ and the $j = l - 1/2$ states) is of order $\alpha^2 E_n \approx 10^{-4} E_n$, since the other factors in Equation (9.29) are all of order unity.

We now introduce a standard, if somewhat arcane, notation to describe the angular momentum states of the hydrogen atom. In this notation, the different $l$ states of the hydrogen atom are written with different capital letters: the $l = 0$ state is called the $S$ state, the $l = 1$ state is called the $P$ state, the $l = 2$ state is called the $D$ state, the $l = 3$ state is called the $F$ state, and then the sequence continues alphabetically $(G, H, I, \ldots)$. (The origin of these abbreviations is buried in the history of spectroscopy; there is nothing particularly logical about them.) The standard way of writing the various $j$ states is to indicate the value of $j$ as a subscript, e.g., $P_{1/2}$ is the notation for $l = 1$, $j = 1/2$. [We will see an additional twist to this notation in Chapter 13.] The different hydrogen energy levels written in this notation are shown in Figure 9.7. The $S$ states $(l = 0)$ do not split since they correspond to a single value of $j$, while the $l \neq 0$ states split into the $j = l + 1/2$ state and the $j = l - 1/2$ state, with the former having higher energy than the latter.

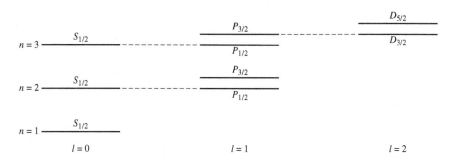

Fig. 9.7    The energy levels of hydrogen showing the fine structure (not drawn to scale).

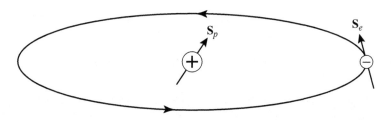

Fig. 9.8    The spin-spin interaction between the proton and electron produces the hyperfine splitting.

Finally, note that we have blithely applied first-order perturbation theory to degenerate states, ignoring the warning in the previous section. However, it is all right to use nondegenerate perturbation theory in this case, since $\langle n \, l \, m_l \, m_s | H_1 | n \, l' \, m_l' \, m_s' \rangle = 0$ if $l \neq l'$ or $m_l \neq m_l'$ or $m_s \neq m_s'$.

There is a second internal magnetic interaction in the hydrogen atom with a much smaller effect. The proton has a spin magnetic moment given by

$$\boldsymbol{\mu}_p = \frac{g_p e}{2m_p} \mathbf{S} \tag{9.30}$$

with $g_p \approx 5.6$. So the electron also feels this magnetic field and is perturbed by it (Figure 9.8). However, a comparison of Equation (9.30) with Equation (9.21) shows that the ratio of the spin magnetic field of the proton to the magnetic field produced by the orbital motion of the electron is roughly $m_e/m_p \approx 6 \times 10^{-4}$. Hence, we expect the splitting from the spin-spin interaction to be much smaller than the effect of the spin-orbit interaction.

Nonetheless, this spin-spin interaction does produce a splitting in energies. Since it is so much smaller than the fine structure, it is called *hyperfine* splitting. In the ground state of hydrogen, for example, the triplet $(S = 1)$ state has a higher energy than the singlet $(S = 0)$ state; the energy difference is $\Delta E = 5.9 \times 10^{-6}$ eV. Although this energy difference is tiny, hyperfine splitting has an importance out of proportion to its magnitude. The universe contains clouds of neutral hydrogen gas; this gas radiates by dropping from the triplet into the singlet state. The frequency of this radiation is $\nu = \Delta E / h = 1420$ MHz, corresponding to a wavelength of $\lambda = 21$ cm: the famous "21-centimeter line."

## *Vacuum Polarization and the Lamb Shift*

The fine-structure calculations in the previous section predict that the hydrogen energy levels do not depend on $l$. Hence, two states with the same $n$ and $j$ quantum numbers but different values of $l$ should be degenerate in energy. In 1947, Willis Lamb and his student, R.C. Retherford, showed experimentally that this was not the case. Specifically, they measured a splitting between the $n = 2$, $S_{1/2}$ state and the $n = 2$, $P_{1/2}$ state.

This splitting, now called the *Lamb shift*, cannot be explained in the context of quantum mechanics, but arises from the more esoteric area of quantum field theory (which was, in part, motivated by Lamb's experimental result). Quantum field theory is beyond the scope of this book; here we will simply use one of the predictions of the theory to explain part of the Lamb shift.

In quantum field theory, the vacuum is no longer simply empty space; it is literally seething with activity. Virtual particles, such as electron-positron pairs, can pop into existence and disappear. As long as the energy of the particles $E$ and their lifetime $t$ satisfy $Et < \hbar/2$, these particle-antiparticle pairs cannot be detected directly. [This is a rather crude explanation, but is made much more precise within the framework of quantum field theory.]

These particle-antiparticle pairs produce an effect called vacuum polarization. Consider a dielectric surrounding a point positive charge. The point charge polarizes the dielectric, attracting negative charge inward and repelling positive charge outward. This tends to cancel the electric field produced by the point charge, leading to a reduced electric field inside the dielectric (Figure 9.9).

Now consider the same positive charge in a vacuum. The production of virtual electron-positron pairs tends to cancel the charge, just as in a physical dielectric. However, unlike a dielectric, we can never remove the

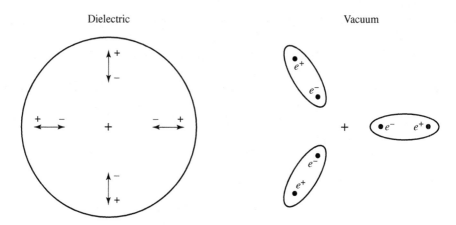

Fig. 9.9   In a dielectric, polarization reduces the electric field produced by a point charge. Vacuum polarization produces the same effect in a vacuum.

positive charge from the vacuum polarization to measure its true charge: the charge we measure has already been cancelled by the effect of the vacuum polarization. This means that the "bare" charge, which cannot be measured directly, is much larger than we thought; in fact, it is mathematically infinite!

The upshot of all of this is that the electric field of a point charge must be modified at the origin (where the charge is "infinite"), but everywhere else in space the charge has already been cancelled by the effects of vacuum polarization, so the electric field is unchanged. The result for $V(r)$, derived from quantum field theory, is

$$V(r) = -\frac{e^2}{4\pi\epsilon_0}\frac{1}{r} - \frac{\alpha e^2}{15\pi\epsilon_0}\frac{\hbar^2}{m_e^2 c^2}\delta^3(\mathbf{r}) \qquad (9.31)$$

where $\delta^3(\mathbf{r})$ is the three-dimensional Dirac delta function discussed in Chapter 7. The second term in Equation (9.31) is the perturbation to the Hamiltonian, so the first-order shift in the energy is

$$E^{(1)} = \langle n\ l\ m_l\ m_s | H_1 | n\ l\ m_l\ m_s \rangle \qquad (9.32)$$

$$= \int d^3\mathbf{r}\ \psi^*_{nlm_l}(\mathbf{r})\left(-\frac{\alpha e^2}{15\pi\epsilon_0}\frac{\hbar^2}{m_e^2 c^2}\delta^3(\mathbf{r})\right)\psi_{nlm_l}(\mathbf{r}) \qquad (9.33)$$

$$= -\frac{\alpha e^2}{15\pi\epsilon_0}\frac{\hbar^2}{m_e^2 c^2}|\psi_{nlm_l}(0)|^2 \qquad (9.34)$$

Recall that the hydrogen wave functions are all identically zero at the origin, except for the $l = 0$ states. Thus, the effect of vacuum polarization is to reduce the energy of the $l = 0$ states relative to the corresponding $l \neq 0$ states. The effect is smaller than the fine-structure splitting, e.g., for the $n = 2$ states, the splitting between the $l = 0$ and $l = 1$ state is about $10^{-7}$ eV. As bizarre as all of this sounds, it is important to remember that this is based on solid experimental evidence.

While all of this is correct as far as it goes, it does not go quite far enough. There are additional contributions to the Lamb shift that arise in quantum field theory. These are actually larger than the effect we have discussed here, and with opposite sign. Thus, the full (experimentally measured) Lamb shift corresponds to the $l = 0$ state having a *higher* energy than the corresponding $l = 1$ state.

### 9.3 The Atom in External Electric or Magnetic Fields

In the previous section, we discussed perturbations which are intrinsic to the atom. We will now examine what happens when the atom is placed in an external electromagnetic field. Since the atom consists of charged particles, and the electrons produce both a spin and orbital magnetic moment, any external electric or magnetic field will perturb the energy levels of the atom. The effect produced by an external electric field is called the *Stark effect*, while the effect of an external magnetic field is the *Zeeman effect*.

#### The Atom in an Electric Field: The Stark Effect

We will first examine the effect of a uniform electric field with magnitude $\mathcal{E}$ on the ground state of hydrogen. Recall that the ground-state wave function is

$$\psi_{100} = \frac{1}{\sqrt{\pi a_0^3}} e^{-r/a_0}$$

where the "100" subscript denotes the $n\ l\ m_l$ quantum numbers. We can ignore the spin state of the electron, since the spin interacts only with magnetic fields through the electron's spin magnetic moment. (Of course, we will have to consider spin in the next section when we discuss external magnetic fields.) We take the electric field to be uniform, static, and pointing in the $z$ direction.

Since the ground state of hydrogen is nondegenerate, we can use the perturbation theory expressions from Section 9.1. Classically, the potential energy of a charge $-e$ in an electric field $\mathcal{E}$ is $V = e\mathcal{E}z$, so the perturbation $H_1$ produced by the electric field is

$$H_1 = e\mathcal{E}z$$

and the first-order change in the energy of the hydrogen atom is, from Equation (9.19),

$$E^{(1)} = \langle \psi_{100} | e\mathcal{E}z | \psi_{100} \rangle \tag{9.35}$$

Taking $z = r\cos\theta$ and writing Equation (9.35) in spherical coordinates, we get

$$E^{(1)} = \int \left( \frac{1}{\sqrt{\pi a_0^3}} e^{-r/a_0} \right) (e\mathcal{E}r\cos\theta) \left( \frac{1}{\sqrt{\pi a_0^3}} e^{-r/a_0} \right) r^2 dr \sin\theta \, d\theta \, d\phi$$

But the integral over $\theta$ vanishes:

$$\int_{\theta=0}^{\pi} \cos\theta \sin\theta \, d\theta = \frac{1}{2} \sin^2\theta \Big|_0^\pi = 0$$

so $E^{(1)} = 0$.

Since the first-order perturbation vanishes, we must use second-order perturbation theory to calculate the change in the energy due to the external electric field. Equation (9.20) gives

$$E^{(2)} = \sum_{n,l,m_l} \frac{|\langle \psi_{100} | e\mathcal{E}z | \psi_{nlm_l} \rangle|^2}{E_1 - E_n} \tag{9.36}$$

where $E_n = 13.6 \text{ eV}/n^2$. Recall from Chapter 6 that every hydrogen wave function can be written as the product of a radial wave function $R_{nl}(r)$ and the spherical harmonic $Y_l^m(\theta, \phi)$. Then the inner product which appears in Equation (9.36) can be written in the form

$$\langle \psi_{100} | e\mathcal{E}z | \psi_{nlm_l} \rangle = e\mathcal{E} \int R_{10}^*(r) Y_0^{0*}(\theta, \phi) \, r\cos\theta \, R_{nl}(r) Y_l^m(\theta, \phi) \, d^3\mathbf{r} \tag{9.37}$$

But now, recall that $Y_0^0 = 1/\sqrt{4\pi}$, and $Y_1^0 = \sqrt{3/4\pi}\cos\theta$, so $Y_0^0 \cos\theta = (1/\sqrt{3})Y_1^0$. This allows us to write Equation (9.37) as

$$\langle \psi_{100} | e\mathcal{E}z | \psi_{nlm_l} \rangle = \frac{e\mathcal{E}}{\sqrt{3}} \int R_{10}^*(r) R_{nl}(r) r^3 \, dr \int Y_1^{0*}(\theta, \phi) Y_l^m(\theta, \phi) \sin\theta \, d\theta \, d\phi$$

Since the $Y_l^m$'s are orthonormal,

$$\int Y_1^{0*}(\theta, \phi) Y_l^m(\theta, \phi) \sin\theta \, d\theta \, d\phi = 1 \quad (l = 1, \ m = 0)$$
$$= 0 \quad (l \neq 1 \text{ or } m \neq 0)$$

Hence, in the sum in Equation (9.36), only the $l = 1$, $m = 0$ terms are nonzero giving

$$E^{(2)} = \sum_n \frac{|(e\mathcal{E}/\sqrt{3}) \int R_{10}^*(r)R_{n1}(r)r^3 \, dr|^2}{E_1 - E_n} \tag{9.38}$$

The integral under the sum in Equation (9.38) can be evaluated exactly for all values of $n$, and the terms in the series decrease rapidly with $n$:

$$E^{(2)} = -(4\pi\epsilon_0)a_0^3\mathcal{E}^2(1.48 + 0.20 + 0.066 + \cdots) = -\frac{9}{4}(4\pi\epsilon_0)a_0^3\mathcal{E}^2$$

The change in energy is negative, since the hydrogen atom becomes polarized and aligns itself so as to partially cancel the external electric field (Figure 9.10; see also Problem 9.2). Since the change in energy is proportional to the square of the applied electric field, this effect is called the *quadratic Stark effect*.

Our use of nondegenerate perturbation theory breaks down for the excited states of hydrogen, since these states are degenerate. Using degenerate perturbation theory, it is possible to show that the change in energy for these excited states is proportional to $\mathcal{E}$ rather than $\mathcal{E}^2$. Hence, the change in energy when an electric field is applied to the excited states of hydrogen is called the *linear Stark effect*.

### The Atom in a Magnetic Field: The Zeeman Effect

Now consider what happens when we apply an external magnetic field $\mathbf{B}$ to the hydrogen atom. Assume that the magnetic field has magnitude $B$ and is pointing in the $z$ direction so that

$$\mathbf{B} = B\hat{z}$$

The potential energy of a magnetic dipole $\boldsymbol{\mu}$ in a magnetic field $\mathbf{B}$ is just $V = -\boldsymbol{\mu}\cdot\mathbf{B}$, so the perturbation produced by the magnetic field is

$$H_1 = -\boldsymbol{\mu}\cdot\mathbf{B} \tag{9.39}$$

In Equation (9.39), there are two contributions to the atomic magnetic moment: the contribution from the orbital magnetic moment $\boldsymbol{\mu}_l$ and the contribution from the spin magnetic moment of the electron $\boldsymbol{\mu}_s$. (In principle, we should also include the spin magnetic moment of the proton, but this is much smaller and can be ignored.) Hence, the total magnetic moment is

$$\boldsymbol{\mu} = \boldsymbol{\mu}_l + \boldsymbol{\mu}_s$$

$$= -\frac{g_l\mu_B}{\hbar}\mathbf{L} - \frac{g_s\mu_B}{\hbar}\mathbf{S}$$

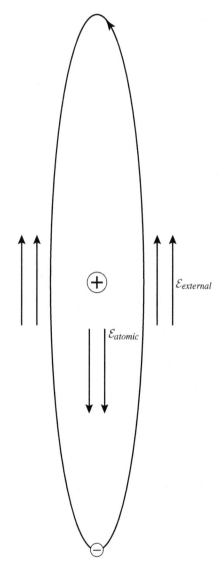

Fig. 9.10   A classical picture of the quadratic Stark effect: the hydrogen atom is polarized by the external electric field, and the field produced by the polarized atom is in the opposite direction to the external field.

Recall that $g_l = 1$ and $g_s \approx 2$, so the expression for $\boldsymbol{\mu}$ becomes

$$\boldsymbol{\mu} = -\frac{\mu_B}{\hbar}[\mathbf{L} + 2\mathbf{S}] \tag{9.40}$$

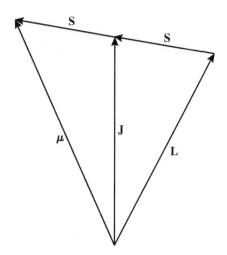

Fig. 9.11   The magnetic moment $\boldsymbol{\mu}$ of the hydrogen atom is proportional to $\mathbf{L} + 2\mathbf{S}$, while the total angular momentum $\mathbf{J}$ is proportional to $\mathbf{L} + \mathbf{S}$, so $\boldsymbol{\mu}$ and $\mathbf{J}$ are not, in general, parallel.

Note that the magnetic moment of the atom is *not* proportional to the total angular momentum operator $\mathbf{J}$, which is $\mathbf{L} + \mathbf{S}$. In classical terms, the angular momentum vector $\mathbf{J}$ and magnetic moment vector $\boldsymbol{\mu}$ are not parallel (Figure 9.11). This has important consequences for the Zeeman effect. (You are already familiar with a much larger classical system in which the angular momentum and magnetic dipole are not parallel; the Earth!)

We can use Equation (9.40) to rewrite the perturbation in Equation (9.39) as

$$H_1 = B\frac{\mu_B}{\hbar}[L_z + 2S_z]$$

Applying this perturbation to the hydrogen state $|n\ l\ m_l\ m_s\rangle$ gives the first-order change in energy,

$$E^{(1)} = \langle n\ l\ m_l\ m_s|(B\mu_B/\hbar)(L_z + 2S_z)|n\ l\ m_l\ m_s\rangle$$
$$= B\mu_B(m_l + 2m_s) \tag{9.41}$$

The problem with this result is that it requires the atom to be in a state of definite $m_l$ and $m_s$ (or, equivalently, an eigenstate of $S_z$ and $L_z$). However, as we have seen in our discussion of fine structure in Section 9.2, the spin-orbit coupling drives the atom into an eigenstate of $J^2$, which does not commute with $S_z$ and $L_z$. Hence, the atom is in a state of definite $j$ and $m_j$ rather than $m_l$ and $m_s$, so our argument would appear to be invalid.

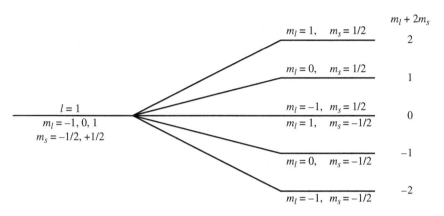

Fig. 9.12   The strong-field Zeeman effect for the energy levels of an $l = 1$ state in hydrogen.

To clarify this situation, we can write the full Hamiltonian as

$$H = H_0 + \frac{e^2}{8\pi\epsilon_0 m_e^2 c^2 r^3}\mathbf{S\cdot L} + B\frac{\mu_B}{\hbar}[L_z + 2S_z] \qquad (9.42)$$

where the second term is the perturbation due to the spin-orbit interaction (given in Equation (9.23)), and the third term is the perturbation from the external magnetic field.

Now consider two possible cases: for very strong magnetic fields ($B \gg 1$ T), the third term in Equation (9.42) dominates the second term, while for weak magnetic fields ($B \ll 1$ T), the second term dominates the third. Consider the case of strong magnetic fields first. For this case we simply ignore the effect of the spin-orbit coupling; the strong magnetic field overwhelms the spin-orbit coupling and drives the atom back into a state of definite $m_l$ and $m_s$. Therefore, for the strong magnetic field case, the expression we derived for $E^{(1)}$ in Equation (9.41) is correct:

$$E^{(1)} = B\mu_B(m_l + 2m_s)$$

This regime of the Zeeman effect is called the *strong-field Zeeman effect* or the *Paschen-Back effect*. An illustration of this perturbation in the energy levels is shown in Figure 9.12 for the case $l = 1$.

Now consider the opposite regime in which spin-orbit coupling dominates the effect of the external magnetic field. In this case the atom is in a state of definite $j$ and $m_j$ rather than $m_l$ and $m_s$, and the perturbation must be written as

$$E^{(1)} = \langle n\, l\, j\, m_j|(B\mu_B/\hbar)(L_z + 2S_z)|n\, l\, j\, m_j\rangle$$

This can be partially simplified by using the fact that $J_z = L_z + S_z$:

$$E^{(1)} = \langle n\,l\,j\,m_j|(B\mu_B/\hbar)(J_z + S_z)|n\,l\,j\,m_j\rangle$$

$$= B\mu_B m_j + \frac{B\mu_B}{\hbar}\langle n\,l\,j\,m_j|S_z|n\,l\,j\,m_j\rangle \qquad (9.43)$$

In order to further simplify this expression, the state $|n\,l\,j\,m_j\rangle$ must be written as a linear combination of the $|n\,l\,m_l\,m_s\rangle$ states. From Chapter 8, we know that $s = 1/2$ and a given value of $l$ can couple to give either $j = l + 1/2$ or $j = l - 1/2$, while $m_j = m_l + m_s$. The actual linear combination is

$$|j = l + 1/2,\ m_j\rangle = \left(\frac{l + 1/2 + m_j}{2l + 1}\right)^{1/2} |m_l = m_j - 1/2,\ m_s = 1/2\rangle$$

$$+ \left(\frac{l + 1/2 - m_j}{2l + 1}\right)^{1/2} |m_l = m_j + 1/2,\ m_s = -1/2\rangle$$

$$|j = l - 1/2,\ m_j\rangle = \left(\frac{l + 1/2 - m_j}{2l + 1}\right)^{1/2} |m_l = m_j - 1/2,\ m_s = 1/2\rangle$$

$$- \left(\frac{l + 1/2 + m_j}{2l + 1}\right)^{1/2} |m_l = m_j + 1/2,\ m_s = -1/2\rangle$$

We can use these equations to solve for $\langle n\,l\,j\,m_j|S_z|n\,l\,j\,m_j\rangle$. For $j = l + 1/2$, we get

$$\langle n\,l\,j\,m_j|S_z|n\,l\,j\,m_j\rangle = \left[\left(\frac{l + 1/2 + m_j}{2l + 1}\right)^{1/2} \langle m_l = m_j - 1/2,\ m_s = 1/2|\right.$$

$$\left. + \left(\frac{l + 1/2 - m_j}{2l + 1}\right)^{1/2} \langle m_l = m_j + 1/2,\ m_s = -1/2|\right]$$

$$\times S_z \left[\left(\frac{l + 1/2 + m_j}{2l + 1}\right)^{1/2} |m_l = m_j - 1/2,\ m_s = 1/2\rangle\right.$$

$$\left. + \left(\frac{l + 1/2 - m_j}{2l + 1}\right)^{1/2} |m_l = m_j + 1/2,\ m_s = -1/2\rangle\right]$$

$$= \frac{\hbar}{2}\left(\frac{l + 1/2 + m_j}{2l + 1}\right) - \frac{\hbar}{2}\left(\frac{l + 1/2 - m_j}{2l + 1}\right)$$

$$= \frac{m_j \hbar}{2l + 1}$$

Similarly, for $j = l - 1/2$, we obtain

$$\langle n \ l \ j \ m_j | S_z | n \ l \ j \ m_j \rangle = -\frac{m_j \hbar}{2l + 1}$$

Combining the results for $j = l + 1/2$ and $j = l - 1/2$, we get

$$\langle n \ l \ j \ m_j | S_z | n \ l \ j \ m_j \rangle = \frac{m_j \hbar}{2l + 1} 2(j - l)$$

and substituting this result into Equation (9.43) yields

$$E^{(1)} = B\mu_B m_j \left( \frac{2j + 1}{2l + 1} \right)$$

In analogy with the $g_s$ factor for the electron spin and $g_l$ for the orbital angular momentum, we can write

$$g = \frac{2j + 1}{2l + 1}$$

where this $g$ is called the *Landé g factor*. In terms of the Landé $g$ factor, the energy shift becomes

$$E^{(1)} = gB\mu_B m_j$$

In contrast to $g_s$ and $g_l$, which are constant, the Landé $g$ factor is not constant, but rather is a function of $j$ and $l$.

The reason for this is the fact, already alluded to, that $\boldsymbol{\mu}$ is not parallel to $\mathbf{J}$, since the operator which determines $\boldsymbol{\mu}$ is $\mathbf{L} + 2\mathbf{S}$ while $\mathbf{J} = \mathbf{L} + \mathbf{S}$. Hence, the ratio between $\boldsymbol{\mu}$ and $\mathbf{J}$ can depend on the relative orientation of $\mathbf{L}$ and $\mathbf{S}$ (Figure 9.11), so $\boldsymbol{\mu}$ is not a fixed multiple of $m_j$. The effect of the weak-field Zeeman effect is to split the energies of the individual $m_j$ levels, with a magnitude which depends on both the magnitude of the magnetic field and the value of the Landé $g$ factor (Figure 9.13).

To summarize, for weak magnetic fields, the hydrogen atom can be taken to be in a state of definite $j$ and $m_j$, and the magnetic field separates the energies of the individual $m_j$ states. As the magnetic field is increased, it eventually becomes stronger than the internal magnetic fields of the atom. In this limit, the magnetic field drives the hydrogen atom into a state of definite $m_l$ and $m_s$, and the perturbation in energy is just proportional to $m_l + 2m_s$.

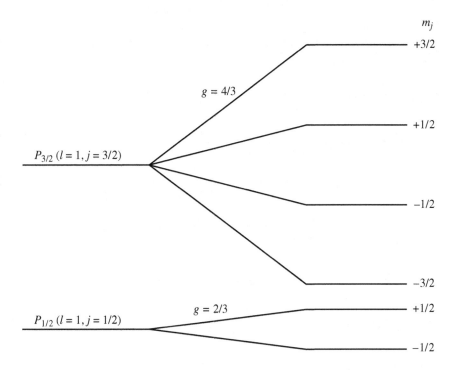

Fig. 9.13   The splitting of the $P_{3/2}$ and $P_{1/2}$ states in the weak-field Zeeman effect.

## PROBLEMS

**9.1** (a) In Example 9.2, the energy of the system can be calculated exactly. Take $\mathbf{B} = B_x \hat{x} + B_z \hat{z}$, and calculate the exact energies. [Hint: Feel free to use a different coordinate system; the energy levels cannot depend on the choice of the coordinate system].
(b) Take the answer in part (a) and expand it out in powers of $B_x$, remembering that $B_x \ll B_z$. Show that the terms proportional to $B_x$ and $B_x^2$ correspond to the answers derived in Example 9.2.

**9.2** A particle is in a potential $V_0$ in its ground state $|\psi_0\rangle$. A small perturbation $H_1$ is applied to the particle. Suppose that the first order perturbation to the energy is zero: $E^{(1)} = \langle \psi_0 | H_1 | \psi_0 \rangle = 0$. Show that the lowest-order effect of $H_1$ is to *decrease* the energy of the ground state.

**9.3** A particle of mass $m$ is confined to move in a one-dimensional square well with infinite potential barriers at $x = 0$ and $x = a$, with $V = 0$ for $0 \leq x \leq a$. The particle is in the ground state. A perturbation $H_1 = \lambda\delta(x - a/2)$ is added, where $\lambda$ is a small constant.

(a) What units does $\lambda$ have?

(b) Calculate the first-order perturbation $E^{(1)}$ due to $H_1$.

(c) Calculate the second-order perturbation $E^{(2)}$. The answer may be expressed as an infinite series.

**9.4** A particle of mass $m$ is confined to move in a narrow, straight tube of length $a$ which is sealed at both ends with $V = 0$ inside the tube. Treat the tube as a one-dimensional infinite square well. The tube is placed at an angle $\theta$ relative to the surface of the earth. The particle experiences the usual gravitational potential $V = mgh$. Calculate the lowest-order change in the energy of the ground state due to the gravitational potential.

**9.5** A particle of mass $m$ is in the ground state in the harmonic oscillator potential

$$V(x) = \frac{1}{2}Kx^2$$

A small perturbation $\beta x^6$ is added to this potential.

(a) What are the units of $\beta$?

(b) How small must $\beta$ be in order for perturbation theory to be valid?

(c) Calculate the first-order change in the energy of the particle.

**9.6** In the hydrogen atom, the proton is not really a point charge but has a finite size. Assume that the proton behaves as a uniformly-charged sphere of radius $R = 10^{-15}$ m. Calculate the shift this produces in the ground-state energy of hydrogen.

**9.7** The photon is normally assumed to have zero rest mass. If the photon had a small mass, this would alter the potential energy which the electron experiences in the electric field of the proton. Instead of

$$V(r) = -\frac{e^2}{4\pi\epsilon_0}\frac{1}{r} \tag{9.44}$$

we would have

$$V(r) = -\frac{e^2}{4\pi\epsilon_0}\frac{e^{-r/r_0}}{r} \tag{9.45}$$

where $r_0$ is a constant with units of length. Assume $r_0$ is large compared to the size of the hydrogen atom, so the potential energy given in Equation (9.45) differs only slightly from the standard one given by Equation (9.44) in the vicinity of the electron. Calculate the change in the ground state energy of hydrogen if the correct potential is given by Equation (9.45) instead of Equation (9.44).

**9.8** Suppose that the proton had spin 0 instead of spin 1/2.
(a) How would this alter the fine structure of the energy levels of the hydrogen atom?
(b) How would this alter the hyperfine structure of the energy levels of the hydrogen atom?

**9.9** We have seen that the spin-orbit interaction splits the $l \neq 0$ states in the hydrogen atom into $j = l + 1/2$ states (with slightly higher energy) and $j = l - 1/2$ states (with slightly lower energy). Suppose that the electron had spin 1. How many different energy levels would the spin-orbit interaction produce, and what would their relative energies be? Be sure to consider how the answer would depend on the value of $l$.

**9.10** Equation (9.29) gives the fine-structure energy shift.
(a) Show that the $j = l + 1/2$ state has a higher energy than the $j = l - 1/2$ state.
(b) Show that the change in energy, $E^{(1)}_{fine\ structure}$, is always negative.

**9.11** An electron is in the ground state in a three-dimensional rectangular box given by $0 \leq x \leq a$, $0 \leq y \leq b$, and $0 \leq z \leq c$, where $V = 0$ inside the box, and there are infinite potential barriers at all of the walls. A homogeneous, static electric field with magnitude $\mathcal{E}$ is applied in the $x$ direction. What is the lowest-order change in the energy of the electron?

**9.12** A hydrogen atom in its ground state is placed in a homogeneous, static electric field with magnitude $\mathcal{E}$ in the $x$ direction.
(a) Show that the first-order perturbation $E^{(1)}$ is 0.
(b) Show that the second-order perturbation $E^{(2)}$ is the same as if the field was pointing in the $z$ direction. [This is obvious from symmetry, but calculate $E^{(2)}$ using perturbation theory and show it explicitly.]

**9.13** A hydrogen atom is in its ground state. A proton is fixed in space a distance $R$ from the nucleus of the hydrogen atom, where $R \gg a_0$. Calculate the perturbation to the energy of the hydrogen atom due to the electric field of this proton.

**9.14** The electron in a hydrogen atom is in a $D$ state. A homogenous, static magnetic field is applied in the $z$ direction.
(a) Draw a diagram showing the splitting of the energy levels in the weak-field limit. Calculate the value of $g$ for each energy level.
(b) Draw a diagram showing the splitting of the energy levels in the strong-field limit.

**9.15** (a) A particle is in a state $|\psi\rangle$ which is an eigenfunction of the Hamiltonian $H_0$ with energy $E$. A perturbation $H_1$ is applied such that $H_1|\psi\rangle = 0$. Show that the energy of the system is completely unchanged by this perturbation.
(b) In the ground state of the helium atom, both electrons are in the $l = 0$ state, and the spin wave function for the two electrons is the singlet spin state ($s = 0$ and $m_s = 0$). [This is a consequence of the Pauli exclusion principle, which will be discussed in Chapter 13.] A homogeneous, static magnetic field is applied in the $z$ direction. Show that the energy of the ground state of helium is completely unaffected by this magnetic field. [Ignore the magnetic moment of the nucleus.] What is the physical reason for this?

Chapter 10

# The variational principle

In the previous chapter, we began with systems for which the Schrödinger equation could be solved exactly, and calculated the change in energy when a small, time-independent perturbation was added to the Hamiltonian. The next logical step is to examine time-dependent perturbations. However, before doing so, we will take a slight detour and develop a technique called the *variational principle*. The variational principle applies to time-independent systems, but it is not a form of perturbation theory; i.e., it does not assume that a small perturbation is applied to a known exact solution. Instead, the variational principle is a technique for estimating the ground-state energy of an arbitrary Hamiltonian for which the Schrödinger equation cannot be solved at all. For example, the Schrödinger equation cannot be solved exactly for atoms with more than one electron, but the variational principle provides a tool to estimate the ground-state energies for such atoms.

The variational principle is based on a simple idea: the expectation value of the Hamiltonian calculated for an arbitrary wave function gives an upper bound on the ground-state energy. By "varying" the wave function used to calculate this expectation value and picking out the smallest resulting value for the expectation value, we obtain an estimate for the ground-state energy. We will derive this result below and then apply it to two different examples.

## 10.1   Variational Principle: Theory

Suppose we have a Hamiltonian $H$ with a set of energies and eigenfunctions:

$$H|\psi_0\rangle = E_0|\psi_0\rangle$$
$$H|\psi_1\rangle = E_1|\psi_1\rangle$$
$$\vdots$$
$$H|\psi_n\rangle = E_n|\psi_n\rangle$$
$$\vdots$$

Assume further that we cannot necessarily solve for any of the energies or eigenfunctions explicitly. Now suppose that we choose a completely arbitrary wave function $|\psi\rangle$, which need not be normalized, and we calculate the expectation value of $H$, $\langle\psi|H|\psi\rangle/\langle\psi|\psi\rangle$. (The $\langle\psi|\psi\rangle$ in the denominator is necessary because we have not assumed that $|\psi\rangle$ is normalized.) The variational principle is based on the fact that this expectation value gives an upper bound on the ground-state energy $E_0$:

$$\frac{\langle\psi|H|\psi\rangle}{\langle\psi|\psi\rangle} \geq E_0 \qquad (10.1)$$

First we will prove the result in Equation (10.1), and then explore the consequences.

Recall that the eigenfunctions of $H$ can be chosen to be an orthonormal basis set, so the arbitrary wave function $|\psi\rangle$ that appears in Equation (10.1) can be expanded as a linear combination of these eigenfunctions:

$$|\psi\rangle = c_0|\psi_0\rangle + c_1|\psi_1\rangle + \cdots + c_n|\psi_n\rangle + \cdots$$

Substituting this expansion into the numerator on the left-hand side of Equation (10.1) gives

$$\langle\psi|H|\psi\rangle = (c_0^*\langle\psi_0| + c_1^*\langle\psi_1| + \cdots + c_n^*\langle\psi_n| + \cdots)H(c_0|\psi_0\rangle + c_1|\psi_1\rangle + \cdots + c_n|\psi_n\rangle + \cdots)$$

But $H$ simply pulls out the appropriate energy when it operates on each eigenfunction, so

$$\langle\psi|H|\psi\rangle = (c_0^*\langle\psi_0| + c_1^*\langle\psi_1| + \cdots + c_n^*\langle\psi_n| + \cdots)(c_0E_0|\psi_0\rangle + c_1E_1|\psi_1\rangle + \cdots + c_nE_n|\psi_n\rangle + \cdots) \qquad (10.2)$$

Because the eigenfunctions are orthonormal, $\langle\psi_0|\psi_0\rangle = 1$, $\langle\psi_1|\psi_1\rangle = 1, \ldots$, $\langle\psi_n|\psi_n\rangle = 1, \ldots$, and all of the terms containing $\langle\psi_m|\psi_n\rangle$ with $m \neq n$ are zero. Therefore, Equation (10.2) simplifies to

$$\langle\psi|H|\psi\rangle = |c_0|^2E_0 + |c_1|^2E_1 + \cdots + |c_n|^2E_n + \cdots$$

Similarly, the denominator in Equation (10.1) can be expressed as

$$\langle \psi | \psi \rangle = (c_0^* \langle \psi_0 | + c_1^* \langle \psi_1 | + \cdots + c_n^* \langle \psi_n | + \cdots)$$
$$(c_0 | \psi_0 \rangle + c_1 | \psi_1 \rangle + \cdots + c_n | \psi_n \rangle + \cdots)$$
$$= |c_0|^2 + |c_1|^2 + \cdots + |c_n|^2 + \cdots$$

so

$$\frac{\langle \psi | H | \psi \rangle}{\langle \psi | \psi \rangle} = \frac{|c_0|^2 E_0 + |c_1|^2 E_1 + \cdots + |c_n|^2 E_n + \cdots}{|c_0|^2 + |c_1|^2 + \cdots + |c_n|^2 + \cdots} \tag{10.3}$$

Because $E_0$ is the ground-state energy, it must be true that

$$E_0 \leq E_1 \leq \cdots \leq E_n \leq \cdots$$

so the numerator in Equation (10.3) satisfies

$$|c_0|^2 E_0 + |c_1|^2 E_1 + \cdots + |c_n|^2 E_n + \cdots$$
$$\geq |c_0|^2 E_0 + |c_1|^2 E_0 + \cdots + |c_n|^2 E_0 + \cdots$$

Substituting this bound into Equation (10.3) yields

$$\frac{\langle \psi | H | \psi \rangle}{\langle \psi | \psi \rangle} \geq \frac{|c_0|^2 E_0 + |c_1|^2 E_0 + \cdots + |c_n|^2 E_0 + \cdots}{|c_0|^2 + |c_1|^2 + \cdots + |c_n|^2 + \cdots}$$

which reduces to Equation (10.1).

Equation (10.1) says that $\langle \psi | H | \psi \rangle / \langle \psi | \psi \rangle$ gives an upper bound on the ground-state energy, but how can it be used to *estimate* the ground-state energy for the Hamiltonian $H$? One possibility would be to simply substitute a large number of different wave functions $|\psi\rangle$ into the left-hand side of Equation (10.1) and pick the one which gives the smallest answer. Since we are guaranteed that this quantity is an upper bound on $E_0$, the smallest answer will give the best approximation to $E_0$. There is, however, a method to sample an *infinite* number of trial wave functions. We can write the trial wave function $|\psi\rangle$ as a function of some continuous parameter $\alpha$, which we can vary. Then the quantity $\langle \psi(\alpha) | H | \psi(\alpha) \rangle / \langle \psi(\alpha) | \psi(\alpha) \rangle$ is a continuous function of $\alpha$, and it is guaranteed to be larger than $E_0$. By choosing $\alpha$ so as to minimize $\langle \psi(\alpha) | H | \psi(\alpha) \rangle / \langle \psi(\alpha) | \psi(\alpha) \rangle$, we obtain the best estimate of $E_0$ (Figure 10.1). What happens if we get lucky and accidently choose a form for $|\psi(\alpha)\rangle$ for which a value of $\alpha$ exists that makes $|\psi(\alpha)\rangle$ exactly equal to the ground-state wave function $|\psi_0\rangle$? In this case it is easy to see that the estimate given by Equation (10.1) is exactly equal to the ground-state energy (Problem 10.1).

$\dfrac{\langle \psi \,|\, H \,|\, \psi \rangle}{\langle \psi \,|\, \psi \rangle}$

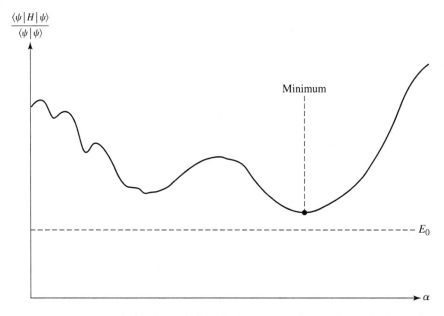

Fig. 10.1   Because $\langle \psi(\alpha)|H|\psi(\alpha)\rangle / \langle \psi(\alpha)|\psi(\alpha)\rangle \geq E_0$, the best estimate for $E_0$ is obtained by minimizing $\langle \psi(\alpha)|H|\psi(\alpha)\rangle / \langle \psi(\alpha)|\psi(\alpha)\rangle$.

## Example 10.1. The Bouncing Ball

Consider a particle subject to the linear potential $V(x) = mgx$ but with an infinite potential barrier at $x = 0$ (Figure 10.2). Estimate the ground-state energy using the variational principle.

First we need to choose a trial wave function $\psi(x)$. There is no "correct" choice for $\psi(x)$, but the more closely $\psi(x)$ can be made to resemble the true ground-state solution of the Schrödinger equation with $V(x) = mgx$, the more accurate our variational estimate of the ground-state energy will be. We first note that the infinite potential barrier at $x = 0$ will give $\psi(0) = 0$ for the true ground-state wave function, so our variational wave function should also have this property. Furthermore, since we are dealing with a bound state, we expect the true wave function to satisfy $\psi(x) \to 0$ as $x \to \infty$. Finally, our experience with solutions of the one-dimensional Schrödinger equation (Chapter 4) indicates that the true ground-state wave function will not cross the $x$-axis for $x > 0$. There are still an infinite number of functions with these desired properties, so our final choice will

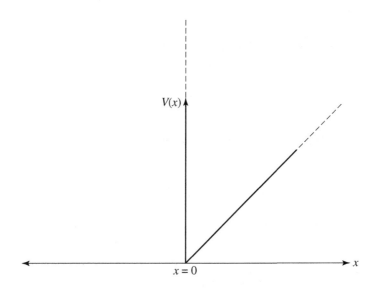

Fig. 10.2   The potential $V(x) = mgx$ with an infinite barrier at $x = 0$.

be somewhat arbitrary. We will use the simple trial wave function

$$\psi(x) = xe^{-\alpha x}$$

Note that this wave function increases from $\psi = 0$ at $x = 0$ up to a maximum at $x = 1/\alpha$, and then $\psi$ decreases exponentially to 0 as $x \to \infty$. The next step is to calculate $\langle\psi|H|\psi\rangle/\langle\psi|\psi\rangle$. The numerator is

$$\langle\psi|H|\psi\rangle = \int_0^\infty (xe^{-\alpha x})\left(-\frac{\hbar^2}{2m}\frac{\partial^2}{\partial x^2} + mgx\right)(xe^{-\alpha x})\,dx$$

$$= \int_0^\infty e^{-2\alpha x}\left(mgx^3 - \frac{\hbar^2}{2m}\alpha^2 x^2 + \frac{\hbar^2}{m}\alpha x\right)dx$$

$$= \frac{3mg}{8\alpha^4} + \frac{\hbar^2}{8m\alpha}$$

and the denominator is

$$\langle\psi|\psi\rangle = \int_0^\infty x^2 e^{-2\alpha x}\,dx$$

$$= \frac{1}{4\alpha^3}$$

so the total expression to be minimized is

$$\frac{\langle\psi|H|\psi\rangle}{\langle\psi|\psi\rangle} = \frac{3mg}{2\alpha} + \frac{\hbar^2}{2m}\alpha^2 \tag{10.4}$$

Taking the derivative of the right-hand side and setting it to zero gives

$$-\frac{3mg}{2\alpha^2} + \frac{\hbar^2}{m}\alpha = 0$$

with the solution

$$\alpha = \left(\frac{3m^2g}{2\hbar^2}\right)^{1/3}$$

Now this value for $\alpha$ must be substituted back into Equation (10.4) to give the estimate for the ground-state energy. We obtain

$$E_{estimated} = \left(\frac{3}{2}\right)^{5/3}(mg^2\hbar^2)^{1/3}$$

To see how our estimate depends on the actual choice of the trial wave function, see Problem 10.5.

---

The variational principle, therefore, is a three-stage process: first, choose a trial wave function that depends on a parameter $\alpha$; next, vary $\alpha$ so as to minimize $\langle\psi|H|\psi\rangle/\langle\psi|\psi\rangle$; and finally, substitute the resulting wave function back into $\langle\psi|H|\psi\rangle/\langle\psi|\psi\rangle$ to obtain the estimate of the ground-state energy $E_0$. It is guaranteed that the answer will provide an upper bound on $E_0$, and it may even provide an excellent approximation to $E_0$, depending on how closely the trial wave function can be made to resemble the true ground-state wave function. Some common mistakes in using the variational principle are forgetting to include the kinetic energy term in $H$ when evaluating $\langle\psi|H|\psi\rangle$, forgetting to divide by $\langle\psi|\psi\rangle$, and forgetting to complete the solution by substituting $\alpha$ back into the expression for $\langle\psi|H|\psi\rangle/\langle\psi|\psi\rangle$.

There is no "correct" choice for the trial wave function, but the form of the potential will often provide a guide as to a reasonable choice. As noted, the goal is to pick something which can be made to resemble the true ground-state wave function as closely as possible. It is also possible to use the variational principle to estimate the first excited-state energy (see Problem 10.2 and Problem 10.6).

## 10.2 Variational Principle: Application to the Helium Atom

As we have seen, quantum mechanics provides an excellent description of the hydrogen atom. The energy levels and wave functions can be calculated via the Schrödinger equation, and perturbation theory can predict the small changes to these energy levels due to magnetic interactions in the atom. Now it is time to reveal an unpleasant fact: the Schrödinger equation cannot be solved in a simple way for any of the other atoms! Even adding a single electron, to produce a helium atom, yields an intractable Schrödinger equation. On the other hand, the calculation of the ground-state energy of helium is the "classic" problem for the variational principle.

The helium atom contains a nucleus of charge $+2e$ and two negatively-charged electrons (Figure 10.3). Labelling the electrons "1" and "2", the Hamiltonian for the helium atom is

$$H = \frac{P_1^2}{2m_e} + \frac{P_2^2}{2m_e} - \frac{2e^2}{4\pi\epsilon_0 r_1} - \frac{2e^2}{4\pi\epsilon_0 r_2} + \frac{e^2}{4\pi\epsilon_0 |\mathbf{r}_1 - \mathbf{r}_2|} \qquad (10.5)$$

where, for simplicity, we use the electron mass $m_e$ rather than the reduced mass. In this equation, $\mathbf{r}_1$ and $\mathbf{r}_2$ are the positions of the two electrons relative to the nucleus with corresponding radial components $r_1$ and $r_2$. The operators $P_1$ and $P_2$ are the momentum operators for the two electrons, given by derivatives with respect to the coordinate of the appropriate electron, e.g., $P_1 = -i\hbar\nabla_1$, where $\nabla_1 = \hat{x}(\partial/\partial x_1) + \hat{y}(\partial/\partial y_1) + \hat{z}(\partial/\partial z_1)$, etc. Then the physical meaning of the various terms in the Hamiltonian is clear: the first and second terms give the kinetic energy of the two electrons, the third and fourth terms are the potentials that each electron feels in the Coulomb field of the nucleus, and the last term gives the change in the energy due to the mutual repulsion of the electrons. It is this last term which causes all the trouble in trying to find an exact solution.

Instead of looking for an exact solution, we will use the variational principle to estimate the ground-state energy. First, consider an arbitrary nucleus of charge $Z$ with only a single electron orbiting it. In this case the

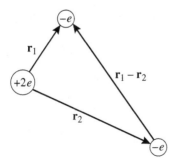

Fig. 10.3   The helium atom contains two electrons at positions $\mathbf{r}_1$ and $\mathbf{r}_2$ relative to the nucleus.

Schrödinger equation can be solved exactly, just as in the case for hydrogen, but now the charge $e$ for the nucleus must be replaced by the charge $Ze$. Recall from Chapter 6 that the ground-state wave function for hydrogen is

$$\psi(\mathbf{r}) = \frac{1}{\sqrt{\pi}} \left( \frac{1}{a_0} \right)^{3/2} e^{-r/a_0}$$

so the ground-state wave function for a single-electron atom with charge $Z$ on the nucleus is

$$\psi(\mathbf{r}) = \frac{1}{\sqrt{\pi}} \left( \frac{Z}{a_0} \right)^{3/2} e^{-Zr/a_0} \tag{10.6}$$

Note that the wave function for helium should be a *single* function of the positions of the two electrons, not two separate wave functions for the two electrons. We will take, as our trial wave function, the product of two wave functions of the form given by Equation (10.6):

$$\psi(\mathbf{r}_1, \mathbf{r}_2) = \left( \frac{1}{\sqrt{\pi}} \left( \frac{Z}{a_0} \right)^{3/2} e^{-Zr_1/a_0} \right) \left( \frac{1}{\sqrt{\pi}} \left( \frac{Z}{a_0} \right)^{3/2} e^{-Zr_2/a_0} \right) \tag{10.7}$$

Rather than setting $Z = 2$ for the charge on the helium nucleus, we will take $Z$ to be our variational free parameter. This choice is reasonable from a physical point of view: each electron partly cancels the positive charge that the other electron feels from the nucleus, so the "effective" value of $Z$ should lie between 1 and 2.

Equation (10.7) simplifies to

$$\psi(\mathbf{r}_1, \mathbf{r}_2) = \frac{1}{\pi} \left( \frac{Z}{a_0} \right)^3 e^{-Z(r_1 + r_2)/a_0}$$

Note that this wave function is already normalized, i.e., $\langle \psi | \psi \rangle = 1$ for any value of $Z$. Then the quantity to be minimized is

$$\langle \psi | H | \psi \rangle = \langle \psi | \frac{P_1^2}{2m_e} | \psi \rangle + \langle \psi | \frac{P_2^2}{2m_e} | \psi \rangle - \langle \psi | \frac{2e^2}{4\pi\epsilon_0 r_1} | \psi \rangle$$

$$- \langle \psi | \frac{2e^2}{4\pi\epsilon_0 r_2} | \psi \rangle + \langle \psi | \frac{e^2}{4\pi\epsilon_0 |\mathbf{r}_1 - \mathbf{r}_2|} | \psi \rangle$$

By symmetry,

$$\langle \psi | \frac{P_1^2}{2m_e} | \psi \rangle = \langle \psi | \frac{P_2^2}{2m_e} | \psi \rangle$$

and

$$\langle \psi | \frac{2e^2}{4\pi\epsilon_0 r_1} | \psi \rangle = \langle \psi | \frac{2e^2}{4\pi\epsilon_0 r_2} | \psi \rangle$$

Using these results, we get

$$\langle\psi|H|\psi\rangle = 2\langle\psi|\frac{P_1^2}{2m_e}|\psi\rangle - 2\langle\psi|\frac{2e^2}{4\pi\epsilon_0 r_1}|\psi\rangle + \langle\psi|\frac{e^2}{4\pi\epsilon_0|\mathbf{r}_1 - \mathbf{r}_2|}|\psi\rangle \quad (10.8)$$

The integrals in the first two terms on the right-hand side of Equation (10.8) are straightforward; they yield

$$2\langle\psi|\frac{P_1^2}{2m_e}|\psi\rangle = \frac{e^2}{4\pi\epsilon_0 a_0}Z^2 \quad (10.9)$$

and

$$-2\langle\psi|\frac{2e^2}{4\pi\epsilon_0 r_1}|\psi\rangle = \frac{e^2}{4\pi\epsilon_0 a_0}(-4Z) \quad (10.10)$$

The third term on the right-hand side of Equation (10.8) is not so easy to evaluate, so we derive it in more detail. Written as an integral, this term is

$$\langle\psi|\frac{e^2}{4\pi\epsilon_0|\mathbf{r}_1 - \mathbf{r}_2|}|\psi\rangle$$

$$= \frac{e^2}{4\pi\epsilon_0}\int \frac{1}{\pi}\left(\frac{Z}{a_0}\right)^3 e^{-Z(r_1+r_2)/a_0}\frac{1}{|\mathbf{r}_1 - \mathbf{r}_2|}\frac{1}{\pi}\left(\frac{Z}{a_0}\right)^3 e^{-Z(r_1+r_2)/a_0}\,d^3\mathbf{r}_1\,d^3\mathbf{r}_2$$

$$= \frac{e^2}{4\pi^3\epsilon_0}\left(\frac{Z}{a_0}\right)^6 \int \frac{e^{-2Z(r_1+r_2)/a_0}}{\sqrt{(\mathbf{r}_1 - \mathbf{r}_2)\cdot(\mathbf{r}_1 - \mathbf{r}_2)}}\,d^3\mathbf{r}_1\,d^3\mathbf{r}_2$$

$$= \frac{e^2}{4\pi^3\epsilon_0}\left(\frac{Z}{a_0}\right)^6 \int \frac{e^{-2Z(r_1+r_2)/a_0}}{\sqrt{r_1^2 + r_2^2 - 2r_1 r_2\cos\theta_{12}}}\,d^3\mathbf{r}_1\,d^3\mathbf{r}_2$$

where $\theta_{12}$ is the angle between the vectors $\mathbf{r}_1$ and $\mathbf{r}_2$. We perform the integral over $\mathbf{r}_2$ first. Since $\mathbf{r}_1$ is treated as a constant for the integration over $\mathbf{r}_2$, we can choose a spherical coordinate system in which $\mathbf{r}_1$ lies on the polar axis. Then $\theta_{12}$ gives the angle between $\mathbf{r}_2$ and the polar axis: it is just $\theta_2$ in the integration over $\mathbf{r}_2$ in polar coordinates. We get

$$\langle\psi|\frac{e^2}{4\pi\epsilon_0|\mathbf{r}_1 - \mathbf{r}_2|}|\psi\rangle = \frac{e^2}{4\pi^3\epsilon_0}\left(\frac{Z}{a_0}\right)^6 \int \frac{e^{-2Z(r_1+r_2)/a_0}}{\sqrt{r_1^2 + r_2^2 - 2r_1 r_2\cos\theta_2}}$$

$$\times r_1^2\,dr_1\sin\theta_1\,d\theta_1\,d\phi_1 r_2^2\,dr_2\sin\theta_2\,d\theta_2\,d\phi_2$$

Performing the integration over $\theta_2$ and $\phi_2$ gives

$$\langle\psi|\frac{e^2}{4\pi\epsilon_0|\mathbf{r_1}-\mathbf{r_2}|}|\psi\rangle = \frac{e^2}{4\pi^3\epsilon_0}\left(\frac{Z}{a_0}\right)^6\int(2\pi)e^{-2Z(r_1+r_2)/a_0}$$

$$\frac{\sqrt{r_1^2+r_2^2+2r_1r_2}-\sqrt{r_1^2+r_2^2-2r_1r_2}}{r_1r_2}$$

$$r_1^2\,dr_1\sin\theta_1\,d\theta_1\,d\phi_1r_2^2\,dr_2$$

$$= \frac{e^2}{2\pi^2\epsilon_0}\left(\frac{Z}{a_0}\right)^6\int e^{-2Z(r_1+r_2)/a_0}$$

$$(r_1+r_2-|r_1-r_2|)r_1\,dr_1\sin\theta_1\,d\theta_1\,d\phi_1r_2\,dr_2$$

Now note that $r_1+r_2-|r_1-r_2| = 2r_2$ when $r_1 > r_2$, and $r_1+r_2-|r_1-r_2| = 2r_1$ when $r_1 < r_2$. This can be expressed as $r_1+r_2-|r_1-r_2| = 2\,\text{Min}(r_1,r_2)$, where the function Min gives the smaller of its two arguments. Furthermore, the integrand is now independent of $\theta_1$ and $\phi_1$, so integration over those variables gives $4\pi$, and we get

$$\langle\psi|\frac{e^2}{4\pi\epsilon_0|\mathbf{r_1}-\mathbf{r_2}|}|\psi\rangle$$

$$= \frac{4e^2}{\pi\epsilon_0}\left(\frac{Z}{a_0}\right)^6\int_{r_1=0}^{\infty}\int_{r_2=0}^{\infty}e^{-2Z(r_1+r_2)/a_0}\text{Min}(r_1,r_2)r_1\,dr_1r_2\,dr_2$$

where we have now written out the limits of integration for $r_1$ and $r_2$ explicitly, in order to explain how to deal with the Min function. The presence of the Min function in the integrand forces us to break the integral over $r_2$ into two pieces: one for $r_2 < r_1$, for which $\text{Min}(r_1,r_2) = r_2$, and the second for $r_2 > r_1$, for which $\text{Min}(r_1,r_2) = r_1$. Then the integral becomes

$$\langle\psi|\frac{e^2}{4\pi\epsilon_0|\mathbf{r_1}-\mathbf{r_2}|}|\psi\rangle = \frac{4e^2}{\pi\epsilon_0}\left(\frac{Z}{a_0}\right)^6\left[\int_{r_1=0}^{\infty}\int_{r_2=0}^{r_1}e^{-2Z(r_1+r_2)/a_0}r_1\,dr_1r_2^2\,dr_2\right.$$

$$\left.+\int_{r_1=0}^{\infty}\int_{r_2=r_1}^{\infty}e^{-2Z(r_1+r_2)/a_0}r_1^2\,dr_1r_2\,dr_2\right]$$

Performing the two integrations over $r_2$ gives

$$\langle\psi|\frac{e^2}{4\pi\epsilon_0|\mathbf{r_1}-\mathbf{r_2}|}|\psi\rangle = \frac{4e^2}{\pi\epsilon_0}\left(\frac{Z}{a_0}\right)^6$$

$$\left[\int_{r_1=0}^{\infty}\left[e^{-4Zr_1/a_0}\left(-\frac{a_0}{2Z}r_1^2-\frac{a_0^2}{2Z^2}r_1-\frac{a_0^3}{4Z^3}\right)+e^{-2Zr_1/a_0}\frac{a_0^3}{4Z^3}\right]r_1\,dr_1\right.$$

$$\left.+\int_{r_1=0}^{\infty}e^{-4Zr_1/a_0}\frac{a_0^2}{4Z^2}\left(\frac{2Zr_1}{a_0}+1\right)r_1^2\,dr_1\right]$$

and integrating over $r_1$ gives the final result:

$$\langle\psi|\frac{e^2}{4\pi\epsilon_0|\mathbf{r}_1 - \mathbf{r}_2|}|\psi\rangle = \frac{e^2}{4\pi\epsilon_0 a_0}\frac{5}{8}Z \qquad (10.11)$$

Combining the results of Equations (10.9), (10.10), and (10.11), and using the fact that $\langle\psi|\psi\rangle = 1$, we get

$$\frac{\langle\psi|H|\psi\rangle}{\langle\psi|\psi\rangle} = \frac{e^2}{4\pi\epsilon_0 a_0}\left(Z^2 - 4Z + \frac{5}{8}Z\right) \qquad (10.12)$$

Taking the derivative of the right-hand side with respect to $Z$ and setting this derivative equal to 0 gives the value $Z_{min}$ for which $\langle\psi|H|\psi\rangle/\langle\psi|\psi\rangle$ is a minimum:

$$2Z_{min} - \frac{27}{8} = 0$$

so $Z_{min} = 27/16$. As expected, $1 < Z_{min} < 2$. Inserting $Z_{min}$ back into the right-hand side of Equation (10.12) gives the estimate of the ground-state energy:

$$E_{estimated} = \frac{\langle\psi|H|\psi\rangle}{\langle\psi|\psi\rangle} = -77.5 \text{ eV}$$

For comparison, the measured ground-state energy of helium (i.e., the energy needed to remove both electrons) is

$$E_0 = -79.0 \text{ eV}$$

As expected, $E_{estimated} > E_0$. However, the error in the value of the ground-state helium energy from the variational principle is only 2%, which is an excellent approximation!

## PROBLEMS

**10.1** Suppose that the trial wave function $|\psi(\alpha)\rangle$ happens to be exactly equal to the true ground-state wave function $|\psi_0\rangle$ for some value of $\alpha$. Show that in this case, the estimate of the ground-state energy given by the variational principle will be equal to the true ground-state energy.

**10.2** Suppose that the trial wave function $|\psi\rangle$ used in the variational principle is orthogonal to the ground-state wave function of the Hamiltonian: $\langle\psi_0|\psi(\alpha)\rangle = 0$ for all values of $\alpha$. Show that in this case

$$\frac{\langle\psi|H|\psi\rangle}{\langle\psi|\psi\rangle} \geq E_1$$

where $E_1$ is the energy of the first excited state of $H$.

**10.3** (a) In order to use the variational principle to estimate the ground-state energy of the one-dimensional potential $V(x) = Kx^4$, where $K$ is a constant, which of the following wave functions would be a better trial wave function?
  (i) $\psi(x) = e^{-\alpha x^2}$
  (ii) $\psi(x) = xe^{-\alpha x^2}$
Explain.
(b) In order to use the variational principle to estimate the ground-state energy of the one-dimensional potential $V(x) = Kx^3$ for $x > 0$ with an infinite potential barrier at $x = 0$, which of the following wave functions would be a better trial wave function?
  (i) $\psi(x) = e^{-\alpha x^2}$
  (ii) $\psi(x) = xe^{-\alpha x^2}$
Explain.

**10.4** A particle of mass $m$ is in the one-dimensional potential given by $V(x) = Kx^3$ for $x \geq 0$, where $K$ is a positive constant. There is an infinite potential barrier at $x = 0$, so $V(0) = \infty$. Use the variational principle with the trial wave function $|\psi\rangle = xe^{-\alpha x}$ to estimate the ground-state energy.

**10.5** Repeat the calculation in Example 10.1 using the trial wave function

$$\psi(x) = xe^{-\alpha x^2}$$

where $\alpha$ is the parameter to be varied. Is the final result a better or a worse approximation to the true ground-state energy than the result of Example 10.1?

**10.6** (a) A particle of mass $m$ is in the one-dimensional potential given by $V(x) = Kx^4$, where $K$ is a positive constant. Use the variational principle with the trial wave function $\psi(x) = e^{-\alpha x^2}$ to estimate the ground-state energy.
(b) The true ground-state wave function for this potential is a symmetric function of $x$, i.e., $\psi_0(-x) = \psi_0(x)$. Use the result of Problem 10.2, along with an appropriately chosen trial wave function, to estimate the energy of the first excited state.

**10.7** A three-dimensional spherically-symmetric harmonic oscillator has the potential $V(r) = (1/2)Kr^2$. The full Hamiltonian is then

$$H = \frac{-\hbar^2}{2m}\left[\frac{1}{r^2}\frac{\partial}{\partial r}r^2\frac{\partial}{\partial r} + \frac{1}{r^2\sin^2\theta}\frac{\partial^2}{\partial\phi^2} + \frac{1}{r^2\sin\theta}\frac{\partial}{\partial\theta}\sin\theta\frac{\partial}{\partial\theta}\right] + \frac{1}{2}Kr^2$$

[Note that the $L^2$ operator has been written out in terms of derivatives.]

(a) Use the trial wave function $\psi(r) = e^{-\alpha r}$ to calculate an approximation to the ground-state energy of the harmonic oscillator.

(b) The exact ground-state energy for the three-dimensional harmonic oscillator is $E = (3/2)\hbar\omega$. What is the relative error in the estimate from part (a)?

**10.8** Here is another approach to solve for the ground-state energy of helium.

(a) Begin with the Hamiltonian of Equation (10.5), but neglect the interaction between the two electrons. Solve the Schrödinger equation in this case to derive the wave function of the two electrons and the energy.

(b) Now add the interaction of the electrons as a perturbation:

$$H_1 = \frac{e^2}{4\pi\epsilon_0|\mathbf{r}_1 - \mathbf{r}_2|}$$

Use first-order perturbation theory to calculate the change in energy, and add this change to the energy derived in part (a) to give an estimate for the total ground-state energy.

(c) Is the estimate in part (b) more accurate or less accurate than the estimate from the variational principle?

**10.9** (a) Singly-ionized lithium has a nucleus of charge $+3e$ and two electrons. Use the variational principle to estimate the ground-state energy.

(b) Now consider a nucleus of charge $Ze$ with two electrons. Use the variational principle to estimate the ground-state energy.

Chapter 11

# Time-dependent perturbation theory

Although we first encountered the Schrödinger equation in full time-dependent form, we have thus far been concerned almost exclusively with solutions to the time-independent Schrödinger equation. There is a good reason for this: time-dependent problems are more difficult! Here we return to the full time-dependent Schrödinger equation. We will not attempt to solve it exactly; instead, we will develop a form of perturbation theory that can be applied to time-dependent problems.

In Chapter 9, we examined what happens if we begin with a Hamiltonian $H_0$ whose eigenfunctions $|\psi_n\rangle$ and energies $E_n$ are known, and then add a small change in the Hamiltonian which is constant in time: $H = H_0 + H_1$. Here we consider what happens if the small change $H_1$ is a function of time:

$$H = H_0 + H_1(t)$$

(Here $H_0$ is still taken to be constant in time.) In practice $H_1(t)$ will normally be produced by a time-dependent change in the potential $V$. Many different types of time dependence are possible for $H_1(t)$. For example, consider an electron in a time-dependent electric field. One could imagine suddenly "turning on" the electric field, applying an oscillating electric field, or producing a very slow change in the electric field (called an *adiabatic* change). These three possibilities give the forms for $H_1(t)$ shown in Figure 11.1. As we shall see, these different time dependences for $H_1(t)$ produce different predictions for the time evolution of the wave function.

The basic idea, as in Chapter 9, is that a small change in the Hamiltonian will produce a small change in the wave function. For a time-dependent perturbation, there can be a nonzero probability that a system initially in a particular eigenstate of $H_0$ will undergo a transition into a different eigenstate of $H_0$. Our goal will be to calculate the probability of such a transition.

$H_1(t)$

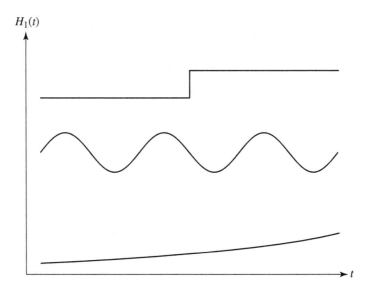

Fig. 11.1    Possible forms for the time dependence of $H_1(t)$ include a sharp change (top), an oscillating perturbation (middle), or an adiabatic change [i.e., a slowly-varying $H_1(t)$] (bottom).

## 11.1    Derivation of Time-Dependent Perturbation Theory

Before deriving the effect of adding a small perturbation to $H_0$, we need to derive some general results about the time evolution of the eigenfunctions of $H_0$ itself. Suppose that at $t = 0$, we have an eigenfunction $|\psi_n\rangle$ of $H_0$ with energy $E_n$. (We use Dirac notation to indicate that these are general eigenfunctions; they could be functions of position or spin states represented in column matrix form.) We will assume for now that $H_0$ is unperturbed. Recall from Chapter 3 that the time evolution of the eigenstates is given by

$$|\psi_n(t)\rangle = |\psi_n\rangle e^{-iE_n t/\hbar}$$

where $|\psi_n\rangle$ represents the state at $t = 0$. (Although we derived this result only for wave functions that are taken to be functions of position, the derivation carries over to the case of abstract states in Dirac notation.)

Now suppose we begin at $t = 0$ in the eigenstate $|\psi_n\rangle$, and we would like to know the probability $P_n(t)$ that we are still in the same eigenstate at some later time $t$. This probability is

$$P_n(t) = |\langle \psi_n | \psi_n(t) \rangle|^2$$
$$= |e^{-iE_n t/\hbar}|^2 \, |\langle \psi_n | \psi_n \rangle|^2$$
$$= 1 \tag{11.1}$$

Thus, a particle in a given eigenstate of $H_0$ remains in that same eigenstate forever.

Now consider a particle in an *arbitrary* state $|\psi(t)\rangle$, not necessarily an eigenfunction of $H_0$. Since the eigenfunctions of $H_0$ at $t = 0$ form a basis set, we can represent the wave function $|\psi(t)\rangle$ at any time $t$ as a linear combination of these eigenfunctions:

$$|\psi(t)\rangle = \sum_m d_m(t)|\psi_m\rangle \tag{11.2}$$

where the coefficients $d_m(t)$ are functions of time since the wave function $|\psi(t)\rangle$ evolves with time, but the $|\psi_m\rangle$'s are *not* functions of time. Suppose a particle is in the state given by $|\psi(t)\rangle$ in Equation (11.2), and a measurement is made to see if it is in a given eigenstate of $H_0$, namely $|\psi_n\rangle$. Equation (11.2) can be used to determine the probability $P_n(t)$ of finding the particle in the state $|\psi_n\rangle$:

$$P_n(t) = |\langle \psi_n | \psi(t) \rangle|^2$$
$$= \left| \sum_m d_m(t) \langle \psi_n | \psi_m \rangle \right|^2$$
$$= |d_n(t)|^2 \tag{11.3}$$

This gives the physical meaning of the expansion in Equation (11.2): the absolute value squared of the coefficient $d_n(t)$ provides the probability that the particle will be in the eigenstate $|\psi_n\rangle$ at some time $t$.

Time-dependent perturbation theory allows us to address the following problem. Suppose a system is initially in some state $|\psi_i\rangle$, which is an eigenstate of $H_0$. At a later time we make a measurement to determine if it is in some other final eigenstate $|\psi_f\rangle$ of $H_0$. From Equation (11.1), we know that this probability is 0. But now we "turn on" a small time-dependent perturbation $H_1(t)$. This means that the full Hamiltonian is now $H = H_0 + H_1(t)$, and the states $|\psi_i\rangle$ and $|\psi_f\rangle$ are no longer eigenstates of $H$. Therefore, Equation (11.1) is no longer valid: now there might be a nonzero probability that the system can begin in the state $|\psi_i\rangle$ and evolve into the state $|\psi_f\rangle$ (where $|\psi_i\rangle$ and $|\psi_f\rangle$ are eigenstates of the original unperturbed

Hamiltonian $H_0$). The calculation of this probability is the main point of time-dependent perturbation theory.

Consider an arbitrary state $|\psi(t)\rangle$, subject to the initial Hamiltonian $H_0$ plus a small time-dependent perturbation $H_1$. The time-dependent Schrödinger equation in this case is

$$[H_0 + H_1(t)]|\psi(t)\rangle = i\hbar \frac{\partial|\psi(t)\rangle}{\partial t} \tag{11.4}$$

We can expand $|\psi(t)\rangle$ in the form given by Equation (11.2). Now, however, we will define a new set of time-dependent coefficients $c_m(t)$ given by

$$d_m(t) = c_m(t)e^{-iE_m t/\hbar}$$

so that $|\psi(t)\rangle$ is given by

$$|\psi(t)\rangle = \sum_m c_m(t)e^{-iE_m t/\hbar}|\psi_m\rangle \tag{11.5}$$

where $E_m$ are the energies of the eigenstates for the original, unperturbed, time-independent Hamiltonian $H_0$. There is no deep significance to this change of variables from $d_m(t)$ to $c_m(t)$; we do it to simplify the algebra. Note that because $|e^{-iE_m t/\hbar}|^2 = 1$, the probability given in Equation (11.3) can be written as

$$P_n(t) = |c_n(t)|^2$$

so we will be interested in determining $c_n(t)$.

Inserting the expansion given by Equation (11.5) into the Schrödinger equation [Equation (11.4)] gives

$$\sum_m c_m(t)e^{-iE_m t/\hbar}E_m|\psi_m\rangle + \sum_m H_1(t)c_m(t)e^{-iE_m t/\hbar}|\psi_m\rangle$$

$$= i\hbar \sum_m \frac{dc_m}{dt}e^{-iE_m t/\hbar}|\psi_m\rangle + i\hbar \sum_m (-iE_m/\hbar)c_m(t)e^{-iE_m t/\hbar}|\psi_m\rangle$$

The first term on the left-hand side and the last term on the right-hand side cancel, since together they just represent the unperturbed Schrödinger equation, and we get

$$\sum_m H_1(t)c_m(t)e^{-iE_m t/\hbar}|\psi_m\rangle = i\hbar \sum_m \frac{dc_m}{dt}e^{-iE_m t/\hbar}|\psi_m\rangle$$

Applying $\langle \psi_n |$ to both sides of this equation and recalling that $\langle \psi_n | \psi_m \rangle = \delta_{mn}$ gives

$$\frac{dc_n}{dt} = \frac{1}{i\hbar} \sum_m \langle \psi_n | H_1(t) | \psi_m \rangle c_m(t) e^{i(E_n - E_m)t/\hbar} \tag{11.6}$$

Note that we have made no approximations so far; Equation (11.6) is exact, and in principle it gives the evolution of all of the coefficients $c_n(t)$. The problem is that it gives the time derivative of $c_n$ as a function of all of the other coefficients $c_m$; in many cases this could be an infinite number!

Making further progress requires using the fact that $H_1(t)$ represents a small perturbation to $H_0$. Suppose that the system begins in an initial eigenstate $|\psi_i\rangle$ for which $c_i = 1$ and $c_m = 0$ for $m \neq i$, and we would like to calculate the probability that it evolves into some final eigenstate $|\psi_f\rangle$. If the perturbation is small, then it is a good approximation to assume that the system does not evolve very far from its initial state, so that at any later time we still have $c_i \approx 1$ on the right-hand side of Equation (11.6), while all of the other $c_m$'s are so small in comparison that they can be ignored. With this approximation (and identifying the state $n$ with the final state $f$), Equation (11.6) becomes

$$\frac{dc_f}{dt} = \frac{1}{i\hbar} \langle \psi_f | H_1(t) | \psi_i \rangle e^{i(E_f - E_i)t/\hbar}$$

which can be integrated to give $c_f$:

$$c_f = \frac{1}{i\hbar} \int_{t_i}^{t_f} dt \langle \psi_f | H_1(t) | \psi_i \rangle e^{i(E_f - E_i)t/\hbar} \tag{11.7}$$

This is the fundamental equation of first-order time-dependent perturbation theory [the time-dependent analog to Equation (9.19)]. Here $|\psi_i\rangle$ and $|\psi_f\rangle$ represent eigenstates of the unperturbed Hamiltonian $H_0$ with $i \neq f$. (Note that $|\psi_i\rangle$ and $|\psi_f\rangle$ are assumed to be orthogonal and must be correctly normalized!) The probability that the system beginning in state $|\psi_i\rangle$ at time $t_i$ will end up in state $|\psi_f\rangle$ at time $t_f$ when a perturbation $H_1(t)$ is applied is

$$P(i \to f) = |c_f|^2 = \frac{1}{\hbar^2} \left| \int_{t_i}^{t_f} dt \langle \psi_f | H_1(t) | \psi_i \rangle e^{i(E_f - E_i)t/\hbar} \right|^2$$

It will often be the case that the perturbation $H_1(t)$ can be factored into a time-independent operator $\mathcal{H}_1$ and a time-dependent piece $f(t)$ which does not operate on the wave functions:

$$H_1(t) = \mathcal{H}_1 f(t)$$

In this case the inner product in Equation (11.7) can be pulled outside of the time integral, giving

$$c_f = \frac{1}{i\hbar}\langle\psi_f|\mathcal{H}_1|\psi_i\rangle \int_{t_i}^{t_f} dt\ f(t)e^{i(E_f-E_i)t/\hbar} \tag{11.8}$$

and the transition probability becomes

$$P(i \to f) = \frac{1}{\hbar^2}|\langle\psi_f|\mathcal{H}_1|\psi_i\rangle|^2 \left|\int_{t_i}^{t_f} dt\ f(t)e^{i(E_f-E_i)t/\hbar}\right|^2 \tag{11.9}$$

This allows us to make general statements about the time evolution of the system even if we only know the time behavior of the perturbation $f(t)$ and know nothing about $\langle\psi_f|\mathcal{H}_1|\psi_i\rangle$.

We now examine how several forms for this time dependence $f(t)$ translate into the time dependence of the transition probability. Consider first the case of a step-function perturbation of the form $H_1 = \mathcal{H}_1 f(t)$, in which the perturbation is "turned on" with constant magnitude at $t = 0$ (Figure 11.1, top). We take $f(t) = 0$ for $t < 0$ and $f(t) = 1$ for $t \geq 0$. Equation (11.8) gives, for this case,

$$c_f = \frac{1}{i\hbar}\langle\psi_f|\mathcal{H}_1|\psi_i\rangle \int_0^{t_f} dt\ e^{i(E_f-E_i)t/\hbar} \tag{11.10}$$

Note that $E/\hbar$ has units of frequency, so it is convenient to define the frequency $\omega_0$ given by

$$\omega_0 = (E_f - E_i)/\hbar$$

Then Equation (11.10) integrates to

$$c_f = \langle\psi_f|\mathcal{H}_1|\psi_i\rangle \frac{1}{\hbar\omega_0}(1 - e^{i\omega_0 t_f})$$

Finally, the probability that the system has evolved into the state $|\psi_f\rangle$ at the time $t_f$ is

$$P(i \to f) = |c_f|^2 = \frac{1}{\hbar^2}|\langle\psi_f|\mathcal{H}_1|\psi_i\rangle|^2 \frac{\sin^2(\omega_0 t_f/2)}{(\omega_0/2)^2} \tag{11.11}$$

This gives the general time dependence of the transition probability for *any* perturbation which is a step function in time. The behavior of $P(i \to f)$ is shown in Figure 11.2. If $t_f$ is small compared with $1/\omega_0$ such that $\omega_0 t_f \ll 1$, the transition probability increases as $t_f^2$. Eventually, however, this probability oscillates as shown.

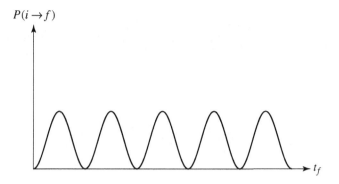

Fig. 11.2   The probability $P(i \to f)$ as a function of time $t_f$ for a transition in the case of a step-function perturbation.

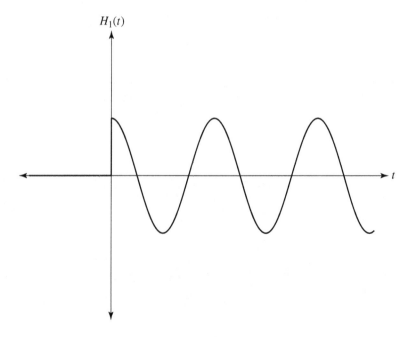

Fig. 11.3   The perturbation $H_1 = \mathcal{H}_1 \cos(\omega t)$ is "turned on" at time $t = 0$.

Second, we consider an oscillatory perturbation of the form $H_1 = \mathcal{H}_1 \cos(\omega t)$ which is "turned on" at time $t = 0$ (Figure 11.3). Equation (11.8) gives, for this case,

$$c_f = \frac{1}{i\hbar} \langle \psi_f | \mathcal{H}_1 | \psi_i \rangle \int_0^{t_f} dt \, \cos(\omega t) e^{i(E_f - E_i)t/\hbar} \qquad (11.12)$$

As in the previous example, define the frequency $\omega_0$ to be given by

$$\omega_0 = (E_f - E_i)/\hbar$$

while $\cos(\omega t)$ can be expanded in exponentials as

$$\cos(\omega t) = \frac{e^{i\omega t} + e^{-i\omega t}}{2}$$

so that Equation (11.12) becomes

$$c_f = \frac{1}{i\hbar}\langle\psi_f|\mathcal{H}_1|\psi_i\rangle \int_0^{t_f} dt\frac{1}{2}\left(e^{i(\omega_0+\omega)t_f} + e^{i(\omega_0-\omega)t_f}\right)$$

which integrates to

$$c_f = \frac{1}{i\hbar}\langle\psi_f|\mathcal{H}_1|\psi_i\rangle\frac{1}{2}\left[\frac{1}{i(\omega_0+\omega)}\left(e^{i(\omega_0+\omega)t_f}-1\right) + \frac{1}{i(\omega_0-\omega)}\left(e^{i(\omega_0-\omega)t_f}-1\right)\right]$$

$$(11.13)$$

This expression is rather complex, but it can be simplified for certain values of $\omega$. Assume first that $E_f > E_i$, so that $\omega_0 > 0$. Then if $\omega$ is close to $\omega_0$ (which is where the transition probability is largest), the second term in Equation (11.13) dominates the first [since, in this case, $1/(\omega_0 - \omega) \gg 1/(\omega_0 + \omega)$]. Dropping the first term and calculating $|c_f|^2$ to derive the transition probability $P(i \to f)$, we get

$$P(i \to f) = |c_f|^2 = \frac{|\langle\psi_f|\mathcal{H}_1|\psi_i\rangle|^2}{4\hbar^2}\frac{\sin^2[(\omega - \omega_0)t_f/2]}{[(\omega - \omega_0)/2]^2} \qquad (11.14)$$

Conversely, if $E_f < E_i$, then $\omega_0 < 0$. In this case the transition probability is largest when $\omega$ is close to $-\omega_0$, in which case the first term in Equation (11.13) dominates the second, and we get

$$P(i \to f) = |c_f|^2 = \frac{|\langle\psi_f|\mathcal{H}_1|\psi_i\rangle|^2}{4\hbar^2}\frac{\sin^2[(\omega + \omega_0)t_f/2]}{[(\omega + \omega_0)/2]^2} \qquad (11.15)$$

These general results are applicable to any sinusoidal perturbation, independent of the value of $\langle\psi_f|\mathcal{H}_1|\psi_i\rangle$. Consider the case $\omega_0 > 0$. If we fix the final time $t_f$ and examine how the transition probability varies with the applied frequency $\omega$, we see that it reaches a maximum at $\omega = \omega_0$; this is similar to the phenomenon of resonance in classical systems. As $\omega$ moves away from $\omega_0$, the transition probability decreases in an oscillatory fashion (Figure 11.4).

$P(i \to f)$

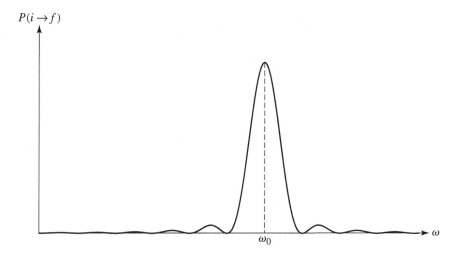

Fig. 11.4 The transition probability $P(i \to f)$ at a fixed time as a function of the applied frequency $\omega$ for a perturbation which varies as $\cos(\omega t)$.

As we have emphasized, Equations (11.11) and (11.14)–(11.15) give the general time evolution for any perturbations of the form $H_1(t) = \mathcal{H}_1 f(t)$, when $f(t)$ is a step function or a sinusoidal function. These cases represent two of the three types of time dependence shown in Figure 11.1. The third case (an adiabatic, i.e., slowly-varying perturbation) is left as an exercise (see Problem 11.4). Of course, these three cases do not exhaust all possibilities; many other forms for the time dependence of $H_1(t)$ are conceivable (see, for example, Problems 11.1 and 11.2).

We now apply these results to two examples.

---

### Example 11.1. A Hydrogen Atom in an Electric Field Turned on at $t = 0$

A hydrogen atom is in its ground state. A uniform electric field with magnitude $\mathcal{E}$ aligned in the positive $z$ direction is turned on at time $t = 0$ and left on. At some later time, $t_f > 0$, what is the probability that the atom will be in each of the following excited states?

(a) $n = 2, \quad l = 1, \quad m_l = -1$

(b) $n = 2, \quad l = 1, \quad m_l = 0$

(c) $n = 2, \quad l = 1, \quad m_l = +1$

This perturbation is of the form $H_1(t) = \mathcal{H}_1 f(t)$ with $f(t)$ a step function, so the transition probability $P(i \to f)$ is given by Equation (11.11).

We need to calculate $\langle \psi_f | \mathcal{H}_1 | \psi_i \rangle$. For the hydrogen atom in an electric field, the time-independent factor in the perturbation, $\mathcal{H}_1$, is

$$\mathcal{H}_1 = e\mathcal{E}z$$

just as in the case of the Stark effect in Chapter 9. Then

$$\langle \psi_f | \mathcal{H}_1 | \psi_i \rangle = e\mathcal{E}\langle \psi_f | z | \psi_i \rangle$$
$$= e\mathcal{E}\langle \psi_f | r \cos\theta | \psi_i \rangle$$

The initial wave function is the ground state of hydrogen:

$$|\psi_i\rangle \rightarrow \psi_{100}(r, \theta, \phi) = \frac{1}{\sqrt{\pi}}\left(\frac{1}{a_0}\right)^{3/2} e^{-r/a_0}$$

In case (a) the final state is

$$|\psi_f\rangle \rightarrow \psi_{21-1}(r, \theta, \phi) = \frac{1}{8\sqrt{\pi}}\left(\frac{1}{a_0}\right)^{3/2}\left(\frac{r}{a_0}\right) e^{-r/2a_0} \sin\theta e^{-i\phi}$$

so that

$$\langle \psi_f | \mathcal{H}_1 | \psi_i \rangle = \frac{e\mathcal{E}}{8\pi a_0^4}\int_{r=0}^{\infty}\int_{\theta=0}^{\pi}\int_{\phi=0}^{2\pi}(r^2 \, dr \sin\theta \, d\theta \, d\phi)(r^2 e^{-3r/2a_0})(\sin\theta\cos\theta)e^{i\phi}$$

The integral over $\phi$ gives zero, so the final transition probability is

$$P(i \rightarrow f) = 0$$

Now consider case (b) for which the final state is

$$|\psi_f\rangle \rightarrow \psi_{210}(r, \theta, \phi) = \frac{1}{4\sqrt{2\pi}}\left(\frac{1}{a_0}\right)^{3/2}\left(\frac{r}{a_0}\right) e^{-r/2a_0} \cos\theta$$

so that

$$\langle \psi_f | \mathcal{H}_1 | \psi_i \rangle = \frac{e\mathcal{E}}{4\sqrt{2}\pi a_0^4}\int_{r=0}^{\infty}\int_{\theta=0}^{\pi}\int_{\phi=0}^{2\pi}(r^2 \, dr \sin\theta \, d\theta \, d\phi)(r^2 e^{-3r/2a_0})(\cos^2\theta)$$

This can be integrated to give

$$\langle \psi_f | \mathcal{H}_1 | \psi_i \rangle = \frac{128\sqrt{2}}{243}e\mathcal{E}a_0$$

Substituting this back into Equation (11.11) gives the final transition probability:

$$P(i \rightarrow f) = |c_f|^2 = \frac{131072}{59049}\frac{e^2\mathcal{E}^2 a_0^2}{(E_2 - E_1)^2}\sin^2\left(\frac{E_2 - E_1}{2\hbar}t\right)$$

Here $E_1$ and $E_2$ are the energies of the ground state and first excited state of hydrogen, respectively.

In case (c) we get the same answer (for the same reason) as in case (a):

$$P(i \rightarrow f) = 0$$

Now consider an example involving spin eigenstates and an oscillating potential.

---

### Example 11.2. Spins in an Oscillating Magnetic Field

An electron is in a strong, uniform, constant magnetic field with magnitude $B_0$ aligned in the $-z$ direction. The electron is initially in the state $|\uparrow\rangle$. A weak, oscillating magnetic field in the $x$ direction of the form

$$\mathbf{B}_1 = B_1 \cos(\omega t)\hat{x}$$

is turned on at $t = 0$. Calculate the probability that the electron is in the state $|\downarrow\rangle$ at some later time $t$, assuming we are near resonance, so Equation (11.14) or (11.15) applies.

Since the constant magnetic field is $\mathbf{B}_0 = -B_0\hat{z}$, the spin-up state $|\uparrow\rangle$ has energy $E = -\mu_B B_0$, while the spin-down state $|\downarrow\rangle$ has energy $E = +\mu_B B_0$, so that the initial and final energies are

$$E_i = -\mu_B B_0$$
$$E_f = +\mu_B B_0$$

Then $\omega_0 > 0$, so Equation (11.14) applies with $\omega_0$ given by

$$\omega_0 = (E_f - E_i)/\hbar = 2\mu_B B_0/\hbar$$

When the perturbation $H_1(t)$ is written as $H_1(t) = \mathcal{H}_1 \cos(\omega t)$, the expression for $\mathcal{H}_1$ is

$$\mathcal{H}_1 = -\boldsymbol{\mu} \cdot \mathbf{B}_1$$
$$= -B_1 \mu_x$$
$$= -B_1 \left(-\frac{2\mu_B}{\hbar}\right) S_x$$
$$= B_1 \mu_B \sigma_x$$

This gives

$$\langle\psi_f|\mathcal{H}_1|\psi_i\rangle = \langle\downarrow|B_1\mu_B\sigma_x|\uparrow\rangle$$

or in matrix form,

$$\langle\psi_f|\mathcal{H}_1|\psi_i\rangle = B_1\mu_B \begin{pmatrix} 0 & 1 \end{pmatrix} \begin{pmatrix} 0 & 1 \\ 1 & 0 \end{pmatrix} \begin{pmatrix} 1 \\ 0 \end{pmatrix}$$
$$= B_1\mu_B$$

Then we obtain, using Equation (11.14), the final result:

$$P(i \to f) = \frac{B_1^2 \mu_B^2}{4\hbar^2} \frac{\sin^2[(\omega - \omega_0)t/2]}{[(\omega - \omega_0)/2]^2}$$

with

$$\omega_0 = 2\mu_B B_0/\hbar$$

## 11.2 Application: Selection Rules for Electromagnetic Radiation

In Chapter 6 we developed a model for the hydrogen atom. In this picture the energy levels $E_1, E_2, \ldots, E_n$ correspond to the principle quantum numbers, and the electron can drop from a higher energy level $E_i$ into a lower energy level $E_f$ by emitting a photon with angular frequency $\omega = (E_i - E_f)/\hbar$ (Figure 11.5, left). In Chapter 9 we showed that various internal interactions and corrections split some of the degenerate states in a given energy level, producing slightly separated energy levels. However, we still expect that transitions will occur between these energy levels with the corresponding emission of a photon with the correct energy (Figure 11.5, right). This picture is essentially complete, in the sense that all observed spectral lines correspond to a predicted pair of hydrogen energy levels. However, the reverse is not true: there are spectral lines predicted by this picture which are very weak or nonexistent. Apparently, something is preventing some transitions from occurring; these are called *forbidden* transitions. For example, the transition from the state $n = 2$, $l = 0$ to the state $n = 1$, $l = 0$ is not seen to occur via emission of a single photon. We will now use time-dependent perturbation theory to understand why. The rules we will derive that determine the allowed and forbidden transitions are called *selection rules*.

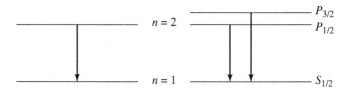

Fig. 11.5  Left: The electron can drop from energy level $E_i$ to energy level $E_f$ with the emission of a photon of angular frequency $\omega = (E_i - E_f)/\hbar$. Right: The actual energy levels display fine structure, but transitions occur in the same way.

We will actually consider photon absorption rather than emission, since the calculation for absorption is more straightforward. Since emission is just the time reverse of absorption, if a given absorption process is allowed, the corresponding emission process will also be allowed, and the same goes for forbidden processes. Consider an electromagnetic wave incident upon the electron in a hydrogen atom. Recall that this wave consists of an oscillating electric field and an oscillating magnetic field, so the obvious question is: which is more important, the interaction of the electron with the magnetic field of the wave, or with its electric field? Classically, recall that the force experienced by an electron in an electric field $\mathcal{E}$ and magnetic field $\mathbf{B}$ is

$$\mathbf{F} = -e[\mathcal{E} + \mathbf{v} \times \mathbf{B}]$$

Further, the magnitudes of the electric and magnetic fields in an electromagnetic wave satisfy the relation

$$\mathcal{E} = Bc$$

Combining these two equations, we see that the force due to the electric field $F_{\mathcal{E}}$ and the force due to the magnetic field $F_B$ satisfy

$$\frac{F_B}{F_{\mathcal{E}}} \approx \frac{v}{c}$$

which is much less than one as long as the electron is nonrelativistic. This is a standard result from classical electromagnetism: in examining the interaction of an electromagnetic wave with matter, it is the electric field of the wave, not its magnetic field, which dominates the interaction.

The electric field due to an electromagnetic wave propagating in the $\mathbf{k}$ direction can be written in the form

$$\mathcal{E} = \mathcal{E}_0 e^{i(\mathbf{k}\cdot\mathbf{r} - \omega t)} = \mathcal{E}_0 e^{i\mathbf{k}\cdot\mathbf{r}} e^{-i\omega t} \tag{11.16}$$

where $\mathcal{E}_0$ points in a direction perpendicular to $\mathbf{k}$.

The dependence of $\mathcal{E}$ on the position $\mathbf{r}$ is sinusoidal, but we can simplify this dependence by noting that we will be interested in frequencies for which $kr \ll 1$. In the hydrogen atom, for example, a transition from one bound state to another corresponds to a photon energy $E < 13.6$ eV, which corresponds to $\lambda > 1.5 \times 10^{-8}$ m. In comparison, the typical "size" of the hydrogen atom over which we need to apply this perturbation is $a_0 \sim 10^{-10}$ m. Thus, it is a good approximation to take $\mathcal{E}$ to be roughly constant over the radius of the hydrogen atom, corresponding to $kr \ll 1$ (Figure 11.6).

Mathematically, this approximation corresponds to expanding out the exponential in Equation (11.16) in the form

$$\mathcal{E} = \mathcal{E}_0 e^{-i\omega t} \left[ 1 + i\mathbf{k}\cdot\mathbf{r} + \frac{1}{2}(i\mathbf{k}\cdot\mathbf{r})^2 + \cdots \right] \tag{11.17}$$

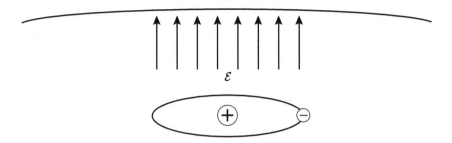

Fig. 11.6   For electromagnetic radiation with wavelength $\lambda \gg a_0$, the electric field can be taken to be constant in space over the radius of the atom.

and retaining only the first term in the expansion:

$$\boldsymbol{\mathcal{E}} = \boldsymbol{\mathcal{E}}_0 e^{-i\omega t}(1) \qquad (11.18)$$

This is called the *dipole approximation*.

Suppose we begin with the hydrogen atom in some initial bound state $|\psi_i\rangle$ and apply the electric field in Equation (11.18). Then the change in the energy of the electron due to this electric field is

$$H_1(t) = e(\boldsymbol{\mathcal{E}}_0 \cdot \mathbf{r})e^{-i\omega t}$$

and the probability that the electron will end up in some other bound state $|\psi_f\rangle$ derived from time-dependent perturbation theory (Equation (11.9)) is

$$P(i \to f) = |c_f|^2 = \frac{e^2}{\hbar^2}|\langle\psi_f|\boldsymbol{\mathcal{E}}_0\cdot\mathbf{r}|\psi_i\rangle|^2 \left| \int e^{i[(E_f-E_i)/\hbar - \omega]t}\, dt \right|^2 \qquad (11.19)$$

The time dependence of this probability is irrelevant (which is why the limits of integration in the time integral have been left unspecified). The important thing is whether or not the transition can occur at all, which is completely determined by whether or not the quantity $\langle\psi_f|\boldsymbol{\mathcal{E}}_0\cdot\mathbf{r}|\psi_i\rangle$ is zero.

We now take the initial state to be the hydrogen wave function with quantum numbers $n_i$, $l_i$, and $m_i$ (all of the $m$'s here will refer to $m_l$; we temporarily drop the $l$ subscript for clarity) and the final state will have quantum numbers $n_f$, $l_f$, and $m_f$. For now we will take $\boldsymbol{\mathcal{E}}_0$ to be in the $z$ direction, so that

$$\boldsymbol{\mathcal{E}}_0 \cdot \mathbf{r} = \mathcal{E}_0 z = \mathcal{E}_0 r \cos\theta$$

This choice for $\boldsymbol{\mathcal{E}}_0$ indicates that the light is polarized in the $z$ direction. Since the value of $l$ does not depend on the choice of coordinate axes, we do not expect the selection rules for $l$ to depend on the choice of polarization. This will *not* be true, however, for the selection rules for $m_l$.

The inner product which determines whether or not the transition can occur becomes

$$\langle \psi_f | \boldsymbol{\mathcal{E}}_0 \cdot \mathbf{r} | \psi_i \rangle$$

$$= \mathcal{E}_0 \int (r^2 \, dr \sin\theta \, d\theta \, d\phi) R^*_{n_f l_f}(r) Y^{m_f *}_{l_f}(\theta, \phi)(r \cos\theta) R_{n_i l_i}(r) Y^{m_i}_{l_i}(\theta, \phi)$$

Consider the integral over $\theta$ and $\phi$:

$$\int (\sin\theta \, d\theta \, d\phi) Y^{m_f *}_{l_f}(\theta, \phi)(\cos\theta) Y^{m_i}_{l_i}(\theta, \phi)$$

It can be shown that the spherical harmonics have the property that

$$Y^m_l(\theta, \phi) \cos\theta = a Y^m_{l+1}(\theta, \phi) + b Y^m_{l-1}(\theta, \phi)$$

where $a$ and $b$ are constants that depend on the particular values of $l$ and $m$. This relation allows us to rewrite the angular integral as

$$\int (\sin\theta \, d\theta \, d\phi) Y^{m_f *}_{l_f}(\theta, \phi)(\cos\theta) Y^{m_i}_{l_i}(\theta, \phi)$$

$$= \int (\sin\theta \, d\theta \, d\phi) Y^{m_f *}_{l_f}(\theta, \phi)[a Y^{m_i}_{l_i+1}(\theta, \phi) + b Y^{m_i}_{l_i-1}(\theta, \phi)] \quad (11.20)$$

Finally, recall the orthogonality relation for the spherical harmonics:

$$\int (\sin\theta \, d\theta \, d\phi) Y^{m' *}_{l'}(\theta, \phi) Y^m_l(\theta, \phi) = 0, \quad \text{unless } l = l' \text{ and } m = m'$$

Applying this orthogonality condition to Equation (11.20), we see that the integral will vanish (and therefore the transition probability in Equation (11.19) will be zero) unless

$$l_f = l_i + 1 \quad \text{or} \quad l_f = l_i - 1 \quad (11.21)$$

and

$$m_f = m_i$$

These are the correct selection rules for light polarized in the $z$ direction, but only the selection rule for $l$ (Equation (11.21)) carries over to arbitrary polarizations, since, as we have argued, anything involving $l$ must be independent of the coordinate system. The selection rule we have derived for $m$ applies only to this particular polarization state; for light polarized in the $x$ or $y$ directions, we obtain instead (Problem 11.8)

$$m_f = m_i + 1 \quad \text{or} \quad m_f = m_i - 1$$

The full selection rules then can be expressed in the succinct form

$$\Delta l = \pm 1$$

and

$$\Delta m_l = 0, \pm 1$$

where $\Delta l = l_f - l_i$ and $\Delta m_l = m_f - m_i$. Finally, there is an additional selection rule for the total angular momentum quantum number $j$:

$$\Delta j = 0, \pm 1 \qquad (11.22)$$

with the single exception that $j_i = 0 \to j_f = 0$ is forbidden. We will not give a rigorous proof of the $j$ selection rules but rather explain their physical origin. The photon has spin 1, so if it is absorbed by an atom with initial total angular momentum quantum number $j_i$, the two angular momenta can couple to give total angular momentum of $j_i - 1$, $j_i$, or $j_i + 1$, from the rules for adding angular momentum in Chapter 8. These, therefore, are the possible values of $j_f$, giving the selection rule in Equation (11.22). The single exception occurs when $j_i = 0$, which can couple to the spin-1 photon only to give $j_f = 1$. This is why the transition $j_i = 0$ to $j_f = 0$ is forbidden. (Of course, $j = 0$ never occurs in the hydrogen atom, but it does occur in multielectron atoms.)

---

**Example 11.3. The Allowed Transitions in Hydrogen from $n = 3$ to $n = 1$**

An electron in the $n = 3$ state of hydrogen emits a photon and drops into the ground state. What are the allowed transitions?

The state $n = 1$ has only one possible value for $l$, namely, $l = 0$ and one possible value for $j$, namely, $j = 1/2$. In the spectroscopic notation introduced in Chapter 9, this state is written $S_{1/2}$.

The state $n = 3$ has the following possible $l$ and $j$ states:

$$l = 0, \quad j = 1/2$$
$$l = 1, \quad j = 1/2, \quad j = 3/2$$
$$l = 2, \quad j = 3/2, \quad j = 5/2$$

The selection rule for $l$ tells us that $\Delta l = \pm 1$. Since the final state has $l = 0$, and $l$ cannot be negative, the only allowed initial state is $l = 1$.

For $l = 1$ in the initial state, we can have either $j = 1/2$ or $j = 3/2$. Since $j = 1/2$ in the final state, the $j = 1/2$ initial state corresponds to $\Delta j = 0$, and the $j = 3/2$ initial state corresponds to $\Delta j = -1$. The selection rule for $j$ is $\Delta j = 0, \pm 1$, so either of these initial $j$ states is allowed.

The allowed transitions, therefore, are

$$l = 1, \ j = 1/2 \to l = 0, \ j = 1/2$$
$$l = 1, \ j = 3/2 \to l = 0, \ j = 1/2$$

or, in spectroscopic notation,

$$P_{1/2} \to S_{1/2}$$
$$P_{3/2} \to S_{1/2}$$

---

Transitions which are not allowed by our selection rules are called *forbidden* transitions, but this is somewhat misleading. Such transitions can sometimes occur but at a much slower rate. There are two possible ways of evading the selection rules: first, a transition may occur through one of the higher-order terms that were dropped in Equation (11.17). For example, a transition occurring through the term linear in **k·r** is called an *electric quadrupole transition*. Second, we have ignored the interaction between the magnetic field of the electromagnetic wave and the electron in the atom, but this interaction can lead to a *magnetic dipole transition*, or a higher order magnetic transition. Therefore, the term "forbidden" transition really means "forbidden to electric dipole radiation."

## PROBLEMS

**11.1** The electron in a hydrogen atom is initially in the ground state. At $t = 0$, a homogeneous electric field aligned in the $z$ direction is turned on. The magnitude of the electric field decreases exponentially:

$$\mathcal{E} = \mathcal{E}_0 \hat{z} e^{-t/\tau}$$

where $\mathcal{E}_0$ and $\tau$ are constants. A measurement is made at $t_f = +\infty$; what is the probability that the electron will be in the first excited state?

**11.2** A system is in an eigenstate $|\psi_i\rangle$ with energy $E_i$. The perturbation

$$H_1(t) = \mathcal{H}_1 e^{-\alpha^2 t^2}$$

is turned on at $t_i = -\infty$ and left on until $t_f = +\infty$. Here $\mathcal{H}_1$ is independent of time, and $\alpha$ is a constant. Show that at $t_f = +\infty$, the probability that the system has evolved into the eigenstate $|\psi_f\rangle$ with energy $E_f$ is

$$P(i \to f) = \frac{\pi}{\hbar^2 \alpha^2} |\langle \psi_f | \mathcal{H}_1 | \psi_i \rangle|^2 e^{-(E_f - E_i)^2 / 2\hbar^2 \alpha^2} \qquad (11.23)$$

**11.3** An electron is in a strong, uniform, constant magnetic field with magnitude $B_0$ aligned in the $+x$ direction. The electron is initially in the state $|\rightarrow\rangle$ with $x$ component of spin equal to $+\hbar/2$. A weak, uniform, constant magnetic field of magnitude $B_1$ (where $B_1 \ll B_0$) in the $+z$ direction is turned on at $t = 0$ and turned off at $t = t_0$. Let $P(i \rightarrow f)$ be the probability that the electron is in the state $|\leftarrow\rangle$ with $x$ component of spin equal to $-\hbar/2$ at a later time $t_f > t_0$. Show that

$$P(i \rightarrow f) = (B_1/B_0)^2 \sin^2(\mu_B B_0 t_0/\hbar)$$

**11.4** Consider a time-dependent perturbation $H_1(t)$ which is adiabatic, i.e., slowly varying. The system is initially in the state $|\psi_i\rangle$ at $t_i = -\infty$. The potential is turned on, and we wish to derive the probability that the system will be in the state $|\psi_f\rangle$ at some later time $t_f$. Write down the standard expression for $c_f$ in this case and use integration by parts to break the expression into two terms, one of which contains $dH_1/dt$. Since $H_1$ is slowly varying, this term may be taken to be 0. Then use the fact that $H_1(-\infty) = 0$ to derive the final expression for the transition probability:

$$P(i \rightarrow f) = \frac{|\langle\psi_f|H_1(t_f)|\psi_i\rangle|^2}{(E_i - E_f)^2}$$

**11.5** A particle with mass $m$ is in a one-dimensional infinite square-well potential of width $a$, so $V(x) = 0$ for $0 \leq x \leq a$, and there are infinite potential barriers at $x = 0$ and $x = a$. Recall that the normalized solutions to the Schrödinger equation are

$$\psi_n(x) = \sqrt{\frac{2}{a}} \sin\left(\frac{n\pi x}{a}\right)$$

with energies

$$E_n = \frac{\hbar^2 \pi^2 n^2}{2ma^2}$$

where $n = 1, 2, 3, \ldots$.
The particle is initially in the ground state. A delta-function perturbation

$$\mathcal{H}_1 = K\delta\left(x - \frac{a}{2}\right)$$

(where $K$ is a constant) is turned on at time $t = -t_1$ and turned off at $t = t_1$. A measurement is made at some later time $t_2$, where $t_2 > t_1$.

(a) What is the probability that the particle will be found to be in the excited state $n = 3$?

(b) There are some excited states $n$ in which the particle will *never* be found, no matter what values are chosen for $t_1$ and $t_2$. Which excited states are these?

**11.6** An electron is in a strong, static, homogeneous magnetic field with magnitude $B_0$ in the $z$ direction. At time $t = 0$, the spin of the electron is in the $+z$ direction. At $t = 0$ a weak, homogenous magnetic field with magnitude $B_1$ (where $B_1 \ll B_0$) is turned on. At $t = 0$ this field is pointing in the $x$ direction, but it rotates counterclockwise in the $x$-$z$ plane with angular frequency $\omega$, so that at any later time $t$ this field is at an angle $\omega t$ relative to the $x$-axis:

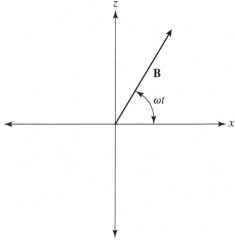

Do not assume anything about the particular value of $\omega$. Calculate the probability that at a later time $t_f$ the electron spin has flipped to the $-z$ direction.

**11.7** A hydrogen atom is in the ground state. At $t = 0$ an electric field with magnitude $\mathcal{E}$ is turned on. At $t = 0$ the electric field points in the $x$ direction, and it rotates counterclockwise in the $x$-$y$ plane with angular frequency $\omega$ (i.e., at any later time $t$ the field is oriented at an angle $\omega t$ relative to the $x$-axis). This rotating field causes the atom to undergo a transition to an $n = 2$ state. Determine which of the $l$, $m_l$ states are possible final states and which are impossible.

**11.8** (a) Consider an electromagnetic wave polarized in the $x$ direction, incident on a hydrogen atom. Show that in this case, the selection rules for $m_l$ are

$$\Delta m_l = \pm 1$$

(b) Repeat this calculation for an electromagnetic wave polarized in the $y$ direction.

**11.9** A hydrogen atom in the $n = 4$ state emits electric dipole radiation and drops into the $n = 3$ state. Determine all possible transitions in terms of their initial and final values for $l$ and $j$. Express the answer in spectroscopic notation.

**11.10** (a) The electron in a hydrogen atom is initially in the state $n = 5$, $l = 0$, $j = 1/2$. The atom emits electric dipole radiation and drops into an $n = 3$ state. Determine all $l$, $j$ states which are possible final states.

(b) An electron in a hydrogen atom is initially in the state $n = 5$, $l = 2$, $j = 5/2$. It emits electric dipole radiation and drops into the state $n = 4$, $l = l_1$, $j = j_1$. From this state, it emits electric dipole radiation again and drops into the hydrogen ground state. Determine $l_1$ and $j_1$.

**11.11** An electron is contained in a three-dimensional rectangular box given by $0 \leq x \leq a$, $0 \leq y \leq b$, and $0 \leq z \leq c$. The solutions of the Schrödinger equation are specified by the quantum numbers $n_x$, $n_y$, and $n_z$. Recall that the normalized wave function is

$$\psi(x, y, z) = \sqrt{\frac{8}{abc}} \sin\left(\frac{n_x \pi x}{a}\right) \sin\left(\frac{n_y \pi y}{b}\right) \sin\left(\frac{n_z \pi z}{c}\right)$$

with energy

$$E = \left(\frac{\hbar^2 \pi^2}{2m}\right)\left(\frac{n_x^2}{a^2} + \frac{n_y^2}{b^2} + \frac{n_z^2}{c^2}\right)$$

where $n_x$, $n_y$, and $n_z$ are positive integers. The electron is initially in the state $n_x, n_y, n_z$. An electromagnetic wave is incident polarized in the $y$ direction, so that the electric field vector is given by:

$$\mathbf{E} = \mathbf{E}_0 e^{-i\omega t}\left[1 + i\mathbf{k}\cdot\mathbf{r} + \frac{1}{2}(i\mathbf{k}\cdot\mathbf{r})^2 + \cdots\right]$$

where $\mathbf{E}_0$ is a constant vector in the $y$ direction. Use the dipole approximation and time-dependent perturbation theory to derive the selection rules for the electron to absorb the radiation and end up in the final state $n'_x$, $n'_y$, $n'_z$.

Chapter 12

# Scattering theory

In this chapter, we return to the theory of scattering, i.e., the behavior of a particle incident on a fixed potential. The one-dimensional case was examined for step-function potentials in Chapter 4. In that case it was possible to solve the Schrödinger equation exactly. Here we extend the discussion to the case of three-dimensional potentials. In most cases the Schrödinger equation for scattering in three dimensions cannot be solved exactly, so approximation methods must be used. This chapter deals with two of these approximation methods: the Born approximation and the method of partial waves. The Born approximation can be applied when the energy of the incident particle is much larger than the potential from which it scatters. While the method of partial waves is applicable to any scattering problem, it takes a particularly simple form in the limit where the energy of the incident particle is low.

## 12.1 Definition of the Cross Section

Before examining these quantum mechanical approximations, we will review some of the concepts of scattering from classical mechanics. Suppose that a region of space is filled with targets having a number density $n$. A particle enters this region and travels a distance $L$ (Figure 12.1). Clearly, the probability $P$ that the particle will strike one of the targets is proportional to the distance $L$ that it travels, and it is also proportional to the number density of targets $n$, so it is possible to write

$$P = nL\sigma \qquad (12.1)$$

where $\sigma$ is a constant of proportionality that depends on the nature of the targets. Since $P$ is a probability, it must be a dimensionless number, while $n$ has units of $1/\text{length}^3$ and $L$ has units of length. Hence, $\sigma$ must have units

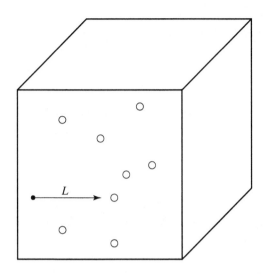

Fig. 12.1   A particle moves a distance $L$ through a region of space having a number density of targets $n$.

of length$^2$, or area. The quantity $\sigma$ is called the *cross section* for scattering. In a classical scattering problem in which the incident particle physically collides with the targets, $\sigma$ is just the cross-sectional area of each target, i.e., the area that the incident particle "sees" head-on. This is not the case (even classically) when the incident particle and the targets interact via a long-range force.

Now suppose that instead of a single particle, a large number of incident particles $N_i$ all travel a distance $L$ through this region of targets, and let $N_s$ be the number of these particles that scatter off of one of the targets. In this case $P$ gives the fraction of particles that scatter, so that

$$P = N_s/N_i$$

which implies that

$$N_s = N_i n L \sigma \tag{12.2}$$

Often, however, it is important to know not only the total probability for scattering but also the probability that the incident particles are scattered in a particular direction. In order to quantify this idea, we set up a spherical coordinate system with the polar axis aligned along the direction of motion of the incident particle (Figure 12.2). Consider a small solid angle of size $d\Omega$ in the $\theta$, $\phi$ direction:

$$d\Omega = \sin\theta \, d\theta \, d\phi$$

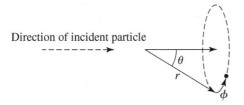

Fig. 12.2  A spherical coordinate system with the polar axis along the direction of motion of the incident particle.

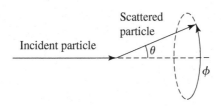

Fig. 12.3  The differential cross section, $d\sigma/d\Omega$, determines the probability that an incident particle will be scattered in the $\theta$, $\phi$ direction.

If $N_s$ is the total number of particles scattered (as in Equation (12.2)), let $dN_s$ be the total number of particles scattered into the small solid angle $d\Omega$ in the $\theta$, $\phi$ direction, where $dN_s$ will be a function of the scattering direction and will be proportional to $d\Omega$. Then the equation analogous to Equation (12.2) is

$$dN_s = N_i n L \frac{d\sigma}{d\Omega} d\Omega$$

Thus, $d\sigma/d\Omega$, which is a function of $\theta$ and $\phi$, determines the probability that the incident particle will be scattered in the $\theta$, $\phi$ direction (Figure 12.3). For instance, $\theta = 0$ corresponds to no scattering, while $\theta = \pi$ represents scattering directly backwards. The quantity $d\sigma/d\Omega$ is called the *differential cross section* and, therefore, $\sigma$ is often called the *total cross section*. Given $d\sigma/d\Omega$, the total cross section is just the integral of the differential cross section over all angles:

$$\sigma = \int_{\theta=0}^{\pi} \int_{\phi=0}^{2\pi} \left( \frac{d\sigma}{d\Omega} \right) \sin\theta \, d\theta \, d\phi$$

It will frequently be the case that the scattering potential is spherically symmetric. In this case the system as a whole (incident particle plus scattering potential) has azimuthal symmetry, i.e., there is no preferred $\phi$ direction. On the other hand, while the potential is symmetric with respect

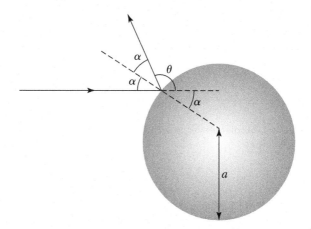

Fig. 12.4 A particle scatters from a hard sphere of radius $a$. The scattering angle is $\theta$, and the angle of incidence and angle of reflection are both $\alpha$.

to $\theta$, the system as a whole is not, since the direction of the incident particle defines a special direction with respect to the $\theta$ coordinate. Thus, for a spherically-symmetric scattering potential, we expect $d\sigma/d\Omega$ to depend on $\theta$ but to be independent of $\phi$.

As an example of how these ideas work, we will now calculate a classical scattering cross section; this is *not* a quantum mechanical calculation!

---

### Example 12.1. Classical Scattering from a Hard Sphere

Suppose that we are shooting particles at a solid sphere of radius $a$, where the size of the incident particles is negligible compared to the size of the sphere. The particles reflect off of the sphere elastically, so that the angle of incidence equals the angle of reflection. We now calculate the differential cross section, $d\sigma/d\Omega$, and the total cross section $\sigma$.

As usual, let $\theta$ be the scattering angle for the particle. When the particle scatters, the angle of reflection equals the angle of incidence; we let $\alpha$ be both of these angles (Figure 12.4). The particle scatters off of a cross-sectional area consisting of a ring of radius $s$ and width $ds$ (Figure 12.5). The cross-sectional area of this ring is

$$d\sigma = 2\pi s\, ds$$

and from Figure 12.5, we have

$$s = a \sin \alpha$$

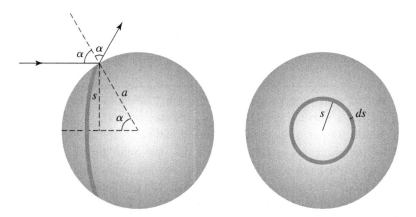

Fig. 12.5   The particle scatters off of a cross-sectional area consisting of a ring of radius $s$, width $ds$: (left) side view, (right) head-on view.

These two equations give $d\sigma$ in terms of $\alpha$:

$$d\sigma = (2\pi)(a\sin\alpha)(a\cos\alpha\,d\alpha) \qquad (12.3)$$

However, we want to express everything in terms of the scattering angle $\theta$. Note from Figure 12.4 that

$$\theta = \pi - 2\alpha$$

Substituting this into Equation (12.3) gives

$$d\sigma = (2\pi)a^2[(\sin\theta)/2](d\theta/2)$$

Note that if the particle strikes this area $d\sigma$, it scatters into the solid angle $d\Omega = (\sin\theta\,d\theta)(2\pi)$, where the $2\pi$ comes from the fact that we have integrated over $\phi$. Then the differential cross section is

$$\frac{d\sigma}{d\Omega} = \frac{2\pi a^2[(\sin\theta)/2](d\theta/2)}{2\pi\sin\theta\,d\theta}$$

$$= \frac{a^2}{4}$$

Thus, the differential cross section in this case is independent of both $\theta$ and $\phi$.

The total cross section is

$$\sigma = \int_{\theta=0}^{\pi}\int_{\phi=0}^{2\pi}\left(\frac{d\sigma}{d\Omega}\right)\sin\theta\,d\theta\,d\phi$$

$$= \int_{\theta=0}^{\pi}\int_{\phi=0}^{2\pi}\frac{a^2}{4}\sin\theta\,d\theta\,d\phi$$

$$= \pi a^2$$

This result makes sense physically because $\pi a^2$ is the cross-sectional area of the sphere, i.e., a particle striking the sphere sees, in projection, a circle of radius $a$ and area $\pi a^2$.

---

The general scattering problem in quantum mechanics is the calculation of $d\sigma/d\Omega$ for a particle incident on an arbitrary potential $V$. In principle, the Schrödinger equation should be solved and the resulting wave function used to find the cross section. In practice, the Schrödinger equation for scattering problems is usually difficult to solve, so approximation methods must be used. There are two cases for which the problem is much simpler. In the limit in which the energy of the particle $E$ is much larger than the potential $V$ we can use a variant of perturbation theory called the Born approximation. In the opposite limit, when $E \ll V$, we utilize the method of partial waves. We now discuss these in turn.

## 12.2   The Born Approximation

Consider an incident particle with energy $E$ scattering off of a potential $V(\mathbf{r})$. (Although the potential will often be taken to have spherical symmetry, this is not essential in using the Born approximation, so we begin with the most general case.) We assume that the potential is "weak" in the sense that $E \gg V(\mathbf{r})$, which allows us to use the results of time-dependent perturbation theory from the previous chapter. We will assume further that $V(\mathbf{r}) \to 0$ when $r \to \infty$ so that the potential can be ignored for sufficiently large $r$.

Assume that the incident particle has wave vector $\mathbf{k}_i$, and it scatters with final wave vector $\mathbf{k}_f$ (Figure 12.6). These correspond to initial and final energies of

$$E_i = \frac{\hbar^2 k_i^2}{2m}$$

$$E_f = \frac{\hbar^2 k_f^2}{2m}$$

We have assumed that the potential vanishes for large $r$, so at locations far from the potential the time-independent wave functions correspond to free particles. These time-independent wave functions (from Chapter 4) are

$$\psi_i = e^{i\mathbf{k}_i \cdot \mathbf{r}}$$

$$\psi_f = e^{i\mathbf{k}_f \cdot \mathbf{r}}$$

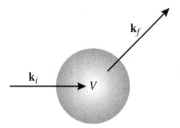

Fig. 12.6   A particle scatters from the potential $V(\mathbf{r})$ with initial wave vector $\mathbf{k}_i$ and final wave vector $\mathbf{k}_f$.

The scattering potential is assumed to be constant in time. However, we can think of this problem in terms of time-dependent perturbation theory. Initially, when the particle is far from the potential, it "sees" $V = 0$. Then, as the particle enters the region where $V \neq 0$, the particle experiences the potential and scatters. Finally, as the particle leaves the region of the potential, it once again "sees" $V = 0$. Hence, we will assume that the potential is "turned on" at some arbitrary time, which we will take to be $t = 0$, and we wish to know the probability that the particle has scattered from the state $\psi_i$ to $\psi_f$ by the time $t$. Formally, since the potential is being treated as a small perturbation, this probability is just given by the result of time-dependent perturbation theory from Chapter 11:

$$
\begin{aligned}
P(i \to f) &= \frac{1}{\hbar^2} \left| \int_0^t dt' \langle \psi_f | V(\mathbf{r}) | \psi_i \rangle e^{i(E_f - E_i)t'/\hbar} \right|^2 \\
&= \frac{1}{\hbar^2} |\langle \psi_f | V(\mathbf{r}) | \psi_i \rangle|^2 \left| \int_0^t dt' \, e^{i(E_f - E_i)t'/\hbar} \right|^2 \\
&= \frac{1}{\hbar^2} |\langle \psi_f | V(\mathbf{r}) | \psi_i \rangle|^2 \left| \frac{\hbar}{i(E_f - E_i)} \left( e^{i(E_f - E_i)t/\hbar} - 1 \right) \right|^2 \\
&= \frac{1}{\hbar^2} |\langle \psi_f | V(\mathbf{r}) | \psi_i \rangle|^2 \frac{4\hbar^2}{(E_f - E_i)^2} \sin^2 \left( \frac{(E_f - E_i)t}{2\hbar} \right) \quad (12.4)
\end{aligned}
$$

Although this expression is formally correct, there are several further steps before this result can be used to derive an actual scattering cross section.

First, the wave functions must be normalized. This represents a problem, since a function of the form $e^{i\mathbf{k} \cdot \mathbf{r}}$ does not have a well-defined integral over all of space. To resolve this, we put a box of volume $\mathcal{V}$ around the incident and scattered wave functions. This may seem like a bit of a fraud, but as long as the volume of the box is much larger than the system under consideration, it cannot affect the final results. In this case the normalized wave functions become

$$\psi_i = \frac{1}{\sqrt{\mathcal{V}}} e^{i\mathbf{k}_i \cdot \mathbf{r}}$$

$$\psi_f = \frac{1}{\sqrt{\mathcal{V}}} e^{i\mathbf{k}_f \cdot \mathbf{r}}$$

since, for example,

$$\int_{\mathcal{V}} \psi_i^* \psi_i \, d^3\mathbf{r} = \int_{\mathcal{V}} \left(\frac{1}{\sqrt{\mathcal{V}}}\right)^2 e^{-i\mathbf{k}_i \cdot \mathbf{r}} e^{i\mathbf{k}_i \cdot \mathbf{r}} \, d^3\mathbf{r} = 1$$

Thus, the inner product in Equation (12.4) can be written as

$$\langle \psi_f | V(\mathbf{r}) | \psi_i \rangle = \int_{\mathcal{V}} d^3\mathbf{r} \frac{1}{\sqrt{\mathcal{V}}} e^{-i\mathbf{k}_f \cdot \mathbf{r}} V(r) \frac{1}{\sqrt{\mathcal{V}}} e^{i\mathbf{k}_i \cdot \mathbf{r}}$$

$$= \frac{1}{\mathcal{V}} \int_{\mathcal{V}} d^3\mathbf{r} \, V(\mathbf{r}) e^{i(\mathbf{k}_i - \mathbf{k}_f) \cdot \mathbf{r}}$$

and the transition probability from Equation (12.4) becomes

$$P(i \rightarrow f) = \frac{4}{(E_f - E_i)^2} \sin^2\left(\frac{(E_f - E_i)t}{2\hbar}\right) \frac{1}{\mathcal{V}^2} \left| \int_{\mathcal{V}} d^3\mathbf{r} \, V(\mathbf{r}) e^{i(\mathbf{k}_i - \mathbf{k}_f) \cdot \mathbf{r}} \right|^2$$

$$(12.5)$$

Now we come to another problem: Equation (12.5) gives the transition probability from a specific initial wave vector $\mathbf{k}_i$ with energy $\hbar^2 k_i^2 / 2m$ to a specific final wave vector $\mathbf{k}_f$ with energy $\hbar^2 k_f^2 / 2m$. In a physical scattering problem, it is certainly possible to control the initial energy, but not the final energy. Hence, Equation (12.5) must be integrated over all possible final energies $E_f$. In doing this, however, there is an additional complication: there are more quantum states corresponding to higher energies than to lower ones. For instance, recall from Chapter 6 that a particle in a cubic box of side $a$ has an energy given by

$$E = \frac{\hbar^2 \pi^2}{2ma^2} [n_x^2 + n_y^2 + n_z^2] \tag{12.6}$$

If the energy is fixed to lie in the small range between $E$ and $E + dE$, then the number of different values of $n_x$, $n_y$, and $n_z$ that produce an energy in this range increases as $E$ increases. To quantify this, consider the three-dimensional "space" defined by $n_x$, $n_y$, and $n_z$ (Figure 12.7). We now fix the energy $E$ and calculate how many states have energies between $E$ and $E + dE$. This number is given by 1/8 of the volume of a thin spherical shell with radius $n$, where

$$n^2 = n_x^2 + n_y^2 + n_z^2$$

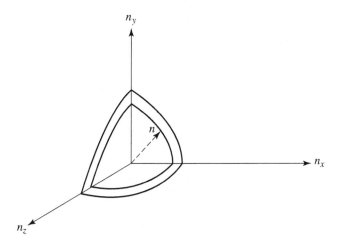

Fig. 12.7 The number of states with energies between $E$ and $E + dE$ is given by $1/8$ of the volume of a thin spherical shell with radius $n$, where $E = (\hbar^2\pi^2/2ma^2)n^2$.

and the factor of $1/8$ arises because the physical states all lie in the octant defined by $n_x > 0$, $n_y > 0$, and $n_z > 0$. If $N(E)\, dE$ gives this number of states, we get

$$N(E)\, dE = \frac{1}{8}(4\pi n^2\, dn) \tag{12.7}$$

Further, Equation (12.6) allows us to express $n$ in terms of $E$:

$$E = \frac{\hbar^2\pi^2}{2ma^2}n^2 \tag{12.8}$$

and using Equation (12.8) to rewrite the right-hand side of Equation (12.7) in terms of $E$ rather than $n$, we get

$$N(E)\, dE = \frac{\mathcal{V}}{4\pi^2}\frac{(2m)^{3/2}}{\hbar^3}\sqrt{E}\, dE \tag{12.9}$$

In this equation, $N(E)$ is called the *density of states*. Physically, Equation (12.9) says that there are more ways for a system to have a large energy than a small energy (and the density of states scales as the square root of the energy).

Now Equation (12.5) can be multiplied by $N(E_f)\, dE_f$ and integrated over the final possible energies. However, we also have to include the fact that the density of final states is proportional to $d\Omega_f/4\pi$. Hence, the

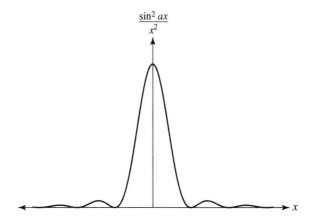

Fig. 12.8   In the limit $a \to \infty$, the function $\sin^2 ax/x^2$ becomes sharply peaked and can be approximated as a multiple of the delta function.

transition probability $dP$ for scattering into a small solid angle $d\Omega_f$ is

$$
dP(i \to \text{any } E) = \int P(i \to f) N(E_f)\, dE_f \frac{d\Omega_f}{4\pi}
$$

$$
= \int \frac{4}{(E_f - E_i)^2} \sin^2 \left( \frac{(E_f - E_i)t}{2\hbar} \right) \frac{1}{\mathcal{V}^2} \left| \int_{\mathcal{V}} d^3\mathbf{r}\, V(\mathbf{r}) e^{i(\mathbf{k}_i - \mathbf{k}_f)\cdot\mathbf{r}} \right|^2
$$

$$
\times \frac{\mathcal{V}}{4\pi^2} \frac{(2m)^{3/2}}{\hbar^3} \sqrt{E_f}\, dE_f \frac{d\Omega_f}{4\pi} \tag{12.10}
$$

Note the factor $\sin^2[(E_f - E_i)t/2\hbar]/(E_f - E_i)^2$ in Equation (12.10). The generic function $\sin^2(ax)/x^2$ becomes arbitrarily sharply peaked as $a \to \infty$ (Figure 12.8). More rigorously,

$$
\frac{\sin^2 ax}{x^2} \to \pi a \delta(x), \quad \text{as } a \to \infty \tag{12.11}
$$

Since we are interested in the behavior of this system in the limit where $t$ becomes large, we can use Equation (12.11) to write

$$
\frac{\sin^2[(E_f - E_i)t/2\hbar]}{(E_f - E_i)^2} \to \pi(t/2\hbar)\delta(E_f - E_i)
$$

When we substitute this delta function into Equation (12.10), the integration over $E_f$ picks out the value $E_f = E_i$. This makes sense on physical grounds; in the limit where perturbation theory is applicable, scattering off of the potential should not change the energy of the incident particle by very much, which implies that $E_f \approx E_i$. After performing this integration and taking $E_i = \hbar^2 k_i^2/2m$, we obtain

$$dP(i \to \text{any } E) = \frac{mtk_i}{\pi\hbar^3\mathcal{V}} \left| \int_{\mathcal{V}} d^3\mathbf{r}\, V(\mathbf{r})e^{i(\mathbf{k}_i - \mathbf{k}_f)\cdot\mathbf{r}} \right|^2 \left(\frac{d\Omega_f}{4\pi}\right) \tag{12.12}$$

with the added restriction that $|\mathbf{k}_f| = |\mathbf{k}_i|$, since the delta function integration gave $E_f = E_i$.

There remains one final task: determining the relationship between the transition probability $dP$ in Equation (12.12) and the differential cross section $d\sigma/d\Omega$. At the beginning of this section, we derived the relationship between the total scattering cross section and the scattering probability in terms of the total distance $L$ travelled by the incident particle (Equation (12.1)). If the incident particle has velocity $v_i$ and travels for a time $t$, we can substitute $L = v_i t$ into Equation (12.1) to obtain an alternative relation between scattering probability and cross section

$$P = nv_i t\sigma$$

For scattering into a small solid angle $d\Omega_f$, the corresponding relation is

$$dP = nv_i t\frac{d\sigma}{d\Omega_f}d\Omega_f$$

so that the differential cross section is

$$\frac{d\sigma}{d\Omega} = \frac{dP}{nv_i t\, d\Omega_f} \tag{12.13}$$

Substituting the expression for $dP$ given by Equation (12.12) into Equation (12.13), using $v_i = p_i/m = \hbar k_i/m$ and assuming one target per volume $\mathcal{V}$ so that $n = 1/\mathcal{V}$, we get a final expression for the differential cross section:

$$\frac{d\sigma}{d\Omega} = \left(\frac{m}{2\pi\hbar^2}\right)^2 \left| \int d^3\mathbf{r}\, V(\mathbf{r})e^{i(\mathbf{k}_i - \mathbf{k}_f)\cdot\mathbf{r}} \right|^2 \tag{12.14}$$

where we have now taken the limit where $\mathcal{V}$ goes to infinity, so the integral is over all space. Equation (12.14) is called the *Born approximation*. Furthermore, when this approximation is valid, it will always be true that the incident wave vector $\mathbf{k}_i$ and the scattered wave vector $\mathbf{k}_f$ satisfy $|\mathbf{k}_i| = |\mathbf{k}_f|$.

The Born approximation can be expressed in a more compact form by defining a new wave vector $\mathbf{K}$ (Figure 12.9):

$$\mathbf{K} = \mathbf{k}_f - \mathbf{k}_i$$

where $\mathbf{K}$ is called the *momentum transfer* because $\hbar\mathbf{K}$ gives the change in the momentum of the particle as it scatters. Taking $\theta$ to be the scattering

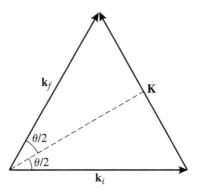

Fig. 12.9  The momentum transfer $\mathbf{K} = \mathbf{k}_f - \mathbf{k}_i$ gives the change in momentum $\hbar\mathbf{K}$ of the particle as it scatters. This figure shows that $K = 2k\sin(\theta/2)$.

angle, i.e., the angle between $\mathbf{k}_f$ and $\mathbf{k}_i$, and taking $k = |\mathbf{k}_i| = |\mathbf{k}_f|$, Figure 12.9 shows that

$$K = 2k\sin(\theta/2) \tag{12.15}$$

In terms of the momentum transfer, the Born approximation is

$$\frac{d\sigma}{d\Omega} = \left(\frac{m}{2\pi\hbar^2}\right)^2 \left| \int d^3\mathbf{r}\, V(\mathbf{r})e^{-i\mathbf{K}\cdot\mathbf{r}} \right|^2 \tag{12.16}$$

When the scattering potential is spherically symmetric $[V(\mathbf{r}) = V(r)]$, the expression for the cross section can be further simplified. In this case we can expand the Born approximation out in spherical coordinates and perform the integrals over $\theta$ and $\phi$. Equation (12.16) becomes

$$\frac{d\sigma}{d\Omega} = \left(\frac{m}{2\pi\hbar^2}\right)^2 \left| \int \sin\theta\, d\theta\, d\phi\, r^2\, dr\, V(r)e^{-i\mathbf{K}\cdot\mathbf{r}} \right|^2$$

The integral over $\phi$ just gives $2\pi$, and the integral over $\theta$ can be simplified by choosing a coordinate system so that the polar axis points in the direction of $\mathbf{K}$; in this case $\mathbf{K}\cdot\mathbf{r} = Kr\cos\theta$ (note that this is *not* the same $\theta$ that appears in Equation (12.15) and Figure 12.9). This gives

$$\frac{d\sigma}{d\Omega} = \left(\frac{m}{2\pi\hbar^2}\right)^2 \left| \int \sin\theta\, d\theta\, d\phi\, r^2\, dr\, V(r)e^{-iKr\cos\theta} \right|^2$$

Integrating over both $\phi$ and $\theta$, the expression for the cross section simplifies to

$$\frac{d\sigma}{d\Omega} = \frac{4m^2}{\hbar^4 K^2} \left( \int_{r=0}^{\infty} \sin(Kr)V(r)r\, dr \right)^2 \tag{12.17}$$

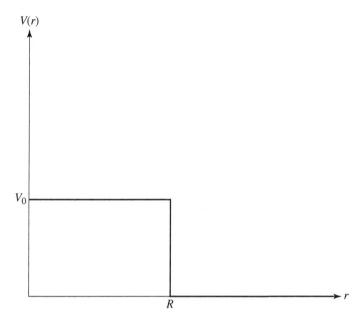

Fig. 12.10  A repulsive spherical well has $V = V_0$ inside a sphere of radius $R$, and $V = 0$ outside of this sphere.

Equation (12.17) is the form of the Born approximation applicable to any spherically-symmetric potential.

---

### Example 12.2. Scattering from a Three-Dimensional Repulsive Spherical Well

A particle scatters from the three-dimensional repulsive spherical well of radius $R$ given by (Figure 12.10):

$$V(r) = V_0, \quad r \leq R$$
$$V(r) = 0, \quad r > R$$

The energy of the particle is much greater than $V_0$. Use the Born approximation to calculate the differential cross section.

This is a spherically-symmetric potential, so we can use the form of the Born approximation given by Equation (12.17). This gives

$$\frac{d\sigma}{d\Omega} = \frac{4m^2}{\hbar^4 K^2} \left( \int_{r=0}^{R} \sin(Kr) V_0 r \, dr \right)^2$$

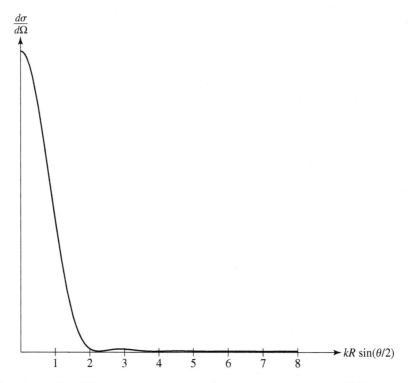

Fig. 12.11　The differential cross section $d\sigma/d\Omega$ as a function of $kR\sin(\theta/2)$ for a repulsive spherical square-well potential of radius $R$, where $k$ is the wave number of the incident particle and $\theta$ is the scattering angle.

Integrating over $r$,

$$\frac{d\sigma}{d\Omega} = \frac{4m^2 V_0^2}{\hbar^4 K^2}\left(\frac{\sin(KR)}{K^2} - \frac{R\cos(KR)}{K}\right)^2$$

Normally, $d\sigma/d\Omega$ is expressed in terms of the scattering angle $\theta$; recall that the magnitude of the momentum transfer $K$ and the scattering angle $\theta$ are related through $K = 2k\sin(\theta/2)$, where $k$ is the magnitude of both the incident and scattered wave vector. Using this expression for $K$,

$$\frac{d\sigma}{d\Omega} = \frac{4m^2}{\hbar^4} V_0^2 R^6 \left(\frac{\sin[2kR\sin(\theta/2)] - 2kR\sin(\theta/2)\cos[2kR\sin(\theta/2)]}{[2kR\sin(\theta/2)]^3}\right)^2$$

A graph of the differential cross section as a function of $kR\sin(\theta/2)$ is given in Figure 12.11. The differential cross section has a large central peak at small values of $kR\sin(\theta/2)$ with tiny oscillations (barely visible on the scale of this figure) at larger values.

Here is another example of the Born approximation, this time with a potential that is not spherically symmetric.

---

**Example 12.3. Scattering from a Delta-Function Potential**
(a) A particle of mass $m$ scatters off of a delta-function potential at the origin:

$$V(x, y, z) = \frac{a\hbar^2}{m}\delta(x)\delta(y)\delta(z)$$

(Here $a$ is a constant with units of length, and the constant in front of the delta functions ensures that $V$ has units of energy.) Use the Born approximation to calculate $d\sigma/d\Omega$.

(b) Repeat this calculation with the delta-function potential located at the point $(b, 0, 0)$, so

$$V(x, y, z) = \frac{a\hbar^2}{m}\delta(x - b)\delta(y)\delta(z)$$

(c) Now suppose that there are delta-function potentials at both $(0, 0, 0)$ and $(b, 0, 0)$, so that

$$V(x, y, z) = \frac{a\hbar^2}{m}[\delta(x - b)\delta(y)\delta(z) + \delta(x)\delta(y)\delta(z)]$$

Calculate $d\sigma/d\Omega$ in this case, and compare it to the sum of the cross sections in (a) and (b).

(a) Since we are dealing with potentials which are not symmetric about the origin, we use the full Born approximation, Equation (12.16):

$$\frac{d\sigma}{d\Omega} = \left(\frac{m}{2\pi\hbar^2}\right)^2 \left| \int d^3\mathbf{r}\, V(\mathbf{r})e^{-i\mathbf{K}\cdot\mathbf{r}} \right|^2$$

Substituting the first delta-function potential from part (a) gives

$$\frac{d\sigma}{d\Omega} = \left(\frac{m}{2\pi\hbar^2}\right)^2 \left| \int dx\, dy\, dz \frac{a\hbar^2}{m}\delta(x)\delta(y)\delta(z)e^{-i(K_x x + K_y y + K_z z)} \right|^2$$

The integral over the delta function picks out the value $x = 0$, $y = 0$, $z = 0$ in the exponential, giving $e^0 = 1$, and the cross section reduces to

$$\frac{d\sigma}{d\Omega} = \frac{a^2}{4\pi^2}$$

Note that the cross section has units of area, as expected.

(b) For the delta function at $(b, 0, 0)$, we have

$$\frac{d\sigma}{d\Omega} = \left(\frac{m}{2\pi\hbar^2}\right)^2 \left| \int dx\, dy\, dz \frac{a\hbar^2}{m} \delta(x - b)\delta(y)\delta(z) e^{-i(K_x x + K_y y + K_z z)} \right|^2$$

$$= \frac{a^2}{4\pi^2} \left| e^{-iK_x b} \right|^2$$

$$= \frac{a^2}{4\pi^2}$$

i.e., the cross section is unchanged when the delta-function potential is moved to a different position.

(c) Now consider the case with both delta functions together:

$$\frac{d\sigma}{d\Omega} = \left(\frac{m}{2\pi\hbar^2}\right)^2 \left| \int dx\, dy\, dz \frac{a\hbar^2}{m} [\delta(x - b)\delta(y)\delta(z) \right.$$

$$\left. + \delta(x)\delta(y)\delta(z)] e^{-i(K_x x + K_y y + K_z z)} \right|^2$$

$$= \frac{a^2}{4\pi^2} \left| e^{-iK_x b} + e^{i(0)} \right|^2$$

Using the identity

$$|1 + e^{ix}|^2 = 4\cos^2(x/2)$$

the cross section reduces to

$$\frac{d\sigma}{d\Omega} = \left(\frac{a^2}{\pi^2}\right) \cos^2\left(\frac{K_x b}{2}\right)$$

Now we see an interesting result: the cross section calculated in part (c) is not the sum of the cross sections in (a) and (b). The reason is the wave nature of the scattering particle: just as for optical scattering, there is *interference* between the waves scattering off of the two delta functions, so the individual cross sections from the two delta functions cannot simply be added together.

---

As noted several times, the Born approximation is valid when the potential can be treated as a small perturbation, i.e., when the energy of the incident particle is much greater than $V$. In the next section we consider the opposite limit of low-energy scattering.

## 12.3 Partial Waves

Although the method of partial waves is simplest when the energy of the incident particle is low, we do not need to introduce this assumption immediately. Suppose we have an incident particle moving in the $z$ direction, and it scatters off of a spherically-symmetric potential $V(r)$ centered at the origin. As in the previous section, we will assume that $V(r) \to 0$ as $r \to \infty$. The time-independent wave function for the incident particle far from the origin is

$$\psi_i = e^{ikz}$$

but what about the wave function $\psi_f$ for the scattered particle? In general, the scattered particle is given by a wave expanding radially outward from the scattering potential with an amplitude that depends on the angular direction.

To construct the wave function corresponding to such a wave, consider first the case of perfect spherical symmetry, i.e., a wave expanding radially outward with equal amplitude in all angular directions. This wave represents a solution to the radial Schrödinger equation (Equation (6.40)):

$$-\frac{\hbar^2}{2m}\frac{\partial^2}{\partial r^2}(rR(r)) + \frac{\hbar^2 l(l+1)}{2mr^2}rR(r) + V(r)rR(r) = ErR(r)$$

Since we are interested in a freely-expanding wave, we take $V = 0$, and the condition that the wave be spherically symmetric implies that $l = 0$. Further, we would like to express the particle momentum in terms of $k$, rather than $E$, where $E = \hbar^2 k^2/2m$. Then the radial Schrödinger equation becomes

$$\frac{\partial^2}{\partial r^2}(rR) + k^2(rR) = 0$$

which has the general solution

$$R = A\frac{e^{ikr}}{r} + B\frac{e^{-ikr}}{r}$$

where $A$ and $B$ are constants to be determined. The first term represents a radially outgoing wave, while the second term represents a radially incoming wave. Thus, for a scattered particle, the second term makes no sense on physical grounds, so $B$ must be zero, giving

$$R = A\frac{e^{ikr}}{r} \tag{12.18}$$

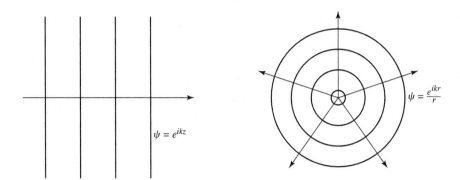

Fig. 12.12   The wave function $\psi = e^{ikz}$ represents a wave moving in the $z$ direction; $\psi = e^{ikr}/r$ represents a spherically-symmetric wave moving radially outward.

Equation (12.18) represents the radial equivalent of a plane wave. Just as $e^{ikz}$ gives a wave moving in the $z$ direction, the quantity $e^{ikr}/r$ represents a spherically-symmetric wave moving radially outward (Figure 12.12).

Now, if a particle scatters from the potential, the outgoing wave need no longer be isotropic. Since we have assumed a spherically-symmetric potential, the amplitude of the scattered wave can be a function of $\theta$, but it should be independent of $\phi$. The most general form we can write for a radially-expanding wave with a dependence on $\theta$ is

$$\psi_f = f(\theta)\frac{e^{ikr}}{r}$$

and the total wave function far from the origin, including both the incident and scattered particle, will be

$$\psi_T = \psi_i + \psi_f$$
$$= e^{ikz} + f(\theta)\frac{e^{ikr}}{r} \tag{12.19}$$

The cross section is a function of $f(\theta)$. To determine this function, we express the cross section in terms of scattered energy rather than discrete particles:

$$\frac{d\sigma}{d\Omega} = \frac{\text{scattered energy/solid angle}}{\text{incident energy}}$$
$$= \frac{r^2|\psi_f|^2}{|\psi_i|^2}$$

Substituting our expressions for $\psi_i$ and $\psi_f$ gives

$$\frac{d\sigma}{d\Omega} = |f(\theta)|^2$$

So the problem of determining the differential cross section reduces to solving the Schrödinger equation in order to find $f(\theta)$.

We will derive such a solution in the region far from the potential, so that $V = 0$. The radial Schrödinger equation for $V = 0$ can be written as

$$\frac{\partial}{\partial r^2}(rR(r)) - \frac{l(l+1)}{r}R(r) + k^2(rR(r)) = 0$$

This equation can be solved exactly for any value of $l$; the solutions for $R(r)$ are called *spherical Bessel functions* and are written as $j_l(kr)$. These solutions, for the first few values of $l$, are

$$l = 0: \quad R = j_0(kr) = \frac{\sin(kr)}{kr}$$

$$l = 1: \quad R = j_1(kr) = \frac{\sin(kr)}{(kr)^2} - \frac{\cos(kr)}{kr}$$

$$l = 2: \quad R = j_2(kr) = \left[\frac{3}{(kr)^3} - \frac{1}{kr}\right]\sin(kr) - \frac{3\cos(kr)}{(kr)^2}$$

As usual, the general solution to the Schrödinger equation is the product of this radial solution and the appropriate spherical harmonic:

$$\psi(r, \theta, \phi) = j_l(kr)Y_l^m(\theta, \phi) \tag{12.20}$$

However, this result leads to a puzzle. We already have a set of solutions to the Schrödinger equation with $V = 0$, namely,

$$\psi = e^{i\mathbf{k}\cdot\mathbf{r}} \tag{12.21}$$

Thus, it would appear that there are two different sets of solutions to the three-dimensional Schrödinger equation for the case of $V = 0$. In fact, both sets of solutions [Equations (12.20)–(12.21)] are valid. They simply represent solutions in spherical and rectangular coordinate systems, respectively. Furthermore, either set of solutions can be used as a basis set. This means, for example, that an incoming wave in the $z$ direction can be written as a sum of spherical waves:

$$e^{ikz} = \sum_{l,m} c_{lm} j_l(kr)Y_l^m(\theta, \phi) \tag{12.22}$$

where $c_{lm}$'s are the constants in the expansion.

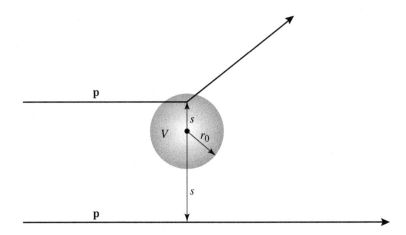

Fig. 12.13    A particle with momentum $p$ scatters off of a potential $V$ of radius $r_0$. The potential will affect the particle only if the point of closest approach $s$ is smaller than $r_0$.

Physically, Equation (12.22) corresponds to expressing a plane wave as the sum of wave functions with different angular momenta. This expansion is useful because we can now show that, in the low-energy limit, it can be a good approximation to retain only the $l = 0$ term. To see the reason for this, consider a particle scattering off of a potential with fixed radius $r_0$, so that $V$ is negligible for $r > r_0$. The spherical Bessel functions have the property that for $kr \ll l$,

$$j_l(kr) \propto (kr)^l$$

Thus, if $k$ is sufficiently small that $kr_0 \ll 1$, all of the spherical Bessel functions are negligible in the vicinity of the potential $(r < r_0)$ except for $l = 0$. In this case only the $l = 0$ component of the incident plane wave "feels" the potential.

This argument has a simple classical analog. Consider a particle with momentum $p$ scattering off of a potential that vanishes for $r > r_0$ (Figure 12.13). Let $s$ be the closest distance that the particle attains relative to the center of the potential. Since the potential is negligible for distances greater than $r_0$, scattering can occur only if $s < r_0$. But this immediately tells us something about the angular momentum of the particle relative to the origin. The angular momentum is $L = ps$, and the requirement that $s < r_0$ translates into the relation

$$L < pr_0$$

Thus, in the classical case, low-energy scattering also translates into low angular momentum scattering.

We can assume then that in the low-energy limit, only the $l = 0$ component of the incoming wave is scattered and the higher $l$ waves are unaffected. Therefore, this approximation is called *s-wave scattering* where $s$ indicates that $l = 0$. In this limit we write the sum in Equation (12.22) as

$$e^{ikz} = c_{00}j_0(kr)Y_0^0(\theta, \phi) + \sum_{l>0} c_{lm}j_l(kr)Y_l^m(\theta, \phi) \qquad (12.23)$$

To find $c_{00}$ we multiply both sides of Equation (12.23) by $Y_0^{0*}$, set $z = r\cos\theta$ on the left-hand side, and integrate over $\theta$ and $\phi$. Because of the orthogonality of the spherical harmonics, only the first term on the right-hand side contributes, and we get

$$\int e^{ikr\cos\theta} Y_0^{0*} \sin\theta\, d\theta\, d\phi = \int c_{00}j_0(kr)|Y_0^0|^2 \sin\theta\, d\theta\, d\phi$$

Taking $Y_0^0 = 1/\sqrt{4\pi}$ and $j_0(kr) = \sin(kr)/kr$ and performing the integrations gives

$$\frac{1}{\sqrt{4\pi}}2\pi\frac{1}{ikr}\left(e^{ikr} - e^{-ikr}\right) = c_{00}\frac{\sin(kr)}{kr} \qquad (12.24)$$

Now we recall that $\sin(x) = (e^{ix} - e^{-ix})/2i$, so Equation (12.24) reduces to

$$c_{00} = \sqrt{4\pi}$$

Using this value for $c_{00}$ and writing $j_0 = \sin(kr)/kr$ in terms of complex exponentials, we can express Equation (12.23) as

$$e^{ikz} = \frac{1}{2i}\left(\frac{e^{ikr}}{kr} - \frac{e^{-ikr}}{kr}\right) + \sum_{l>0}\text{terms}$$

Note that this represents the *incident* wave $\psi_i$. What does the scattered wave look like? Since we have assumed that only the $l = 0$ part of the wave is actually affected by the potential, the only contribution of the scattering will be an outgoing spherical wave with $l = 0$. Thus, the total wave function $\psi_T = \psi_i + \psi_f$ will be

$$\psi_T = \frac{1}{2i}\left(\eta_0\frac{e^{ikr}}{kr} - \frac{e^{-ikr}}{kr}\right) \qquad (12.25)$$

where we have dropped the $l > 0$ terms, and the effect of the scattering has been to add a term proportional to $e^{ikr}/kr$. We do not yet know the amplitude of this additional term, so we have absorbed it into a new unknown complex number $\eta_0$, which we need to calculate.

Conservation of particle probability means that $|\eta_0|^2 = 1$, so it is conventional to re-express $\eta_0$ in terms of a function with unit magnitude

$$\eta_0 = e^{2i\delta_0} \tag{12.26}$$

where $\delta_0$, defined by Equation (12.26), is called the *s-wave phase shift*. In order to calculate a cross section, we need to write $\psi_T$ in Equation (12.25) in the form of an incident plane wave and a scattered spherical wave (as in Equation (12.19)). Pulling out the terms that correspond to the incident plane wave $e^{ikz}$ and expressing $\eta_0$ in terms of $\delta_0$, we get

$$\psi_T = e^{ikz} + \left(\frac{e^{2i\delta_0} - 1}{2ik}\right)\frac{e^{ikr}}{r}$$

Then the cross section is

$$\frac{d\sigma}{d\Omega} = |f(\theta)|^2$$
$$= \left|\frac{e^{2i\delta_0} - 1}{2ik}\right|^2$$
$$= \frac{\sin^2\delta_0}{k^2}$$

Note that the differential cross section in this case is completely isotropic, i.e., independent of the scattering angle. This is because of our assumption of *s*-wave scattering; in this limit only the $l = 0$ part of the wave undergoes scattering, and the $l = 0$ wave is isotropic. The total cross section is then just

$$\sigma = 4\pi\frac{\sin^2\delta_0}{k^2} \tag{12.27}$$

Of course, the problem now is to find the phase shift $\delta_0$; this calculation is performed by solving the Schrödinger equation for the scattering potential, as shown here.

---

### Example 12.4. Low-Energy Scattering from an Infinitely Hard Sphere

A particle is incident on a central potential $V(r)$ which is infinitely high at $r \leq a$ with $V = 0$ for $r > a$. The energy of the particle is sufficiently low that *s*-wave scattering is a good approximation. Find the total cross section.

The wave function outside the potential is given by Equation (12.25):

$$\psi_T = \frac{1}{2i} \left( \eta_0 \frac{e^{ikr}}{kr} - \frac{e^{-ikr}}{kr} \right)$$

$$= \frac{1}{2i} \left( \frac{e^{ikr+2i\delta_0}}{kr} - \frac{e^{-ikr}}{kr} \right)$$

At the surface of the potential, $r = a$, the infinite potential forces the wave function to zero, so $\psi_T(r = a) = 0$. This means that

$$e^{ika+2i\delta_0} - e^{-ika} = 0$$

which has the solution $\delta_0 = -ka$. Then the $s$-wave expression for the cross section, Equation (12.27), gives

$$\sigma = 4\pi \frac{\sin^2 ka}{k^2}$$

In the low-energy limit, $k \to 0$, and we get

$$\sigma = 4\pi a^2$$

Note that this is four times the classical scattering cross section from a hard sphere (Example 12.1). In the classical case, the incident particle "sees" the geometrical cross-sectional area of the sphere, which is just $\pi a^2$. In the quantum mechanical system, the incident particle acts like a wave and diffracts around the target, giving a larger cross section.

---

The idea of $s$-wave scattering can be extended by summing over all of the $l$'s in the spherical wave expansion and finding the phase shift $\delta_l$ for each partial wave. The result is a total cross section which looks like

$$\sigma = \frac{4\pi}{k^2} \sum_{l=0}^{\infty} (2l + 1) \sin^2 \delta_l$$

The case of $s$-wave scattering, which is the only one we have examined in detail, can be obtained from this result by taking only the $l = 0$ term in the series.

## PROBLEMS

**12.1** Show that in the Born approximation, the differential cross section obtained from the negative of a given potential $-V(\mathbf{r})$ is exactly the same as that obtained from the potential $V(\mathbf{r})$.

**12.2** An incident particle with mass $m$, velocity $v$, and charge $ze$ scatters off of a charge $Ze$ at the origin. Use the Born approximation to calculate the differential scattering cross section for the screened Coulomb potential

$$V(r) = (zZe^2/4\pi\epsilon_0 r)e^{-r/d}$$

Then let $d \to \infty$, so that $V(r)$ approaches the normal Coulomb potential, and show that $d\sigma/d\Omega$ approaches the Rutherford scattering differential cross section

$$\frac{d\sigma}{d\Omega} = \left(\frac{1}{4\pi\epsilon_0}\right)^2 \left(\frac{zZe^2}{2mv^2}\right)^2 \frac{1}{\sin^4(\theta/2)}$$

**12.3** (a) A particle with charge $+e$ is incident on an electric dipole consisting of a charge of $+e$ and a charge of $-e$ separated by the vector $\mathbf{d}$ (which runs from $-e$ to $+e$). The energy of the incident particle is sufficiently large to treat the dipole as a small perturbation. Calculate the differential scattering cross section $d\sigma/d\Omega$ as a function of the initial wave vector $\mathbf{k}_i$, the scattered wave vector $\mathbf{k}_f$, and the standard Rutherford scattering cross section $(d\sigma/d\Omega)_R$, given by

$$\left(\frac{d\sigma}{d\Omega}\right)_R = \left| -\frac{m}{2\pi\hbar^2} \int d^3\mathbf{r}\, V_c(\mathbf{r}) e^{-i\mathbf{K}\cdot\mathbf{r}} \right|^2$$

where $\mathbf{K} = \mathbf{k}_f - \mathbf{k}_i$ and $V_c(\mathbf{r})$ is the Coulomb potential.
(b) In the limits $k_i d \ll 1$ and $k_i d \gg 1$, determine whether the dipole differential cross section is larger or smaller than the Rutherford differential cross section. Explain the physical reason for these results.

**12.4** A particle which is travelling in the $+z$ direction scatters off of a potential consisting of four delta functions at the vertices of a square in the $x$-$y$ plane at the points $(-a, 0, 0)$, $(+a, 0, 0)$, $(0, -a, 0)$, and $(0, +a, 0)$.

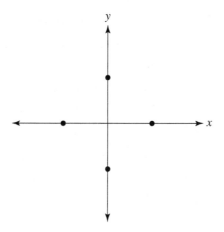

The potential is

$$V = A\delta(x+a)\delta(y)\delta(z) + A\delta(x-a)\delta(y)\delta(z)$$
$$+ A\delta(x)\delta(y+a)\delta(z) + A\delta(x)\delta(y-a)\delta(z)$$

where $A$ is a constant. Use the Born approximation to calculate the differential scattering cross section. Express the answer in terms of the magnitude of the incident wave vector $k$ and the scattering angles $\theta$ and $\phi$. (This is an example where the cross section *does* depend on $\phi$.)

**12.5** (a) A particle of mass $m$ and energy $E$ scatters off of the central potential $V(r) = Ar^{-2}$, where $A$ is a constant. Use the Born approximation to calculate the differential cross section $d\sigma/d\Omega$ as a function of $E$ and the scattering angle $\theta$.
(b) Show that the total cross section $\sigma$ is infinite.

**12.6** A particle of mass $m$ is incident on the potential $V(r) = V_0 e^{-r/r_0}$, where $V_0$ and $r_0$ are constants with units of energy and length, respectively. The potential is independent of $\theta$ and $\phi$. The energy of the particle is large, so that the potential can be treated as a small perturbation. Calculate the differential scattering cross section $d\sigma/d\Omega$. Express the final answer as a function of the scattering angle $\theta$ and the energy of the particle $E$.

**12.7** (a) A particle with mass $m$ scatters off of the potential

$$V = A\delta(z), \quad \text{for } -a \leq x \leq a \text{ and } -a \leq y \leq a$$
$$V = 0, \quad \text{otherwise}$$

where $A$ is a constant. In other words, this potential forms a square in the $x$-$y$ plane and is infinitesimally thin in the $z$ direction. Use the Born approximation to calculate the differential scattering cross section. (Assume an arbitrary direction for the incident particle, and express the answer in terms of the momentum transfer vector $\mathbf{K} = \mathbf{k}_f - \mathbf{k}_i$.)

(b) Show that in the limit where $a \to \infty$ (so that the scattering potential occupies the entire $x$-$y$ plane), the resulting differential cross section corresponds to only two possible results: either the particle will pass through without any scattering or else it will scatter with the angle of incidence equal to the angle of reflection.

**12.8** Suppose a particle with mass $m$ scatters off of a finite spherical potential of radius $R_0$ given by

$$V(r) = V_0 \quad (r \leq R_0)$$
$$V(r) = 0 \quad (r > R_0)$$

In the limit where the energy of the incident particle is small, show that the total cross section is

$$\sigma = 4\pi R_0^2 \left( \frac{\tanh(kR_0)}{kR_0} - 1 \right)^2$$

where $k = \sqrt{2m(V_0 - E)/\hbar^2}$.

Chapter 13

# Multiparticle Schrödinger equation

Thus far, we have examined the solution of the Schrödinger equation for a single particle. However, the real world consists of systems composed of many particles. While it is straightforward to generalize the Schrödinger equation to systems of many particles, something very interesting happens when dealing with systems of *identical* particles: the requirement that two particles be treated as identical imposes certain restrictions on the properties of the wave function.

When dealing with systems containing multiple particles, we will not write down a separate wave function for each particle. Instead, we will have a single wave function that encodes the information about all of the particles together. We will show that for systems of identical particles, the wave function is either unchanged or multiplied by $-1$ when any two of the particles are exchanged. This, in turn, leads to an important result called the *Pauli Exclusion Principle*, which turns out to be crucial to the very existence of matter as we know it.

## 13.1 Wave Function for Identical Particles

For a single particle with definite energy $E$, the Schrödinger equation has the familiar form

$$-\frac{\hbar^2}{2m}\nabla^2\psi(\mathbf{r}) + V(\mathbf{r})\psi(\mathbf{r}) = E\psi(\mathbf{r})$$

Now suppose that we have two particles, not necessarily identical. In order to treat these two particles as a single system, we must have a single wave function which combines the information for both particles (i.e., instead of separate wave functions for each particle). This wave function is written as $\psi(\mathbf{r}_1, \mathbf{r}_2)$. The physical interpretation of this wave function in terms of

313

probabilities is similar to the one-particle wave function. Recall that for a single particle, $|\psi(\mathbf{r})|^2\, d^3\mathbf{r}$ gives the probability that the particle can be found in a small volume $d^3\mathbf{r}$ near $\mathbf{r}$. For the two-particle wave function, $|\psi(\mathbf{r}_1, \mathbf{r}_2)|^2\, d^3\mathbf{r}_1\, d^3\mathbf{r}_2$ gives the probability that particle 1 is located in the small volume $d^3\mathbf{r}_1$ near $\mathbf{r}_1$ and particle 2 is located in the small volume $d^3\mathbf{r}_2$ near $\mathbf{r}_2$.

---

**Example 13.1. Interpretation of the Two-Particle Wave Function**
Two particles are confined in an infinite one-dimensional square well with width $a$. They are in a state with a wave function given by

$$\psi(x_1, x_2) = \frac{2}{a} \sin\left(\frac{\pi x_1}{a}\right) \sin\left(\frac{\pi x_2}{a}\right)$$

A measurement is made of the positions of both particles. Calculate the probability that particle 1 is on the left-hand side of the potential well $(x < a/2)$ and particle 2 is on the right-hand side $(x > a/2)$.

This probability is

$$P = \int_{x_1=0}^{a/2} \int_{x_2=a/2}^{a} |\psi(x_1, x_2)|^2\, dx_1\, dx_2$$

$$= \int_{x_1=0}^{a/2} \int_{x_2=a/2}^{a} \frac{4}{a^2} \sin^2\left(\frac{\pi x_1}{a}\right) \sin^2\left(\frac{\pi x_2}{a}\right) dx_1\, dx_2$$

$$= \frac{4}{a^2} \left[\frac{x_1}{2} - \frac{a}{4\pi} \sin\left(\frac{2\pi x_1}{a}\right)\right]_{x_1=0}^{a/2} \left[\frac{x_2}{2} - \frac{a}{4\pi} \sin\left(\frac{2\pi x_2}{a}\right)\right]_{x_2=a/2}^{a}$$

$$= \frac{1}{4}$$

---

The two-particle wave function $\psi(\mathbf{r}_1, \mathbf{r}_2)$ obeys the Schrödinger equation in the form

$$-\frac{\hbar^2}{2m} \nabla_1^2 \psi(\mathbf{r}_1, \mathbf{r}_2) - \frac{\hbar^2}{2m} \nabla_2^2 \psi(\mathbf{r}_1, \mathbf{r}_2) + V(\mathbf{r}_1, \mathbf{r}_2)\psi(\mathbf{r}_1, \mathbf{r}_2) = E\psi(\mathbf{r}_1, \mathbf{r}_2)$$

$$\tag{13.1}$$

where the symbol $\nabla_1^2$ is the sum of second derivatives taken with respect to $\mathbf{r}_1$:

$$\nabla_1^2 = \frac{\partial^2}{\partial x_1^2} + \frac{\partial^2}{\partial y_1^2} + \frac{\partial^2}{\partial z_1^2}$$

and, similarly, for $\nabla_2^2$ the derivatives are taken with respect to $\mathbf{r}_2$:

$$\nabla_2^2 = \frac{\partial^2}{\partial x_2^2} + \frac{\partial^2}{\partial y_2^2} + \frac{\partial^2}{\partial z_2^2}$$

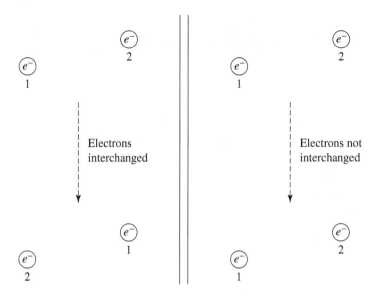

Fig. 13.1 Electron 1 and electron 2 are indistinguishable. It is therefore impossible to determine whether or not they have been interchanged.

and the potential $V$ is now a function of the positions of both particles. So far, this is a straightforward generalization of the one-particle wave function and one-particle Schrödinger equation. Now, however, we introduce a twist. Subatomic particles have the property that any two of the same kind of particle (two protons, two electrons, etc.) are *indistinguishable*. This means that they have exactly the same mass, the same spin, etc. In practical terms, consider the following experiment: two electrons are placed at two different locations in a laboratory (Figure 13.1). You turn your back, and the laboratory assistant either leaves the electrons in place or interchanges the two particles. When you again observe the electrons, there is literally *no way* to determine whether or not the electrons have been interchanged, since they are absolutely identical. This is a practical definition of indistinguishable particles. This property of indistinguishability has profound consequences.

To determine these consequences, define a new operator called the *exchange operator*, $\mathcal{E}_{12}$ which has the effect of interchanging particle 1 and particle 2. Thus, the effect of the exchange operator on the wave function is

$$\mathcal{E}_{12}\psi(\mathbf{r}_1, \mathbf{r}_2) = \psi(\mathbf{r}_2, \mathbf{r}_1)$$

It is possible to show (Problem 13.1) that the exchange operator commutes with the Hamiltonian as long as the two-particle potential has the property that

$$V(\mathbf{r}_1, \mathbf{r}_2) = V(\mathbf{r}_2, \mathbf{r}_1) \tag{13.2}$$

In fact, most reasonable potentials have this property; the most common situation is for two particles each to be subject to the same external potential, and to interact with each other via a potential that is symmetric under the interchange of the two particles. For example, each electron in the helium atom experiences the Coulomb potential of the nucleus, and the electrons also repel each other via a Coulomb potential, leading to the potential

$$V(\mathbf{r}_1, \mathbf{r}_2) = -\frac{2e^2}{4\pi\epsilon_0}\frac{1}{r_1} - \frac{2e^2}{4\pi\epsilon_0}\frac{1}{r_2} + \frac{e^2}{4\pi\epsilon_0}\frac{1}{|\mathbf{r}_1 - \mathbf{r}_2|} \tag{13.3}$$

Clearly, this potential satisfies Equation (13.2). We will assume throughout the remainder of this chapter that we are dealing exclusively with Hamiltonians that commute with the exchange operator.

Therefore, we can take the two-particle wave function to be an eigenfunction of $\mathcal{E}_{12}$ and attempt to determine the possible eigenvalues. Let $\gamma$ be an eigenvalue of $\mathcal{E}_{12}$:

$$\mathcal{E}_{12}\psi(\mathbf{r}_1, \mathbf{r}_2) = \gamma\psi(\mathbf{r}_1, \mathbf{r}_2)$$

Now note that for any wave function, applying $\mathcal{E}_{12}$ twice simply restores the original wave function; the particles are interchanged, then interchanged back to their original positions. Mathematically,

$$\mathcal{E}_{12}^2\psi(\mathbf{r}_1, \mathbf{r}_2) = \mathcal{E}_{12}\psi(\mathbf{r}_2, \mathbf{r}_1) = \psi(\mathbf{r}_1, \mathbf{r}_2) \tag{13.4}$$

But if $\psi(\mathbf{r}_1, \mathbf{r}_2)$ is an eigenfunction of $\mathcal{E}_{12}$ with eigenvalue $\gamma$, then applying $\mathcal{E}_{12}$ twice will pull out a factor of $\gamma^2$:

$$\mathcal{E}_{12}^2\psi(\mathbf{r}_1, \mathbf{r}_2) = \gamma^2\psi(\mathbf{r}_1, \mathbf{r}_2) \tag{13.5}$$

Combining Equations (13.4) and (13.5) gives the possible values for $\gamma$:

$$\gamma^2 = 1$$

so

$$\gamma = \pm 1$$

Therefore, the exchange operator can produce only two possible results when applied to the wave function. If $\gamma = 1$, then the eigenfunction equation gives $\mathcal{E}_{12}\psi(\mathbf{r}_1,\mathbf{r}_2) = \psi(\mathbf{r}_1,\mathbf{r}_2)$, while the definition of $\mathcal{E}_{12}$ gives $\mathcal{E}_{12}\psi(\mathbf{r}_1,\mathbf{r}_2) = \psi(\mathbf{r}_2,\mathbf{r}_1)$ so that

$$\psi(\mathbf{r}_1,\mathbf{r}_2) = \psi(\mathbf{r}_2,\mathbf{r}_1) \tag{13.6}$$

and the wave function is *symmetric* under the interchange of the two particles. Conversely, if $\gamma = -1$, then $\mathcal{E}_{12}\psi(\mathbf{r}_1,\mathbf{r}_2) = -\psi(\mathbf{r}_1,\mathbf{r}_2)$, and $\mathcal{E}_{12}\psi(\mathbf{r}_1,\mathbf{r}_2) = \psi(\mathbf{r}_2,\mathbf{r}_1)$ so

$$\psi(\mathbf{r}_1,\mathbf{r}_2) = -\psi(\mathbf{r}_2,\mathbf{r}_1) \tag{13.7}$$

and the wave function is *antisymmetric* under the interchange of the two particles.

It is observed in nature that any given particle obeys either Equation (13.6) or Equation (13.7) but not both. Particles with wave functions that are symmetric under particle interchange are said to obey *Bose–Einstein statistics* and are called *bosons*, while particles with wave functions that antisymmetric under particle interchange obey *Fermi–Dirac statistics* and are called *fermions*. It is further observed that the category that a given particle belongs to is entirely determined by its spin. Particles with integer spin ($s = 0,1,2,\ldots$) behave as bosons, while particles with half-integer spin ($s = 1/2,3/2,\ldots$) behave as fermions. Hence, the electron, proton, and neutron, each with spin $1/2$, are fermions, while the photon, with spin 1, is a boson.

This result leads immediately to the *Pauli exclusion principle*, which states that two fermions cannot occupy the same quantum state. For example, suppose that the two electrons in the helium atom have all of the same quantum numbers, and represent the two-particle wave function in Dirac notation as $|1\ \ 2\ \rangle$. If the two electrons have exactly the same quantum numbers, then the wave function is invariant when the two electrons are exchanged, i.e., $|1\ \ 2\ \rangle = |2\ \ 1\ \rangle$. However, this contradicts the fact that the wave function of two fermions must be antisymmetric under their interchange: $|1\ \ 2\ \rangle = -|2\ \ 1\ \rangle$. Hence, two fermions cannot occupy the exact same quantum state. If it were not for the exclusion principle, all of the electrons in a multielectron atom would simply drop down into the ground state, $n = 1$. It is the exclusion principle which forces the electrons into higher-energy states, making atoms and chemistry as we know it possible. (This is discussed in more detail in Section 13.2.)

More generally, the symmetry or antisymmetry of the wave function restricts the allowed solutions of the Schrödinger equation. As an example, consider a potential of the form

$$V(\mathbf{r}_1,\mathbf{r}_2) = V_0(\mathbf{r}_1) + V_0(\mathbf{r}_2) + V_1(|\mathbf{r}_1 - \mathbf{r}_2|)$$

for which both particles experience the same external potential $V_0$ while interacting with each other via the potential $V_1$. The Schrödinger equation is difficult to solve with a general potential of this type, but it is instructive to take the limit where $V_1 \ll V_0$, so that as a first approximation, the interaction potential can be ignored. Even in this limit, the particles still affect each other through the requirement that the wave function be either symmetric or antisymmetric. When $V_1 \ll V_0$, the two-particle Schrödinger equation (Equation (13.1)) takes the form

$$-\frac{\hbar^2}{2m}\nabla_1^2\psi(\mathbf{r}_1,\mathbf{r}_2) + V(\mathbf{r}_1)\psi(\mathbf{r}_1,\mathbf{r}_2) - \frac{\hbar^2}{2m}\nabla_2^2\psi(\mathbf{r}_1,\mathbf{r}_2) + V(\mathbf{r}_2)\psi(\mathbf{r}_1,\mathbf{r}_2)$$
$$= E\psi(\mathbf{r}_1,\mathbf{r}_2) \tag{13.8}$$

where we have dropped the "0" subscript on the potential for simplicity. Equation (13.8) is applicable whenever two particles experience the same external potential but do not interact directly with each other.

This equation resembles the sum of two single-particle Schrödinger equations, and this property can be exploited to find a solution of the two-particle equation. Suppose that the single-particle Schrödinger equation with potential $V$,

$$-\frac{\hbar^2}{2m}\nabla^2\psi(\mathbf{r}) + V(\mathbf{r})\psi(\mathbf{r}) = E\psi(\mathbf{r})$$

can be solved exactly, yielding single-particle wave functions and corresponding energies $\psi_n(\mathbf{r})$ and $E_n$, respectively. Then we can verify by direction substitution into Equation (13.8) that the product of any two of these solutions for the two particles, $\psi_m(\mathbf{r}_1)\psi_n(\mathbf{r}_2)$, is a solution of Equation (13.8):

$$-\frac{\hbar^2}{2m}\nabla_1^2[\psi_m(\mathbf{r}_1)\psi_n(\mathbf{r}_2)] + V(\mathbf{r}_1)[\psi_m(\mathbf{r}_1)\psi_n(\mathbf{r}_2)]$$
$$-\frac{\hbar^2}{2m}\nabla_2^2[\psi_m(\mathbf{r}_1)\psi_n(\mathbf{r}_2)] + V(\mathbf{r}_2)[\psi_m(\mathbf{r}_1)\psi_n(\mathbf{r}_2)]$$
$$= \psi_n(\mathbf{r}_2)\left[-\frac{\hbar^2}{2m}\nabla_1^2\psi_m(\mathbf{r}_1) + V(\mathbf{r}_1)\psi_m(\mathbf{r}_1)\right]$$
$$+ \psi_m(\mathbf{r}_1)\left[-\frac{\hbar^2}{2m}\nabla_2^2\psi_n(\mathbf{r}_2) + V(\mathbf{r}_2)\psi_n(\mathbf{r}_2)\right]$$
$$= \psi_n(\mathbf{r}_2)E_m\psi_m(\mathbf{r}_1) + \psi_m(\mathbf{r}_1)E_n\psi_n(\mathbf{r}_2)$$
$$= (E_m + E_n)\psi_m(\mathbf{r}_1)\psi_n(\mathbf{r}_2)$$

Thus, a general solution to Equation (13.8) is the wave function $\psi_{mn}(\mathbf{r}_1,\mathbf{r}_2)$ given by

$$\psi_{mn}(\mathbf{r}_1,\mathbf{r}_2) = \psi_m(\mathbf{r}_1)\psi_n(\mathbf{r}_2) \tag{13.9}$$

where $\psi_m$ and $\psi_n$ are the eigenfunctions of the one-particle Schrödinger equation with potential $V$, and the energy corresponding to $\psi_{mn}$ is

$$E_{mn} = E_m + E_n$$

Although $\psi_{mn}$ satisfies the two-particle Schrödinger equation, this solution is neither symmetric nor antisymmetric under the exchange of the two particles. The way to resolve this problem is to note that there are actually two different solutions which have the same energy $E_{mn}$; in addition to the solution in Equation (13.9), there is another solution obtained by switching $m$ and $n$:

$$\psi_{nm}(\mathbf{r}_1, \mathbf{r}_2) = \psi_n(\mathbf{r}_1)\psi_m(\mathbf{r}_2)$$

Since both of these wave functions correspond to the same energy, $E_m + E_n$, any linear combination of them will also satisfy the Schrödinger equation and have energy $E_m + E_n$. In particular, these two solutions can be combined to produce a wave function that is symmetric under interchange of the two particles:

$$\psi_{mn}(\mathbf{r}_1, \mathbf{r}_2)_{symm} = \frac{1}{\sqrt{2}}[\psi_m(\mathbf{r}_1)\psi_n(\mathbf{r}_2) + \psi_n(\mathbf{r}_1)\psi_m(\mathbf{r}_2)] \qquad (13.10)$$

and a wave function that is antisymmetric under interchange of the two particles:

$$\psi_{mn}(\mathbf{r}_1, \mathbf{r}_2)_{antisymm} = \frac{1}{\sqrt{2}}[\psi_m(\mathbf{r}_1)\psi_n(\mathbf{r}_2) - \psi_n(\mathbf{r}_1)\psi_m(\mathbf{r}_2)] \qquad (13.11)$$

where the $1/\sqrt{2}$ factor insures that the two-particle wave functions are normalized as long as the individual one-particle wave functions are correctly normalized. These, then, are the symmetric and antisymmetric solutions to the two-particle Schrödinger equation.

One special case must be treated separately: the wave functions for which $n = m$. For the case of antisymmetric wave functions, Equation (13.11) gives $\psi_{nn} = 0$, so such wave functions are not allowed. (This is another corollary of the exclusion principle.) Thus, if the allowed single-particle wave functions correspond to the quantum number $n$, where $n = 1, 2, 3, \ldots$, then for antisymmetric wave functions, the lowest-energy state is $n = 1$, $m = 2$. For symmetric wave functions, there is no similar restriction; the state $m = n$ gives a perfectly acceptable wave function, and $m = 1$, $n = 1$ is the lowest-energy state. However, in this case the normalizing factor $1/\sqrt{2}$, which is derived based on the assumption that

$\psi_m(\mathbf{r}_1)\psi_n(\mathbf{r}_2)$ and $\psi_n(\mathbf{r}_1)\psi_m(\mathbf{r}_2)$ are orthogonal wave functions, is no longer correct. Instead, the normalized wave function is simply

$$\psi_{nn}(\mathbf{r}_1, \mathbf{r}_2) = \psi_n(\mathbf{r}_1)\psi_n(\mathbf{r}_2) \tag{13.12}$$

The spin states of the particles produce an additional complication, but before adding this complication, consider a solution for spinless particles.

---

## Example 13.2. Two Identical Spin-0 Particles in an Infinite One-Dimensional Square Well

Two identical spin-0 particles are confined in an infinite one-dimensional square well with width $a$. Find the energy levels and corresponding wave functions, and determine the ground state.

The individual wave functions for the infinite one-dimensional square well are

$$\psi(x) = \sqrt{\frac{2}{a}} \sin\left(\frac{n\pi x}{a}\right)$$

with energy

$$E_n = \frac{\pi^2 \hbar^2}{2ma^2} n^2$$

Since the particles have spin 0, they are bosons, and their wave function must be symmetric as in Equations (13.10) and (13.12). Therefore, the total wave function is

$$\psi_{mn}(x_1, x_2) = \frac{1}{\sqrt{2}} \frac{2}{a} \left[ \sin\left(\frac{m\pi x_1}{a}\right) \sin\left(\frac{n\pi x_2}{a}\right) + \sin\left(\frac{n\pi x_1}{a}\right) \sin\left(\frac{m\pi x_2}{a}\right) \right]$$

for $m \neq n$, and for $m = n$ the solution is

$$\psi_{nn} = \frac{2}{a} \sin\left(\frac{n\pi x_1}{a}\right) \sin\left(\frac{n\pi x_2}{a}\right)$$

with, in either case, a corresponding energy of

$$E_{mn} = E_m + E_n$$
$$= \frac{\pi^2 \hbar^2}{2ma^2} [m^2 + n^2]$$

The ground-state energy is then

$$E_{11} = \frac{\pi^2 \hbar^2}{ma^2}$$

and the ground-state wave function is

$$\psi_{11} = \frac{2}{a} \sin\left(\frac{\pi x_1}{a}\right) \sin\left(\frac{\pi x_2}{a}\right)$$

This was the wave function used in Example 13.1.

What happens when the particles have spin? As a specific example, consider the case of two spin-1/2 particles. Recall from Chapter 8 that the spin states can be expressed in terms of the total spin quantum number for the two particles $s$ and the $z$ component of this total spin $m_s$. These states, in turn, are related to the "spin up" and "spin down" states of the individual particles through the relations:

$$| 1 \ 1 \rangle = | \uparrow \uparrow \rangle$$

$$| 1 \ 0 \rangle = \frac{1}{\sqrt{2}} | \uparrow \downarrow \rangle + \frac{1}{\sqrt{2}} | \downarrow \uparrow \rangle$$

$$| 1 \ -1 \rangle = | \downarrow \downarrow \rangle$$

$$| 0 \ 0 \rangle = \frac{1}{\sqrt{2}} | \uparrow \downarrow \rangle - \frac{1}{\sqrt{2}} | \downarrow \uparrow \rangle$$

where the states on the left-hand side of these equations are the $|s \ m_s\rangle$ states, and the states on the right-hand side are the spin up or spin down states of the two individual particles.

Now note an important point: All three of the triplet states (i.e., the states with $s = 1$) are symmetric under interchange of the two particles, while the singlet state ($s = 0$) is antisymmetric under interchange of the two particles. When the exchange operator is applied to two particles, it interchanges *both* their spatial positions and their spins (Figure 13.2). Hence, in considering whether a wave function is symmetric or antisymmetric under particle exchange, the full wave function must include both the spatial wave function and the spin wave function. For two spin-1/2 particles, for example, we write the full wave function $|1 \ 2\rangle$ as the product of the spatial part of the wave function and spin part of the wave function:

$$|1 \ 2\rangle \Rightarrow \psi(\mathbf{r}_1, \mathbf{r}_2)|s \ m_s\rangle$$

and we require that this *total* wave function be symmetric for bosons and antisymmetric for fermions. Since the spatial part of the wave function can be either symmetric or antisymmetric, and the spin part of the wave function can be either symmetric or antisymmetric, there are only four possibilities:

$\psi(\mathbf{r}_1, \mathbf{r}_2)$ is symmetric, $|s \ m_s\rangle$ is symmetric → total wave function is symmetric

$\psi(\mathbf{r}_1, \mathbf{r}_2)$ is antisymmetric, $|s \ m_s\rangle$ is symmetric → total wave function is antisymmetric

$\psi(\mathbf{r}_1, \mathbf{r}_2)$ is symmetric, $|s \ m_s\rangle$ is antisymmetric → total wave function is antisymmetric

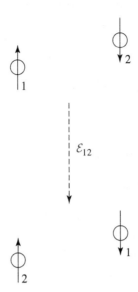

Fig. 13.2   The exchange operator interchanges both the positions and spins of the two particles.

$\psi(\mathbf{r}_1, \mathbf{r}_2)$ is antisymmetric, $|s \ m_s\rangle$ is antisymmetric $\rightarrow$ total wave function is symmetric.

Thus, for bosons the spatial part of the wave function and the spin part of the wave function must either be both symmetric or both antisymmetric. For fermions, a symmetric spin state implies an antisymmetric spatial wave function, and an antisymmetric spin state implies a symmetric spatial wave function. Thus, two spin-1/2 particles in the triplet state will have an antisymmetric spatial wave function, while two spin-1/2 particles in the singlet state will have a symmetric spatial wave function. It is impossible to determine the spatial wave function and spin state in isolation; the knowledge of one is needed to determine the possible choices for the other.

---

### Example 13.3. Two Identical Spin-1/2 Particles in an Infinite One-Dimensional Square Well

Two identical spin-1/2 particles are confined in an infinite one-dimensional square well with width $a$. Find the spatial wave function and energy for the lowest-energy singlet state and lowest-energy triplet state, respectively.

The singlet state has an antisymmetric spin state and, therefore, a symmetric spatial wave function. Thus, the lowest energy state is the same as in Example 13.2: the state $m = n = 1$:

$$\psi_{11} = \frac{2}{a} \sin\left(\frac{\pi x_1}{a}\right) \sin\left(\frac{\pi x_2}{a}\right)$$

with energy

$$E_{11} = \frac{\pi^2 \hbar^2}{ma^2}$$

For the triplet state, the spin state is symmetric, so the spatial wave function must be antisymmetric. For this case, $m = n$ is not allowed, and the lowest energy state corresponds to the antisymmetric wave function with $m = 1$, $n = 2$:

$$\psi_{12}(x_1, x_2) = \frac{1}{\sqrt{2}} \frac{2}{a} \left[ \sin\left(\frac{\pi x_1}{a}\right) \sin\left(\frac{2\pi x_2}{a}\right) - \sin\left(\frac{2\pi x_1}{a}\right) \sin\left(\frac{\pi x_2}{a}\right) \right]$$

with energy

$$E_{12} = \frac{\pi^2 \hbar^2}{2ma^2}[1^2 + 2^2]$$

$$= \frac{5\pi^2 \hbar^2}{2ma^2}$$

---

The requirement that the fermion states must be antisymmetric has a profound effect on the behavior of electrons in atoms. This will be explored in more detail in the next section; here we examine the simplest multielectron atom: the helium atom.

---

## Example 13.4. The Ground State of the Helium Atom

The full potential for the two electrons in a helium atom is given by Equation (13.3):

$$V(\mathbf{r}_1, \mathbf{r}_2) = -\frac{2e^2}{4\pi\epsilon_0} \frac{1}{r_1} - \frac{2e^2}{4\pi\epsilon_0} \frac{1}{r_2} + \frac{e^2}{4\pi\epsilon_0} \frac{1}{|\mathbf{r}_1 - \mathbf{r}_2|} \tag{13.13}$$

As a first approximation, we neglect the final term which represents the mutual repulsion between the two electrons. In this limit the electrons are treated as independent particles which each feel the central Coulomb potential produced by the nucleus with charge $+2e$.

Writing the spatial part of the one-particle wave function in the standard form $\psi_{nlm_l}$, the lowest-energy wave function for a single electron in the potential $V = -2e^2/4\pi\epsilon_0 r$ is

$$\psi_{100}^{He} = \frac{1}{\sqrt{\pi}} \left(\frac{2}{a_0}\right)^{3/2} e^{-2r/a_0}$$

with energy

$$E_1^{He} = 4E_1^H$$

where the "$He$" superscript refers to the wave function and energy of a single electron in the Coulomb potential of the helium nucleus, and the "$H$" superscript refers to the corresponding quantities in hydrogen. Since $E_1^H$ is the ground-state energy of hydrogen, we have $E_1^H = -13.6$ eV, so $E_1^{He} = -54.4$ eV.

For the helium atom to be in the lowest possible energy state, both electron wave functions must correspond to $\psi_{100}^{He}$, so the spatial part of the wave function is symmetric:

$$\psi(\mathbf{r}_1, \mathbf{r}_2) = \psi_{100}^{He}(\mathbf{r}_1)\psi_{100}^{He}(\mathbf{r}_2)$$
$$= \frac{8}{\pi a_0^3} e^{-2r_1/a_0} e^{-2r_2/a_0} \tag{13.14}$$

Since the spatial part of the wave function is symmetric, and the electrons are fermions, which must have an antisymmetric total wave function, the spin part of the wave function must be antisymmetric. Thus, the two electrons in the helium atom must be in the singlet state:

$$|0\ 0\rangle = \frac{1}{\sqrt{2}}|\uparrow \downarrow\rangle - \frac{1}{\sqrt{2}}|\downarrow \uparrow\rangle \tag{13.15}$$

In elementary explanations of atomic structure, it is often stated that the atomic electrons fill the lowest possible energy states (in this case, the two states with $n = 1$, $l = 0$, $m_l = 0$) but then to fulfill the Pauli exclusion principle, the electrons must have opposite spins. However, Equation (13.15) shows that this explanation is an oversimplification: the two electrons are not forced into a state, for instance, where electron 1 has spin up and electron 2 has spin down. Rather, they are forced into the singlet spin state with total spin $s = 0$ and $m_s = 0$; the spins of the electrons are a mixture of $|\uparrow \downarrow\rangle$ and $|\downarrow \uparrow\rangle$.

The total energy corresponding to the spatial wave function in Equation (13.14) is $E = 2E_1^{He} = -108.8$ eV, while the true ground-state energy of helium is $-79.0$ eV. Thus, our estimate of the ground-state energy is rather far off the mark. However, this estimate can be improved by treating the electron-electron repulsion (the last term in Equation 13.13) as a perturbation and applying first-order perturbation theory.

The change in the energy due to this repulsion is then

$$E^{(1)} = \int d^3\mathbf{r}_1 \, d^3\mathbf{r}_2 \left( \frac{8}{\pi a_0^3} e^{-2r_1/a_0} e^{-2r_2/a_0} \right) \left( \frac{e^2}{4\pi\epsilon_0} \frac{1}{|\mathbf{r}_1 - \mathbf{r}_2|} \right)$$
$$\left( \frac{8}{\pi a_0^3} e^{-2r_1/a_0} e^{-2r_2/a_0} \right)$$
$$= \frac{5}{4} \frac{e^2}{4\pi\epsilon_0 a_0}$$
$$= 34.0 \text{ eV}$$

As expected, this change in the energy is positive since it represents a repulsion between the two electrons. Adding it to the energy of the unperturbed two-particle wave function, we get an estimate of the total ground-state energy: $E = -108.8$ eV $+ 34.0$ eV $= -74.8$ eV. This differs from the true ground-state energy by about 5%, which is not bad agreement (although the variational principle, Chapter 10, does provide a better estimate for this energy).

---

These arguments that we have derived for the two-particle wave function can be extended to systems of more particles. For instance, the total wave function for three fermions must be antisymmetric under interchange of any two of the particles. If the spin part is symmetric, then this implies, for example,

$$\psi(\mathbf{r}_1, \mathbf{r}_2, \mathbf{r}_3) = -\psi(\mathbf{r}_2, \mathbf{r}_1, \mathbf{r}_3) = -\psi(\mathbf{r}_1, \mathbf{r}_3, \mathbf{r}_2)$$

and so on. If $\psi_1(\mathbf{r})$, $\psi_2(\mathbf{r})$, $\psi_3(\mathbf{r})$ are three different solutions to the one-particle Schrödinger equation, then the fully antisymmetric spatial wave function is given by

$$\psi(\mathbf{r}_1, \mathbf{r}_2, \mathbf{r}_3) = \frac{1}{\sqrt{6}} [\psi_1(\mathbf{r}_1)\psi_2(\mathbf{r}_2)\psi_3(\mathbf{r}_3) - \psi_1(\mathbf{r}_2)\psi_2(\mathbf{r}_1)\psi_3(\mathbf{r}_3)$$
$$- \psi_1(\mathbf{r}_1)\psi_2(\mathbf{r}_3)\psi_3(\mathbf{r}_2) - \psi_1(\mathbf{r}_3)\psi_2(\mathbf{r}_2)\psi_3(\mathbf{r}_1)$$
$$+ \psi_1(\mathbf{r}_2)\psi_2(\mathbf{r}_3)\psi_3(\mathbf{r}_1) + \psi_1(\mathbf{r}_3)\psi_2(\mathbf{r}_1)\psi_3(\mathbf{r}_2)]$$

This can be written in a particularly compact form as a determinant:

$$\psi(\mathbf{r}_1, \mathbf{r}_2, \mathbf{r}_3) = \frac{1}{\sqrt{6}} \begin{vmatrix} \psi_1(\mathbf{r}_1) & \psi_1(\mathbf{r}_2) & \psi_1(\mathbf{r}_3) \\ \psi_2(\mathbf{r}_1) & \psi_2(\mathbf{r}_2) & \psi_2(\mathbf{r}_3) \\ \psi_3(\mathbf{r}_1) & \psi_3(\mathbf{r}_2) & \psi_3(\mathbf{r}_3) \end{vmatrix}$$

which is called a *Slater determinant*. Of course, the usefulness of this approach diminishes as the number of particles increases. For the uranium atom, for instance, the Slater determinant for the electrons is a $92 \times 92$ determinant, which expands out into $92! \sim 10^{142}$ terms. (This expansion written out would far exceed the length of the entire visible universe.) Clearly, other approaches are called for.

## 13.2   Multielectron Atoms

The electron in hydrogen is characterized by four quantum numbers: $n$, $l$, $m_l$, and $m_s$. For electrons in multielectron atoms, the electron-electron interactions alter the wave functions and energy levels, and it is not possible to solve the Schrödinger equation analytically. On the other hand, the electrons in a multielectron atom can still be characterized by the same set of quantum numbers, and the Pauli exclusion principle prevents any two electrons from sharing the same complete set of quantum numbers.

Recall that for the hydrogen atom, the energy of a given state is entirely determined by the principle quantum number $n$; all $l$, $m_l$, and $m_s$ states for a given $n$ have the same energy (aside from the small perturbations which were discussed in Chapter 9). In multielectron atoms, the electron-electron interactions break this degeneracy so that the energy of a given electron depends on *both* $n$ and $l$. However, the various $m_l$ and $m_s$ states for a given $n$ and $l$ are still degenerate in energy (since they represent quantities which depend on the orientation of the coordinate system). For this reason, the set of all possible $m_l$ and $m_s$ states for a fixed $n$ and $l$ is called a *subshell*; each subshell corresponds to a distinct energy level. Since $m_l$ ranges from $-l$ to $+l$, and there are two spin states for each value of $n$, $l$, and $m_l$, a given subshell can hold $2(2l + 1)$ electrons. For a given value of $n$, the set of all possible $l$, $m_l$, and $m_s$ states is called a *shell*; each shell can hold $2n^2$ electrons (Problem 13.7).

The notation which indicates the number of electrons occupying each subshell in a multielectron atom is rather arcane. Unfortunately, it is also rather standard, so we shall use it here. A given $n$, $l$ subshell is denoted by a number (which gives $n$) and a letter (which gives $l$, in accordance with the previously-used notation: $l = 0$ is denoted by $s$, $l = 1$ is denoted by $p$, $l = 2$ is denoted by $d$, and so on). So the first few subshells in an atom are:

$$1s \rightarrow n = 1, \quad l = 0$$
$$2s \rightarrow n = 2, \quad l = 0$$
$$2p \rightarrow n = 2, \quad l = 1$$
$$3s \rightarrow n = 3, \quad l = 0$$
$$3p \rightarrow n = 3, \quad l = 1$$
$$3d \rightarrow n = 3, \quad l = 2$$

The number of electrons occupying each subshell is indicated by a superscript. For instance, the ground state of boron is written as

$$1s^2 \, 2s^2 \, 2p^1 \tag{13.16}$$

indicating that two electrons have $n = 1$ and $l = 0$, two electrons have $n = 2$ and $l = 0$, and one electron has $n = 2$ and $l = 1$.

Just as in hydrogen, the higher $n$ states have higher energy, but what about the $l$ states? In general, states with higher $l$ have wave functions which are peaked further from the nucleus (see, for example, Figure 6.7). Thus, the lower $l$ states "feel" a stronger Coulomb attraction and have lower energy. For example, in the ground state of the lithium atom, which has three electrons, the first two electrons fall into the $1s$ subshell, and the third electron can have either $n = 2$, $l = 0$ or $n = 2$, $l = 1$. Our argument indicates that the $n = 2$, $l = 0$ state has the lower energy, so the ground state of lithium is

$$1s^2 \, 2s^1$$

In general then, subshells are filled first from lower $n$ to higher $n$, and for a given value of $n$, from lower $l$ to higher $l$; this is called the *Aufbau* principle. Thus, we expect the subshells to be filled in the order: $1s$, $2s$, $2p$, $3s$, $3p$, $3d$, $4s$, .... This simple rule works well for the lightest atoms, but it breaks down at the $3d$ state. At this point screening by the inner electrons becomes so important that the $n = 4$, $l = 0$ state is pulled below the $n = 3$, $l = 2$ state, so that the order in which the subshells are actually filled is: $1s$, $2s$, $2p$, $3s$, $3p$, $4s$, $3d$, .... For heavier elements, the order in which the subshells are filled becomes even more complicated.

This filling of subshells and shells is the basis for all of chemistry. A shell with a full set of electrons, called a *closed shell*, plays no role in binding to other atoms; it is the electrons in a partially filled shell which determine the chemical behavior of an atom. (Note that it is a filled *shell*, not a filled *subshell*, which renders the electrons inactive as far as chemical activity is concerned.) Atoms consisting entirely of closed shells are chemically inert. Because these atoms cannot bind even to each other, the corresponding elements are all gasses at room temperature: the "noble gasses." The first few such atoms with their electron configurations are:

$$\text{helium} \quad 1s^2$$

$$\text{neon} \quad 1s^2\,2s^2\,2p^6$$

$$\text{argon} \quad 1s^2\,2s^2\,2p^6\,3s^2\,3p^6$$

Note that the $n = 3$ shell for argon is considered closed despite the fact that there are no electrons in the $3d$ subshell. As noted earlier this is the point at which our simple rules for filling subshells break down, with the $3d$ states having higher energy than the $4s$ states.

The alkali metals (lithium, sodium, potassium, etc.) have a single electron in an unfilled shell, making them very reactive as electron donors:

$$\text{lithium} \quad 1s^2\,2s$$

$$\text{sodium} \quad 1s^2\,2s^2\,2p^6\,3s$$

$$\text{potassium} \quad 1s^2\,2s^2\,2p^6\,3s^2\,3p^6\,4s$$

Similarly, the halogens (fluorine, chlorine, bromine, ...) are one electron short of a filled shell, rendering them prone to grabbing electrons from other atoms. This shell filling, of course, is what produces the periodic table.

A given atom can also be characterized by its angular momentum states. These states are labeled by the quantum numbers $L$, $S$, and $J$, which correspond to the total orbital angular momentum for the electrons in the atom $L$, the total spin angular momentum quantum number for the atomic electrons $S$, and the total angular momentum $J$. (We use uppercase letters to denote total angular momenta, and lowercase letters to denote angular momenta of individual electrons. Note that the angular momentum of the nucleus is ignored here.) The way in which the individual spins and orbital angular momenta of the electrons in an atom couple to give $L$, $S$, and $J$ depends on the number of electrons in the atom. For the lightest elements, it is a good approximation to assume that all of the individual orbital angular momenta couple to give the value for $L$, and all of the individual spin angular momenta couple to give the value for $S$. These total values for $L$ and $S$ then couple, in accordance with the relation

$$|L - S| \le J \le L + S$$

to give the total angular momentum $J$. This coupling scheme is called *L-S coupling* or *Russell–Saunders coupling*, and it is the only one we will consider in detail here. For some heavy elements, it is a better approximation to assume that the individual $l$ and $s$ of each electron couple to give a distinct total angular momentum $j$ for each electron; all of these $j$'s then couple to give the total $J$. This is called *j-j coupling* and will not be considered further. While a given atom can have a variety of values for $L$, $S$, and $J$, the ground state of a given atom always has a unique set of these quantum numbers. (Note that the values of $L$, $S$, and $J$ for the ground state of an atom have no particular importance for its chemical behavior.)

Consider first the simplest case, hydrogen. The ground state has a single electron with $l = 0$ and $s = 1/2$, which means that the *total* orbital angular momentum and *total* spin angular momentum must also be $L = 0$ and $S = 1/2$. These can couple to give only a single value for $J$, namely, $J = 1/2$. This is written in the rather unusual (but standard!) notation introduced in Chapter 9; the value for $L$ is written as a letter (in this case $S$), and the value of $J$ is written as a subscript to give $S_{1/2}$. Now we must add a superscript to denote the value of $S$, but instead of writing the value of $S$, we write the value of $2S + 1$ as an upper left-hand superscript (don't blame me; I didn't invent this). Thus, the full set of values for $L$, $S$, and $J$ are given as

$$^{2S+1}L_J$$

In this notation, the ground state of hydrogen is written as

$$^2S_{1/2}$$

(Note that capital letters $S, P, D, \ldots$ denote the *total* orbital angular momentum of an atom, while small letters $s, p, d, \ldots$ such as we used in describing the electron configuration, refer to the angular momentum states of individual electrons.)

Moving on to helium, recall from the previous section that the electrons in the ground state must be in the singlet state, so that $S = 0$. Further, both electrons occupy $l = 0$ states, so they can only couple to give $L = 0$. Finally $S = 0$ and $L = 0$ can only give $J = 0$, so we get

$$^1S_0$$

as the ground state of helium. This result can be generalized: the electrons in a closed subshell always pair off to give $L = 0$, $S = 0$, and $J = 0$; it is the electrons in partially-filled subshells which then determine the total angular momentum state. (Note the difference from chemistry, where electrons become chemically irrelevant only if the full shell is filled, not a subshell.)

Hydrogen and helium are relatively straightforward examples in the sense that the angular momenta can couple to give only a single unique set of quantum numbers. Now consider an example where this is not the case. Boron has 5 electrons, in the configuration shown in Equation (13.16). The closed subshells contribute nothing to the angular momentum, which is determined entirely by the single electron with $l = 1$ and spin $1/2$. Thus, we must have $L = 1$ and $S = 1/2$. However, this leads to two possible values for $J$: $J = 1/2$ or $J = 3/2$. Which is the lowest energy state?

To determine the ground state for boron (and other atoms), a set of empirical rules (called *Hund's rules*) provides a guide to choosing the values of $S$, $L$, and $J$ that give the lowest energy state. These rules are applied in the following order:

Hund's Rule #1: Given more than one allowed value for $S$, choose the largest possible value.

Hund's Rule #2: Given more than one allowed value for $L$, choose the largest possible value.

Hund's Rule #3: Given more than one allowed value for $J$, choose the smallest possible value for $J$ if the subshell under consideration is less than half full, and choose the largest possible value for $J$ if the subshell is more than half full.

Applying these rules to the case of boron, we see that Rule #1 and Rule #2 are irrelevant; we have only a single possible value for $S$ and for $L$. It is Rule #3 which determines the ground state. The $2p$ subshell can hold six electrons, so with only a single electron, it is clearly less than half full. Therefore, Rule #3 tells us to choose the smallest possible value for $J$, which in this case is $J = 1/2$. This gives the ground state of boron:

$$^2P_{1/2}$$

When there is more than one electron in an unfilled shell, things become much more complex as illustrated in the next example.

---

**Example 13.5. The Ground State of Carbon**

The electron configuration for carbon is

$$1s^2\ 2s^2\ 2p^2$$

The unfilled subshell contains two electrons, each with spin $1/2$ and $l = 1$. Thus, they can couple to give

$$L = 0, 1, 2$$

and

$$S = 0, 1$$

Together, these yield six possible pairs for $L$ and $S$:

$$L = 0, \quad S = 0 \tag{13.17}$$
$$L = 1, \quad S = 0 \tag{13.18}$$
$$L = 2, \quad S = 0 \tag{13.19}$$
$$L = 0, \quad S = 1 \tag{13.20}$$
$$L = 1, \quad S = 1 \tag{13.21}$$
$$L = 2, \quad S = 1 \tag{13.22}$$

However, not all of these possible states are allowed; some of them can only be obtained by putting electrons into the same $m_l$ and $m_s$ states, violating the exclusion principle. This is a subtle point which requires further exploration. Each electron can have $m_l = -1$, 0, or 1, and $m_s = +1/2$ or $-1/2$. We write the $m_l$, $m_s$ state for the two electrons in Dirac notation as $|m_l(1)\ m_s(1)\ m_l(2)\ m_s(2)\rangle$, where (1) and (2) denote the first and second electron, respectively. To avoid confusion, we use the numerical value for $m_l$ and arrow notation for $m_s$. In this notation, for instance, a state for which the first electron has $m_l = 1, m_s = 1/2$ and the second electron has $m_l = 0$, $m_s = -1/2$ would be written as

$$|1 \uparrow\ 0 \downarrow\rangle$$

We can now catalog all possible states for the two electrons in terms of their $m_l$ and $m_s$ values. In doing so we recall that states for which the two electrons have the same values for $m_l$ and $m_s$ are not allowed by the exclusion principle. For example, the state $|0 \uparrow\ 0 \uparrow\rangle$ is excluded. Further, we don't want to double-count states which are obtained by interchanging the two electrons; e.g., the states $|1 \uparrow\ 0 \downarrow\rangle$ and $|0 \downarrow\ 1 \uparrow\rangle$ are identical, so we will list only one of them. With these cautionary notes, we proceed to list all of the allowed states for $m_l$ and $m_s$ for the two electrons:

$$|1 \uparrow \quad 1 \downarrow\rangle$$
$$|1 \uparrow \quad 0 \uparrow\rangle$$
$$|1 \uparrow \quad 0 \downarrow\rangle$$
$$|1 \uparrow \quad -1 \uparrow\rangle$$
$$|1 \uparrow \quad -1 \downarrow\rangle$$
$$|1 \downarrow \quad 0 \uparrow\rangle$$
$$|1 \downarrow \quad 0 \downarrow\rangle$$
$$|1 \downarrow \quad -1 \uparrow\rangle$$
$$|1 \downarrow \quad -1 \downarrow\rangle$$
$$|0 \uparrow \quad 0 \downarrow\rangle$$
$$|0 \uparrow \quad -1 \uparrow\rangle$$
$$|0 \uparrow \quad -1 \downarrow\rangle$$
$$|0 \downarrow \quad -1 \uparrow\rangle$$
$$|0 \downarrow \quad -1 \downarrow\rangle$$
$$|-1 \uparrow \quad -1 \downarrow\rangle$$

For each of these states, we then determine the *total* values $M_L$ and $M_S$, which are simply the sums $M_L = m_l(1) + m_l(2)$ and $M_S = m_s(1) + m_s(2)$:

| | | | | |
|---|---|---|---|---|
| $\|1 \uparrow$ | $1 \downarrow\rangle$ | $\rightarrow M_L = 2,$ | $M_S = 0$ | (13.23) |
| $\|1 \uparrow$ | $0 \uparrow\rangle$ | $\rightarrow M_L = 1,$ | $M_S = 1$ | (13.24) |
| $\|1 \uparrow$ | $0 \downarrow\rangle$ | $\rightarrow M_L = 1,$ | $M_S = 0$ | (13.25) |
| $\|1 \uparrow$ | $-1 \uparrow\rangle$ | $\rightarrow M_L = 0,$ | $M_S = 1$ | (13.26) |
| $\|1 \uparrow$ | $-1 \downarrow\rangle$ | $\rightarrow M_L = 0,$ | $M_S = 0$ | (13.27) |
| $\|1 \downarrow$ | $0 \uparrow\rangle$ | $\rightarrow M_L = 1,$ | $M_S = 0$ | (13.28) |
| $\|1 \downarrow$ | $0 \downarrow\rangle$ | $\rightarrow M_L = 1,$ | $M_S = -1$ | (13.29) |
| $\|1 \downarrow$ | $-1 \uparrow\rangle$ | $\rightarrow M_L = 0,$ | $M_S = 0$ | (13.30) |
| $\|1 \downarrow$ | $-1 \downarrow\rangle$ | $\rightarrow M_L = 0,$ | $M_S = -1$ | (13.31) |
| $\|0 \uparrow$ | $0 \downarrow\rangle$ | $\rightarrow M_L = 0,$ | $M_S = 0$ | (13.32) |
| $\|0 \uparrow$ | $-1 \uparrow\rangle$ | $\rightarrow M_L = -1,$ | $M_S = 1$ | (13.33) |
| $\|0 \uparrow$ | $-1 \downarrow\rangle$ | $\rightarrow M_L = -1,$ | $M_S = 0$ | (13.34) |
| $\|0 \downarrow$ | $-1 \uparrow\rangle$ | $\rightarrow M_L = -1,$ | $M_S = 0$ | (13.35) |
| $\|0 \downarrow$ | $-1 \downarrow\rangle$ | $\rightarrow M_L = -1,$ | $M_S = -1$ | (13.36) |
| $\|-1 \uparrow$ | $-1 \downarrow\rangle$ | $\rightarrow M_L = -2,$ | $M_S = 0$ | (13.37) |

Now note an important point: some of the $L$, $S$ states listed in Equations (13.17)–(13.22) will lead to values of $M_L$ and $M_S$ which do not appear in Equations (13.23)–(13.37). These values of $L$ and $S$ must be excluded; the physical reason they cannot exist is that they violate the exclusion principle. For instance, one of our six $L$, $S$ pairs is $L = 2$, $S = 1$. If this was a possible state for the electrons, then it would lead to all possible pairs of the states $M_L = -2, -1, 0, 1, 2$ and $M_S = -1, 0, 1$ in Equations (13.23)–(13.37). While some of these states are observed, others are not. For instance, we do not see $M_L = 2$, $M_S = 1$. The reason for this is that this state can arise only if *both* electrons have $m_l = 1$ and $m_s = 1/2$; however, the exclusion principle prevents both electrons from being in this same state. Thus, the state $L = 2$, $S = 1$ is ruled out. The task is then to find a subset of the $L$, $S$ states from the six given in Equations (13.17)–(13.22) which produces all of the $M_L$ and $M_S$ states in Equations (13.23)–(13.37), but which does not produce any "extra" $M_L$, $M_S$ states not on this list. It's a bit like solving a puzzle.

Having ruled *out* the state $L = 2$, $S = 1$, we can now rule *in* several other states. The state $L = 1$, $S = 1$ must be allowed, since this is the only other $L$, $S$ state which can give us $M_L = 1$, $M_S = 1$, which is seen in Equation (13.24). This $L$, $S$ state then produces all of the following values of $M_L$ and $M_S$:

$$M_L = -1, \quad M_S = -1 \tag{13.38}$$
$$M_L = -1, \quad M_S = 0 \tag{13.39}$$
$$M_L = -1, \quad M_S = 1 \tag{13.40}$$
$$M_L = 0, \quad M_S = -1 \tag{13.41}$$
$$M_L = 0, \quad M_S = 0 \tag{13.42}$$
$$M_L = 0, \quad M_S = 1 \tag{13.43}$$
$$M_L = 1, \quad M_S = -1 \tag{13.44}$$
$$M_L = 1, \quad M_S = 0 \tag{13.45}$$
$$M_L = 1, \quad M_S = 1 \tag{13.46}$$

Similarly, the state $L = 2$, $S = 0$ is the only remaining state which can give $M_L = 2$, $S = 0$, seen in Equation (13.23). This state produces the following five $M_L$, $M_S$ states:

$$M_L = -2, \quad M_S = 0 \tag{13.47}$$
$$M_L = -1, \quad M_S = 0 \tag{13.48}$$
$$M_L = 0, \quad M_S = 0 \tag{13.49}$$
$$M_L = 1, \quad M_S = 0 \tag{13.50}$$
$$M_L = 2, \quad M_S = 0 \tag{13.51}$$

When we remove the 14 states given by Equations (13.38)–(13.51) from the list of states given by Equations (13.23)–(13.37), we are left with only a single state to account for: $M_L = 0$, $M_S = 0$. We can produce this state (and only this state) by taking $L = 0$, $S = 0$.

Therefore, of the six possible pairs of $L$, $S$ states identified in Equations (13.17)–(13.22), only three are actually allowed:

$$L = 2, \quad S = 0$$

which can give only $J = 2$,

$$L = 0, \quad S = 0$$

which can give only $J = 0$, and

$$L = 1, \quad S = 1$$

which can give $J = 0$, 1, or 2.

Now Hund's rules can be used to find which of these states is the ground state. Hund's Rule #1 instructs us to choose the state with the largest $S$, which is $L = 1$, $S = 1$. Hund's Rule #2 is irrelevant, since we have only a single $L$ value ($L = 1$) corresponding to $S = 1$. Finally, the $2p$ subshell can hold 6 electrons, and it contains only 2 in this case, so it is less than half full. Thus, Hund's Rule #3 indicates that the ground state is the lowest allowed $J$ state: $J = 0$. Therefore, the ground state is $L = 1$, $S = 1$, $J = 0$, written as

$$^3P_0$$

---

Clearly, complex calculations of the sort given in Example 13.5 arise only when there are two are more electrons in an unfilled subshell. Atoms in which all of the subshells are filled have $L = 0$, $S = 0$, and $J = 0$. Note that this includes atoms which are not chemically "inert", since an atom can have all of its subshells filled but still have an unfilled shell. For instance, beryllium (with four electrons) has the electron configuration

$$1s^2 \, 2s^2$$

Both of its subshells are filled, giving a ground-state angular momentum of $^1S_0$. However, the $n = 2$ shell is unfilled (it can hold 8 electrons) so beryllium is quite chemically reactive. Similarly, any atom with a single electron in an unfilled subshell will have the angular momentum quantum numbers of that electron, as in the case of boron considered above.

We have barely scratched the surface of the study of multielectron atoms. Atomic physics remains an active area of research with many problems still being explored. However, almost this entire field traces back to fundamental concepts from quantum mechanics.

## PROBLEMS

**13.1** Consider the two-particle Hamiltonian given by

$$H = -\frac{\hbar^2}{2m}\nabla_1^2 - \frac{\hbar^2}{2m}\nabla_2^2 + V(\mathbf{r}_1, \mathbf{r}_2)$$

Show that the exchange operator $\mathcal{E}_{12}$ commutes with $H$ as long as the two-particle potential has the property that $V(\mathbf{r}_1, \mathbf{r}_2) = V(\mathbf{r}_2, \mathbf{r}_1)$.

**13.2** Two identical spin-1/2 particles with mass $m$ are in a one-dimensional infinite square-well potential with width $a$, so $V(x) = 0$ for $0 \le x \le a$, and there are infinite potential barriers at $x = 0$ and $x = a$. The particles do not interact with each other; they see only the infinite square-well potential.

(a) Calculate the energies of the three lowest-energy singlet states.

(b) Calculate the energies of the three lowest-energy triplet states.

(c) Suppose that the particles are in a state with wave function

$$\psi(x_1, x_2) = \frac{1}{\sqrt{2}}\frac{2}{a}\left[\sin\frac{\pi x_1}{a}\sin\frac{7\pi x_2}{a} + \sin\frac{\pi x_2}{a}\sin\frac{7\pi x_1}{a}\right]$$

where $x_1$ is the position of particle 1 and $x_2$ is the position of particle 2. Are the particles in a triplet spin state or a singlet spin state? Explain.

**13.3** Two identical spin-0 particles with mass $m$ are confined inside a three-dimensional rectangular box given by $0 \le x \le a$, $0 \le y \le b$, and $0 \le z \le c$, where $a < b < c$, and the potential barriers at the walls of the box are infinitely high. The particles do not interact with each other; they see only the potential of the box. Write down the normalized wave function $\psi(x_1, y_1, z_1, x_2, y_2, z_2)$ for the ground state, and indicate the energy of the ground state.

**13.4** (a) Two identical spin-1/2 particles are confined inside of the rectangular box from Problem 13.3. The particles do not interact with each other; they see only the potential of the box. Write down the normalized spatial part of the lowest-energy singlet wave function. What is the energy of this state?

(b) Write down the normalized spatial part of the lowest-energy triplet wave function. What is the energy of this state?

(c) What is the total spin $s$ of the ground-state wave function for the system?

**13.5** Two identical spin-1/2 particles are confined to an infinite one-dimensional square well of width $a$ with infinite potential barriers at $x = 0$ and $x = a$. The potential is $V(x) = 0$ for $0 \leq x \leq a$. Suppose that the particles interact weakly by the potential $V_1(x) = K\delta(x_1 - x_2)$, where $x_1$ and $x_2$ are the positions of the two particles, $K$ is a constant, and $\delta$ is the Dirac delta function. This represents a very short-range weak force between the two particles.

(a) Using first-order perturbation theory, find the perturbation to the energy of the lowest-energy singlet state.

(b) Show that the first-order perturbation to the energy of the lowest-energy triplet state is zero.

(c) What is the physical reason for the answer in part (b)?

**13.6** Two identical spin-1/2 particles are in the one-dimensional simple harmonic oscillator potential $V(x) = (1/2)Kx^2$. The particles do not interact with each other; they see only the harmonic oscillator potential. The particles are in the lowest-energy triplet state, with $s = 1$.

(a) Write down the normalized spatial part of the wave function.

(b) Calculate the energy of this state.

(c) If the positions of both particles are measured, what is the probability that both particles will be located on the right-hand side of the minimum in the potential (i.e., the probability that both particles have $x > 0$)?

**13.7** Show that the $n$ shell in an atom can hold $2n^2$ electrons.

**13.8** Sodium has $Z = 11$. Determine the ground-state electron configuration.

**In Problems 13.9–13.12, express all angular momentum states in the notation $^{2S+1}L_J$.**

**13.9** Determine the ground-state $L$, $S$, and $J$ values for

(a) calcium which has the electron configuration

$$1s^2 \; 2s^2 \; 2p^6 \; 3s^2 \; 3p^6 \; 4s^2$$

(b) yttrium which has the electron configuration

$$1s^2 \; 2s^2 \; 2p^6 \; 3s^2 \; 3p^6 \; 3d^{10} \; 4s^2 \; 4p^6 \; 4d^1 \; 5s^2$$

**13.10** Consider the excited state of beryllium with the electron configuration

$$1s^2 \, 3p^1 \, 3d^1$$

Determine all possible $L$, $S$, and $J$ values. Note that the Pauli exclusion principle for the two $n = 3$ electrons can be ignored; why?

**13.11** Zirconium $(Z = 40)$ consists of closed subshells plus 2 electrons in an unfilled $d$ subshell. Derive the set of allowed $L$, $S$, and $J$ values, and determine which state has the lowest energy.

**13.12** The electron configuration for nitrogen is

$$1s^2 \, 2s^2 \, 2p^3$$

Calculate $L$, $S$, and $J$ for the ground state.

Chapter 14

# Modern applications of quantum mechanics

The first half of the 20th century saw two revolutions in physics, culminating in the theories of quantum mechanics and general relativity. Of these two, quantum mechanics has certainly had a more profound impact on the subsequent development of physics. Quantum mechanics provides the basis for nuclear physics, particle physics, and solid state physics, with applications in almost all other fields of physics. Furthermore, it has yielded a surprising number of practical applications. In this chapter we examine just two of these, in which the use of quantum mechanics is particularly straightforward: magnetic resonance imaging and quantum computing. Magnetic resonance imaging is already a well-developed technology, while quantum computing is still very much in the formative stage. These are presented merely as representative examples; there are many others.

## 14.1 Magnetic Resonance Imaging

Imagine that a patient needs to undergo a magnetic resonance imaging (MRI) scan (Figure 14.1). The patient first removes any objects that can affect or be affected by magnetic fields (jewelry, credit cards, metallic objects), and then lies on a padded bench which slides into a long tube. The only sounds are occasional loud thumping noises. After 30–90 minutes, the patient slides back out again, and the exam is finished.

The development of magnetic resonance imaging technology represents an enormous advancement in diagnostic medicine. MRI scans are among the least invasive and safest of all medical tests. Unlike X-rays, they appear to have absolutely no harmful effects, and they allow physicians to image soft tissues, especially brain and nerve tissue, which are difficult to examine with conventional X-rays. Real-time imagining of the brain has opened

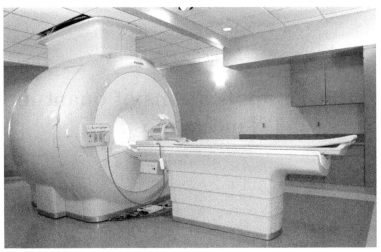

Courtesy of John Gore, Vanderbilt University

Fig. 14.1   A magnetic resonance imaging (MRI) machine.

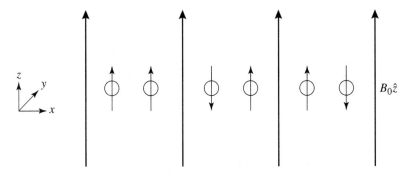

Fig. 14.2   The protons in hydrogen placed in a magnetic field $\mathbf{B}_0 = B_0\hat{z}$ will line up parallel or antiparallel to the field.

up new areas of research in neuroscience, as investigators are able to see different parts of the brain "light up" in response to different stimuli. And it's all derived from quantum mechanics.

Magnetic resonance imaging involves the interaction of external magnetic fields with the protons in hydrogen atoms. Consider a hydrogen atom placed in a strong magnetic field (Figure 14.2). The proton magnetic moment $\boldsymbol{\mu}_p$ is given by an expression similar to that for the electron magnetic moment in Equation (8.7):

$$\boldsymbol{\mu}_p = \frac{g_p \mu_B}{\hbar} \frac{m_e}{m_p} \mathbf{S} \tag{14.1}$$

where $g_p = 5.59$ is determined experimentally. The factor multiplying $\mathbf{S}$ is positive in this equation and negative in Equation (8.7) because the proton and electron have opposite charges, and the ratio of $m_e$ to $m_p$ in Equation (14.1) arises from the fact that the magnetic moment is inversely proportional to the mass of the particle (Equation (8.5)). As before, $\mu_B$ is the Bohr magneton, $\mu_B = 9.3 \times 10^{-24}$ A·m², and $\mathbf{S}$ is the spin.

Recall from Chapter 8 that the potential experienced by the proton in an external magnetic field $\mathbf{B}$ is

$$V = -\boldsymbol{\mu}_p \cdot \mathbf{B}$$

If, for instance, $\mathbf{B}$ is a static field in the $z$ direction with magnitude $B_0$,

$$\mathbf{B}_0 = B_0 \hat{z}$$

then the protons will be forced into eigenstates of the Hamiltonian and will line up with spins either in the $+z$ direction with energy $E = -(g_p \mu_B / 2)(m_e/m_p)B_0$, or in the $-z$ direction with energy $E = (g_p \mu_B / 2)(m_e/m_p)B_0$ (Figure 14.2).

Now suppose that we add a perturbation in the form of an electromagnetic wave. The magnetic field generated by this wave will have the form $\mathbf{B}_1 \cos(\omega t)$. This applied magnetic field can be chosen to be perpendicular to the $z$-axis; for simplicity we will take it to be in the $x$ direction, so that

$$\mathbf{B}_1 = B_1 \cos(\omega t)\hat{x}$$

What happens to the proton spins when this oscillating magnetic field is applied? This problem was previously discussed in Example 11.2 for the case of an electron. It was found that the applied field produces a nonzero probability for the electron to flip into the opposite $z$ spin state (Figure 14.3). Using the results of Example 11.2, but replacing the electron magnetic moment with the proton magnetic moment, gives

$$P(i \to f) = \frac{B_1^2 \mu_p^2}{4\hbar^2} \frac{\sin^2[(\omega - \omega_0)t/2]}{[(\omega - \omega_0)/2]^2} \tag{14.2}$$

where $P(i \to f)$ is the transition probability for a proton to go from a lower energy (spin up) state to a higher energy (spin down) state. The frequency $\omega_0$ in this case is

$$\omega_0 = 2\mu_B(g_p/2)(m_e/m_p)B_0/\hbar$$

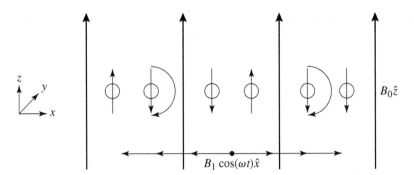

Fig. 14.3    An oscillating magnetic field in the $\hat{x}$ direction causes some of the proton spins to flip.

The important point is that the transition probability in Equation (14.2) is sharply peaked around $\omega \approx \omega_0$. In order to drive a large number of transitions, the applied frequency of the radiation should be close to this resonance frequency. A typical MRI field strength is $B_0 = 1.5$ T; using this value, we obtain $\omega_0 = 4.0 \times 10^8$ sec$^{-1}$, corresponding to a frequency of

$$\nu = \omega/2\pi = 6.4 \times 10^7 \text{ Hz}$$

Thus, the applied frequency needs to be in the MHz region, corresponding to radio frequencies.

After the spins flip into the high-energy state, they will relax back into the low-energy state. As they do so, they will develop a spin component (and therefore a magnetic moment component) perpendicular to the strong static magnetic field $B_0\hat{z}$, causing them to precess about this field (Figure 14.4). We examined spin precession in Chapter 8, where we found that the angular frequency of precession for an electron with spin perpendicular to the magnetic field $\mathbf{B}$ was $\omega = 2\mu_B B/\hbar$. Again, we must change the electron magnetic moment $\mu_B$ to the proton magnetic moment $\mu_p$, so that the precession frequency for the protons is

$$\omega = 2\mu_p B_0/\hbar = 2\mu_B(g_p/2)(m_e/m_p)B_0/\hbar$$

In fact, this is exactly the same as the applied frequency that maximized the probability for the protons to flip! As the protons precess, they emit radiation at the precession frequency. As noted, this frequency is in the MHz range, so that the precessing protons give off electromagnetic radiation at radio frequencies. This radiation can then be detected and used to map the emitting protons. Thus, MRI allows the direct imaging of hydrogen atoms

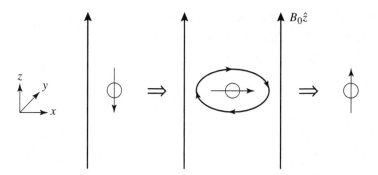

Fig. 14.4   As the protons relax back into the lower-energy spin state, they precess about the static magnetic field.

in the body (Figure 14.5). Tuning MRI to detect hydrogen atoms makes sense, since hydrogen is the most abundant element (by number) in the human body. (This picture is an oversimplification, since the protons actually precess about the combined total magnetic field, $\mathbf{B} = \mathbf{B}_0 + \mathbf{B}_1 \cos(\omega t)$, but it gives a reasonably accurate picture of how MRI works.)

MRI has several advantages over X-rays. Perhaps the biggest advantage is one of safety: X-rays ionize atoms in the body, leading to the danger of cancer at low doses and actual destruction of tissue at high doses. In contrast, no health risk from strong magnetic fields has ever been demonstrated. MRI images soft tissues more effectively than X-rays, which scatter less effectively off of low-density tissues than off of high-density materials such as bone. Furthermore, since MRI couples to hydrogen atoms, it can actually provide information on the chemical content of body tissues. The major drawback of MRI (compared with conventional X-rays) is its high cost.

It is amusing to note that MRI was originally (and accurately) called "nuclear magnetic resonance," since it relies on the resonant flip of the atomic nuclei in magnetic fields, with subsequent precession and re-emission of radiation at the resonant frequency. However, patients were disturbed by the use of the word "nuclear," which conjured images of nuclear weapons, nuclear waste, etc. This was particularly unfortunate given that the process itself is so noninvasive and benign. Hence it was repackaged as "magnetic resonance imaging," avoiding the negative connotations of the word "nuclear."

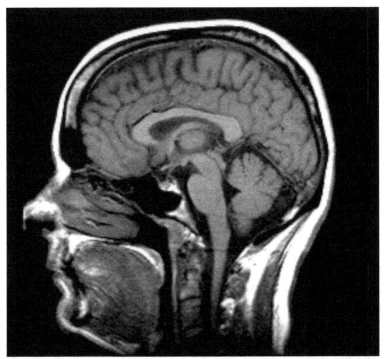

Courtesy of the Department of Diagnostic Radiology, Yale University School of Medicine

Fig. 14.5 An example of the image produced by an MRI machine.

## 14.2 Quantum Computing

We now examine a more recent, and completely different, application of quantum mechanics: quantum computing. Classical computing is based on binary digits or bits. For instance, a two-bit computer can be in one of the following four states:

$$0 \quad 0$$
$$0 \quad 1$$
$$1 \quad 0$$
$$1 \quad 1$$

where 0 and 1 can represent, electronically, circuits which are "on" or "off" or, logically, states that are "true" or "false." A sequence of $n$ bits can be in $2^n$ possible states; this provides a measure of the information storage in such a system. Note, however, that the system can be in only one of these states at any given time, so the computer can operate on only one state at a time.

Now consider a quantum analog: a system of two spin-1/2 particles. Just as in the case of the two-bit circuit, each particle can be in one of two states, so there are four possible states all together:

$$|\downarrow \downarrow\rangle$$
$$|\downarrow \uparrow\rangle$$
$$|\uparrow \downarrow\rangle$$
$$|\uparrow \uparrow\rangle$$

Now "spin down" and "spin up" represent the "0" and "1" states, respectively, of a classical computer. These quantum bits are called *qubits*. Note, however, that the system need not be in a single state; it can be in a superposition of all four states, e.g.,

$$|\psi\rangle = a_1 |\downarrow \downarrow\rangle + a_2 |\downarrow \uparrow\rangle + a_3 |\uparrow \downarrow\rangle + a_4 |\uparrow \uparrow\rangle$$

It is possible then to access all four of these spin states simultaneously, since each of them contributes to $|\psi\rangle$. One can also imagine operating on this linear combination of spin states to obtain a second linear combination of the spin states for the two particles. This represents a radical degree of parallel computing: all four two-qubit states can be manipulated simultaneously. This parallelism becomes more pronounced as the number of qubits increases; with ten qubits, for example, a million spin states can be accessed simultaneously!

It is possible to carry this analogy further and actually perform logical operations on the qubits. First, recall the sort of logic gates that are possible in a classical computer. Binary gates such as AND and OR take two bits and produce a single output bit (Figure 14.6). If 1 and 0 represent "true" and "false," respectively, then the AND gate produces an output of 1 if and only if both input bits are 1; otherwise it produces 0. Similarly, the OR gate produces 1 if either bit is 1; it produces 0 only if both bits are 0. A third example is the XOR, or exclusive OR gate, which is identical to the OR gate with the exception that it produces a "false" output if both inputs are "true."

In order to produce the quantum analog of these logic gates, we first need a slightly different representation of our four two-particle spin states.

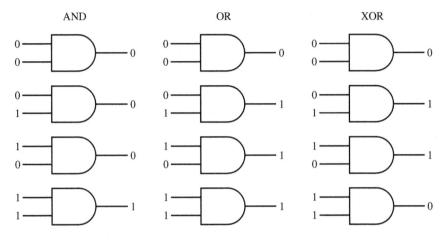

Fig. 14.6    The classical AND, OR, and XOR gates take in two bits and produce a one-bit output.

We represent the four states as column vectors in the following way:

$$|\downarrow \downarrow\rangle \Leftrightarrow \begin{pmatrix} 1 \\ 0 \\ 0 \\ 0 \end{pmatrix} \qquad (14.3)$$

$$|\downarrow \uparrow\rangle \Leftrightarrow \begin{pmatrix} 0 \\ 1 \\ 0 \\ 0 \end{pmatrix} \qquad (14.4)$$

$$|\uparrow \downarrow\rangle \Leftrightarrow \begin{pmatrix} 0 \\ 0 \\ 1 \\ 0 \end{pmatrix} \qquad (14.5)$$

$$|\uparrow \uparrow\rangle \Leftrightarrow \begin{pmatrix} 0 \\ 0 \\ 0 \\ 1 \end{pmatrix} \qquad (14.6)$$

As noted, the spin-down state represents a "0" or "false" bit, and the spin-up state represents a "1" or "true" state. A logic gate can then be represented as multiplication by a $4 \times 4$ matrix. As an example, the matrix representing the XOR gate is

$$U_{\text{XOR}} = \begin{pmatrix} 1 & 0 & 0 & 0 \\ 0 & 1 & 0 & 0 \\ 0 & 0 & 0 & 1 \\ 0 & 0 & 1 & 0 \end{pmatrix}$$

Using this matrix representation as well as the column vector representations in Equations (14.3)–(14.6), we can derive how $U_{\text{XOR}}$ operates on our four spin states:

$$U_{\text{XOR}}|\downarrow\downarrow\rangle = |\downarrow\downarrow\rangle$$
$$U_{\text{XOR}}|\downarrow\uparrow\rangle = |\downarrow\uparrow\rangle$$
$$U_{\text{XOR}}|\uparrow\downarrow\rangle = |\uparrow\uparrow\rangle \tag{14.7}$$
$$U_{\text{XOR}}|\uparrow\uparrow\rangle = |\uparrow\downarrow\rangle$$

In what sense can this result be treated as an XOR logic gate? The two inputs are simply the spin states of the two particles. The output of the logic gate is read off of the spin state of the second particle, while the spin state of the first particle is left unchanged. For instance, in Equation (14.7), the input spins of the two particles represent "true" and "false," respectively, so the XOR output should be "true." Hence, the spin of the second particle is set to spin up, while the spin of the first particle remains unchanged by the operator. Clearly, this differs from a classical logic gate in that there are two outputs rather than one: the spin of the first particle (which never changes) and the spin of the second particle (which corresponds to the output of the classical XOR gate). There is a reason for this: quantum mechanics is invariant under time reversal, so any quantum mechanical operation must be reversible. An operation such as a classical logic gate, which takes two inputs and produces only a single output, reduces the total information in the system; one cannot, in general, reconstruct the input values that go into a classical logic gate if only the output is known. Hence, quantum mechanical logic gates must have two outputs. Note, as emphasized earlier, that the quantum XOR gate can operate on a *linear superposition* of spin states, producing a linear superposition of outputs.

It is also possible to produce quantum logic circuits with no classical analog. Consider, instead of a two-qubit system, a single-qubit system with the two possible states

$$|\uparrow\rangle \Leftrightarrow \begin{pmatrix} 1 \\ 0 \end{pmatrix} \rightarrow \text{TRUE}$$

$$|\downarrow\rangle \Leftrightarrow \begin{pmatrix} 0 \\ 1 \end{pmatrix} \rightarrow \text{FALSE}$$

The NOT gate gives an output of TRUE if the input is FALSE and FALSE if the input is TRUE. In matrix form, this operator can be written as

$$U_{\text{NOT}} = \begin{pmatrix} 0 & 1 \\ 1 & 0 \end{pmatrix}$$

The NOT gate is a well-behaved classical logic gate. Now, however, consider the $\sqrt{\text{NOT}}$ gate, given by

$$U_{\sqrt{\text{NOT}}} = \frac{1}{2} \begin{pmatrix} 1-i & 1+i \\ 1+i & 1-i \end{pmatrix}$$

It is possible to show (Problem 14.7) that $U_{\sqrt{\text{NOT}}} U_{\sqrt{\text{NOT}}} = U_{\text{NOT}}$, so that applying the operation $\sqrt{\text{NOT}}$ twice yields NOT. There is, however, no classical logic gate with this property.

This all sounds fine in theory, but is it actually possible to design a quantum computing system to perform useful calculations? A breakthrough in this area was achieved in 1994 at Bell Labs by Peter Shor, who devised a quantum algorithm for factoring prime numbers. This algorithm was successfully implemented in 2001 at the IBM Almaden Research Center. The IBM scientists constructed a molecule consisting of five fluorine-19 atoms and two carbon-13 atoms, with the spins of the atomic nuclei serving as the qubits. These nuclear spins were manipulated using nuclear magnetic resonance technology similar to that described in the previous section. This quantum computer succeeded in factoring the number 15 (into 3 and 5). Since then, work in this field has accelerated enormously, becoming an area of intense current research.

## PROBLEMS

**14.1** Oxygen is the second most abundant element in the human body (by number) and the most abundant by mass. However, MRI detection of oxygen atoms is not practical. Why?

**14.2** Electrons in hydrogen have a much larger magnetic moment (both orbital and spin) than the magnetic moment of the proton. Why then are MRI machines not tuned, for example, to the resonant frequency of the spin magnetic moment of the electron rather than the proton?

**14.3** Consider an MRI machine with a 1.5 T static field. How far from the resonant frequency would one have to be in order for the

spin-flip probability to decrease from its maximum value to each of the following?
(a) one-half of the maximum value
(b) zero

**14.4** The random thermal energy of each molecule in the human body is roughly $E \approx kT$, where $k$ is Boltzmann's constant ($k = 1.38 \times 10^{-23}$ J K$^{-1}$) and $T$ is the temperature. Compare this thermal energy to the potential energy experienced by a proton in a 1.5 T MRI machine. What does the answer say about the efficiency with which protons will align into lower energy states in such a magnetic field?

**14.5** (a) Verify that the matrix corresponding to $U_{\text{NOT}}$ produces the correct output when applied to the $|\downarrow\rangle$ and $|\uparrow\rangle$ states.
(b) Compute the matrix corresponding to $U_{\text{NOT}}^2$, and calculate the result when it is applied to $|\downarrow\rangle$ and $|\uparrow\rangle$. What logical operation does $U_{\text{NOT}}^2$ correspond to?

**14.6** (a) Show explicitly that the classical XOR gate cannot be inverted.
(b) Calculate the matrix corresponding to the inverse of the quantum XOR gate.

**14.7** Show that $U_{\sqrt{\text{NOT}}} U_{\sqrt{\text{NOT}}} = U_{\text{NOT}}$.

**14.8** Apply $U_{\text{XOR}}$ to the mixed state $(1/\sqrt{2})(|\downarrow\downarrow\rangle + |\uparrow\uparrow\rangle)$. Explain what the result means.

Chapter 15

# Relativistic quantum mechanics

The theory of quantum mechanics that we have developed thus far is based on the nonrelativistic definition of energy, namely,

$$E = \frac{p^2}{2m} + V \tag{15.1}$$

The replacement of $p$, $V$, and $E$ with the appropriate operators leads to the Schrödinger equation. However, special relativity, which predates the Schrödinger equation by 20 years, indicates that Equation (15.1) is only an approximation valid at low velocities. When particle velocities become comparable to the speed of light, this equation breaks down. In this chapter we will examine what happens when we attempt to incorporate special relativity into quantum mechanics; the result is called *relativistic quantum mechanics*.

## 15.1 The Klein–Gordon Equation

*Derivation of the Klein–Gordon Equation*

As noted, the relationship between momentum and energy given by Equation (15.1) is valid only in the limit of low velocities, $v \ll c$, where $c$ is the speed of light. In special relativity, Einstein generalized this equation to give the correct relation between $p$ and $E$ at all velocities:

$$E^2 = p^2 c^2 + m^2 c^4 \tag{15.2}$$

where we have assumed a free particle with no potential (we will assume $V = 0$ throughout this chapter), and $m$ is the mass of the particle at rest, which is a constant. (In dealing with relativistic quantities, it is possible to simplify the equations considerably by setting $c = 1$. We will resist the urge to do that here, but it is important to be aware that it is often done.) In the limit where $v \ll c$, it is possible to show (Problem 15.1) that Equation (15.2) reduces to

$$E = \frac{p^2}{2m} + mc^2 \qquad (15.3)$$

This equation is similar to Equation (15.1) for the case $V = 0$, but there is an extra term on the right-hand side of Equation (15.3), corresponding to an extra contribution to the energy: $E = mc^2$. In special relativity, this is called the *rest energy* of the particle, and it must be included in the total energy. Note that nonrelativistic quantum mechanics ignores this rest energy, but it *is* included in the equations of relativistic quantum mechanics.

In order to derive an equation for the wave function that corresponds to Equation (15.2), we follow a procedure very similar to our "derivation" of the Schrödinger equation in Chapter 3. Assume that we have a wave function $\phi(\mathbf{r}, t)$ that is an eigenfunction of the energy operator $i\hbar(\partial/\partial t)$ with eigenvalue $E$, and also an eigenfunction of the momentum operator $-i\hbar\nabla$ with eigenvalue $\mathbf{p}$. In that case, we have

$$\left(i\hbar\frac{\partial}{\partial t}\right)^2 \phi = E^2\phi$$

and

$$\left(-i\hbar\nabla\right)^2 \phi = p^2\phi$$

and we can reproduce Equation (15.2) by writing

$$\left(i\hbar\frac{\partial}{\partial t}\right)^2 \phi = (-i\hbar\nabla)^2 c^2\phi + m^2c^4\phi \qquad (15.4)$$

As in the derivation of the Schrödinger equation, we now make the assumption that Equation (15.4) is always valid, regardless of whether or not $\phi$ is an eigenfunction of energy or momentum. Simplifying Equation (15.4) gives

$$\frac{1}{c^2}\frac{\partial^2\phi}{\partial t^2} - \nabla^2\phi + \frac{m^2c^2}{\hbar^2}\phi = 0 \qquad (15.5)$$

Equation (15.5) is called the *Klein–Gordon* equation.

The Klein–Gordon equation describes a particle with spin 0, which limits its usefulness, since the particles of greatest interest (e.g., the electron, proton, etc.) all have spin 1/2. Further, this equation leads to some problems connected with the interpretation of probabilities. To see this we need to digress.

## Probability Densities and Currents

For the Schrödinger equation, the probability density is given by $|\psi|^2$. However, it is *not* true that the corresponding quantity for the Klein–Gordon equation is $|\phi|^2$. To find the probability density in this case, we need to introduce a new quantity called the *probability current*.

We argue in analogy to a classical fluid. For a fluid with density $\rho$ and velocity $\mathbf{v}$, the rate $\partial\rho/\partial t$ at which the density changes at a fixed point is given by

$$\frac{\partial\rho}{\partial t} + \nabla\cdot(\rho\mathbf{v}) = 0 \tag{15.6}$$

The validity of this equation can be seen by integrating it over a closed volume $V$, and using the divergence theorem to transform the integral of $\nabla\cdot(\rho\mathbf{v})$ into an integral over the surface:

$$\int_V \frac{\partial\rho}{\partial t} + \int_A (\rho\mathbf{v})\cdot d\mathbf{A} = 0 \tag{15.7}$$

The first term in this equation is just the rate at which the total mass inside the closed surface changes; the second term is the rate at which mass is crossing the surface. Thus, Equation (15.7) (or equivalently, Equation (15.6)) simply says that the rate at which mass increases or decreases inside of a bounded region (the first term) is just given by the rate at which mass crosses the boundary of the region (the second term). Therefore, Equation (15.6) is called the *continuity equation*.

In quantum mechanics, probability can be treated in exactly the same way. The quantity equivalent to the density $\rho$ is just the probability density (which for the Schrödinger equation is $\rho = |\psi|^2$), and we can define a *probability current* $\mathbf{J}$ that satisfies the continuity equation for probabilities:

$$\frac{\partial\rho}{\partial t} + \nabla\cdot\mathbf{J} = 0 \tag{15.8}$$

As an example, we determine the expression corresponding to $\mathbf{J}$ for the Schrödinger wave function.

---

### Example 15.1. The Nonrelativistic Probability Current
What is the value for $\mathbf{J}$ arising from the nonrelativistic Schrödinger equation?

Substituting $\rho = \psi^*\psi$ into Equation (15.8) gives

$$\psi^*\frac{\partial\psi}{\partial t} + \psi\frac{\partial\psi^*}{\partial t} + \nabla\cdot\mathbf{J} = 0 \tag{15.9}$$

The first two terms can be simplified using the nonrelativistic Schrödinger equation, which can be written, in the absence of a potential, as

$$i\hbar \frac{\partial \psi}{\partial t} = -\frac{\hbar^2}{2m}\nabla^2\psi$$

Multiplying by $-i\psi^*/\hbar$ gives the first term in Equation (15.9):

$$\psi^* \frac{\partial \psi}{\partial t} = \frac{i\hbar}{2m}\psi^*\nabla^2\psi$$

and the complex conjugate of this equation gives the second term in Equation (15.9):

$$\psi \frac{\partial \psi^*}{\partial t} = \frac{-i\hbar}{2m}\psi\nabla^2\psi^*$$

Substituting these expressions into Equation (15.9) gives

$$\nabla \cdot \mathbf{J} = \frac{i\hbar}{2m}(\psi\nabla^2\psi^* - \psi^*\nabla^2\psi)$$

which can be integrated to give the expression for $J$:

$$\mathbf{J} = \frac{i\hbar}{2m}(\psi\nabla\psi^* - \psi^*\nabla\psi) \tag{15.10}$$

---

Taking the same expression for $\mathbf{J}$ as given in Equation (15.10), but using the Klein–Gordon equation instead of the Schrödinger equation, we can derive an expression for $\rho$ that satisfies the continuity equation (see Problem 15.4):

$$\rho = \frac{i\hbar}{2mc^2}\left(\phi^* \frac{\partial \phi}{\partial t} - \phi \frac{\partial \phi^*}{\partial t}\right) \tag{15.11}$$

Note that this expression for $\rho$ is quite different from our familiar expression $\rho = \psi^*\psi$. In particular, the expression for $\rho$ given by Equation (15.11) can be negative, which is clearly a "bad thing." Note further that in using Equation (15.2), we have inadvertently introduced negative energy solutions! Equation (15.2) corresponds to both $E = \sqrt{p^2c^2 + m^2c^4}$ and $E = -\sqrt{p^2c^2 + m^2c^4}$.

---

### Example 15.2. Solution to the Klein–Gordon Equation for a Particle at Rest

Consider a particle at rest for which $\mathbf{p} = 0$. We will solve the Klein–Gordon equation for this particle.

Since the momentum is zero, we have $-i\hbar\nabla\phi = 0$, and Equation (15.5) becomes

$$\frac{1}{c^2}\frac{\partial^2\phi}{\partial t^2} + \frac{m^2c^2}{\hbar^2}\phi = 0$$

The most general solution to this equation is

$$\phi = A_1 e^{imc^2t/\hbar} + A_2 e^{-imc^2t/\hbar} \tag{15.12}$$

where $A_1$ and $A_2$ are unknown constants. Applying the energy operator $i\hbar(\partial/\partial t)$ to each term in Equation (15.12), we see that the first term corresponds to a state with negative energy, $E = -mc^2$, while the second term corresponds to a state with positive energy, $E = mc^2$.

---

Because the Klein–Gordon equation contains a second derivative with respect to time, two boundary conditions are necessary to determine the solution: both $\phi$ and $\partial\phi/\partial t$ must be specified. This is clear, for example, in the solution given by Equation (15.12); two boundary conditions are necessary to determine the two unknown constants. This represents an additional degree of freedom not present in the Schrödinger equation. It is reasonable, therefore, to see if it is possible to find an equation corresponding to relativistic dynamics that contains only first derivatives with respect to time; the result will be the Dirac equation.

## 15.2 The Dirac Equation

Consider what happens if we try to construct an operator equation corresponding to Equation (15.2) but restrict the equation to be first order in all of the operators. As a first attempt, we write

$$\left(i\hbar\frac{\partial}{\partial t}\right)\psi = \left[-i\hbar c\left(\alpha_1\frac{\partial}{\partial x} + \alpha_2\frac{\partial}{\partial y} + \alpha_3\frac{\partial}{\partial z}\right) + \beta mc^2\right]\psi$$

where $\alpha_1$, $\alpha_2$, $\alpha_3$, and $\beta$ are constants to be determined. In order to find the values of these constants, we square the operator on both sides of the equation, and require the final result to reduce to the Klein–Gordon equation:

$$-\hbar^2\frac{\partial^2\psi}{\partial t^2} = \left(-i\hbar c\sum_{j=1}^{3}\alpha_j\nabla_j + \beta mc^2\right)\left(-i\hbar c\sum_{k=1}^{3}\alpha_k\nabla_k + \beta mc^2\right)\psi$$

$$= \left(-\hbar^2 c^2\sum_{j=1}^{3}\sum_{k=1}^{3}\alpha_j\alpha_k\nabla_j\nabla_k - i\hbar mc^3\sum_{j}(\beta\alpha_j + \alpha_j\beta)\nabla_j + \beta^2 m^2 c^4\right)\psi$$

$$\tag{15.13}$$

where we define $\nabla_1 = \partial/\partial x$, $\nabla_2 = \partial/\partial y$, and $\nabla_3 = \partial/\partial z$. We want this to reduce to the Klein–Gordon equation, which can be written as

$$-\hbar^2 \frac{\partial^2 \phi}{\partial t^2} = -\hbar^2 c^2 \nabla^2 \phi + m^2 c^4 \phi$$

In order for Equation (15.13) to reduce to the Klein–Gordon equation, the first term on the right-hand side of Equation (15.13) must simplify to $-\hbar^2 c^2 \nabla^2 \psi$, which requires

$$\alpha_j \alpha_j = 1 \tag{15.14}$$

and

$$\alpha_j \alpha_k + \alpha_k \alpha_j = 0, \quad \text{for } j \neq k \tag{15.15}$$

The second term on the right-hand side of Equation (15.13) must vanish, which gives

$$\beta \alpha_j + \alpha_j \beta = 0 \tag{15.16}$$

Finally, the last term on the right-hand side of Equation (15.13) must reduce to $m^2 c^4 \psi$, so that

$$\beta^2 = 1 \tag{15.17}$$

To summarize, each of the four constants $\alpha_1$, $\alpha_2$, $\alpha_3$, and $\beta$ must square to give 1, but the four constants all anticommute with each other ($\alpha_i$ and $\alpha_j$ are said to *anticommute* if $\alpha_i \alpha_j = -\alpha_j \alpha_i$). In fact, it is impossible to find four numbers that satisfy these relations! On the other hand, it *is* possible to satisfy these relations if we take $\alpha_1$, $\alpha_2$, $\alpha_3$, and $\beta$ to be *matrices*. For example, the Pauli spin matrices from Chapter 8 satisfy exactly the desired relations: $\sigma_x^2 = \sigma_y^2 = \sigma_z^2 = I$, and $\sigma_x \sigma_y + \sigma_y \sigma_x = 0$, $\sigma_x \sigma_z + \sigma_z \sigma_x = 0$, $\sigma_y \sigma_z + \sigma_z \sigma_y = 0$. The problem is that there are only three of these matrices and we need four. In order to find four mutually anticommuting matrices that all square to give the identity matrix, the minimum size of the matrices must be $4 \times 4$. Thus, the wave function in our differential equation is no longer a one-component object (as it is in the Schrödinger and Klein–Gordon equations). Instead, it is a four-component column vector. There are an infinite number of different choices for matrices satisfying Equations (15.14)–(15.17), but the choice of which ones to use doesn't change any physical calculations. The conventional choice is the following:

$$\alpha_j = \begin{pmatrix} 0 & \sigma_j \\ \sigma_j & 0 \end{pmatrix}$$

and

$$\beta = \begin{pmatrix} I & 0 \\ 0 & -I \end{pmatrix}$$

where each symbol stands for a $2 \times 2$ matrix. Here the $\sigma_j$'s are the $2 \times 2$ Pauli spin matrices, and $I$ is the $2 \times 2$ identity matrix. For example,

$$\alpha_3 = \begin{pmatrix} 0 & 0 & 1 & 0 \\ 0 & 0 & 0 & -1 \\ 1 & 0 & 0 & 0 \\ 0 & -1 & 0 & 0 \end{pmatrix}$$

and

$$\beta = \begin{pmatrix} 1 & 0 & 0 & 0 \\ 0 & 1 & 0 & 0 \\ 0 & 0 & -1 & 0 \\ 0 & 0 & 0 & -1 \end{pmatrix}$$

With these values for $\alpha_j$ and $\beta$, we can go back to our original equation and write

$$i\hbar\frac{\partial\psi}{\partial t} = -i\hbar c(\boldsymbol{\alpha}\cdot\nabla)\psi + \beta mc^2\psi \tag{15.18}$$

where $\boldsymbol{\alpha}\cdot\nabla$ is shorthand for $\alpha_1(\partial/\partial x) + \alpha_2(\partial/\partial y) + \alpha_3(\partial/\partial z)$; hence, $\boldsymbol{\alpha}$ is a three-component object whose three components are each $4 \times 4$ matrices! Equation (15.18) is called the *Dirac equation*. It forms the basis of relativistic quantum mechanics, and is perhaps the second most important equation in this book after the Schrödinger equation itself. Of course, as we have emphasized, the $\psi$ which appears in Equation (15.18) is really a four-component column vector:

$$\psi = \begin{pmatrix} \psi_1 \\ \psi_2 \\ \psi_3 \\ \psi_4 \end{pmatrix}$$

where each of the four components $\psi_1, \ldots, \psi_4$, is a function of position and time. If all of the matrices in the Dirac equation are multiplied out, the Dirac equation breaks into a set of four differential equations relating these four components and their derivatives (see Problem 15.6).

To find the probability density and probability current for the Dirac wave function, we begin with the Dirac equation and derive a relation which looks like the continuity equation. Starting with Equation (15.18), we first take the adjoint (i.e., the conjugate transpose) of this equation to obtain

$$-i\hbar\frac{\partial \psi^\dagger}{\partial t} = i\hbar c(\nabla\psi^\dagger)\cdot\boldsymbol{\alpha} + \psi^\dagger\beta mc^2 \tag{15.19}$$

where we have used the fact that all of the $\alpha_j$, $\beta$ matrices are Hermitian. Using Equations (15.18) and (15.19) for the time derivatives of $\psi$ and $\psi^\dagger$,

$$\frac{\partial}{\partial t}(\psi^\dagger\psi) = \psi^\dagger\frac{\partial \psi}{\partial t} + \frac{\partial \psi^\dagger}{\partial t}\psi$$
$$= -c\nabla\cdot(\psi^\dagger\boldsymbol{\alpha}\psi) \tag{15.20}$$

Equation (15.20) looks just like the continuity equation if we take $\rho$ and $\mathbf{J}$ to be given by

$$\rho = \psi^\dagger\psi$$

and

$$\mathbf{J} = c\psi^\dagger\boldsymbol{\alpha}\psi$$

These then are the probability density and probability current for the Dirac equation. Note that, unlike the case for the Klein–Gordon equation, the Dirac probability density is always nonnegative, since

$$\rho = \psi^\dagger\psi = |\psi_1|^2 + |\psi_2|^2 + |\psi_3|^2 + |\psi_4|^2$$

and each of these four terms is nonnegative. This is a good thing, since probabilities should be positive.

Now consider the solutions of the Dirac equation for a particle at rest. (The solutions of the Dirac equation for nonzero momentum are derived in Problem 15.8.) For a particle at rest, $p = 0$ so the term $-i\hbar c(\boldsymbol{\alpha}\cdot\nabla)\psi$ is zero, and the Dirac equation becomes

$$i\hbar\frac{\partial \psi}{\partial t} = \beta mc^2\psi$$

Multiplying out $\beta\psi$ on the right-hand side gives four ordinary differential equations:

$$i\hbar\frac{d\psi_1}{dt} = mc^2\psi_1$$
$$i\hbar\frac{d\psi_2}{dt} = mc^2\psi_2$$
$$i\hbar\frac{d\psi_3}{dt} = -mc^2\psi_3$$
$$i\hbar\frac{d\psi_4}{dt} = -mc^2\psi_4$$

Solving for $\psi_1, \psi_2, \psi_3$, and $\psi_4$ and recombining the solutions back into the form of a column vector gives four linearly-independent solutions. The first two are

$$\psi = \begin{pmatrix} 1 \\ 0 \\ 0 \\ 0 \end{pmatrix} e^{-imc^2t/\hbar} \tag{15.21}$$

and

$$\psi = \begin{pmatrix} 0 \\ 1 \\ 0 \\ 0 \end{pmatrix} e^{-imc^2t/\hbar} \tag{15.22}$$

The energies corresponding to these two solutions can be determined by applying the energy operator $i\hbar(\partial/\partial t)$; we obtain

$$E = mc^2$$

for both solutions. This is precisely the expected result for the energy of a particle at rest. But why are there *two* linearly-independent solutions? These solutions apparently describe a two-component object. But we are already familiar with such an object: a spin-1/2 particle! The Dirac equation describes an elementary particle with spin of $1/2$, and the wave function $\psi$ combines both the spatial and the spin information into a single quantity called a *spinor*. Thus, the wave function in Equation (15.21) represents a particle at rest with $m_s = +1/2$, and the wave function in Equation (15.22) represents a particle at rest with $m_s = -1/2$. These two solutions can be combined linearly to yield any other spin state for a spin-1/2 particle.

The other two linearly-independent solutions are

$$\psi = \begin{pmatrix} 0 \\ 0 \\ 1 \\ 0 \end{pmatrix} e^{imc^2t/\hbar}$$

and

$$\psi = \begin{pmatrix} 0 \\ 0 \\ 0 \\ 1 \end{pmatrix} e^{imc^2t/\hbar}$$

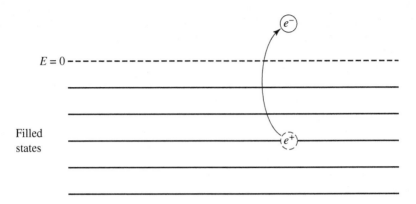

Fig. 15.1   In a vacuum, all of the negative-energy states are filled with electrons, and all of the positive-energy states are empty. Moving an electron into a positive-energy state leaves behind a "hole" in the negative-energy states, corresponding to a positron. Electron-positron annihilation corresponds to the electron dropping back down and filling the hole.

which both have *negative* energy:

$$E = -mc^2$$

It is tempting to dismiss these as "spurious" solutions with no physical significance, but Dirac did not do so. Rather, he assumed that these solutions were also valid. If this is the case, where are the negative-energy electrons? Dirac argued that in an ordinary vacuum, all of the positive-energy states are empty, while the negative-energy states are filled by a sea of electrons (Figure 15.1). If an electron is removed from a negative-energy state and boosted into a positive-energy state, it leaves behind a "hole" in the negative-energy states. The energy of this "hole" is the negative of the corresponding negative-energy electron state, so the "hole" has energy $E = -(-mc^2) = mc^2$. In this way Dirac postulated the existence of antimatter. The holes in the negative-energy states are antielectrons, called *positrons*. Note that it is *not* the electrons in the negative-energy states that correspond to positrons, but rather the *holes* in the negative-energy states generated by removing electrons from them. A positron can annihilate with an electron; this corresponds to the electron dropping back down and filling the vacant negative-energy state, thus eliminating both the electron and the hole. This prediction of Dirac was confirmed experimentally in 1932 with the discovery, by Carl Anderson, of the positron.

This is only a small subset of the important results to which the Dirac equation leads. It is possible, for example, to show that the Dirac equation predicts that $g_s$ for the spin magnetic moment of the electron should be exactly 2 (a prediction which must be modified, by a small amount, using the more advanced results of quantum field theory). The Dirac equation forms one of the main pathways leading from classical quantum mechanics into our modern theories of particle physics.

## PROBLEMS

**15.1** Show that the relativistic relation between energy and momentum (Equation (15.2)) reduces to

$$E = \frac{p^2}{2m} + mc^2$$

for the case when $v \ll c$.

**15.2** If $\phi$ is an eigenfunction of both energy and momentum, then another differential equation corresponding to Equation (15.2) is

$$\frac{1}{c^2} \left( \frac{\partial \phi}{\partial t} \right)^2 - (\nabla \phi)^2 + \frac{m^2 c^2}{\hbar^2} \phi^2 = 0$$

Why is this a less desirable equation than the Klein–Gordon equation?

**15.3** (a) If **J** is the Schrödinger probability current, show that

$$\int \mathbf{J} \, d^3\mathbf{r} = \langle \mathbf{v} \rangle$$

(b) What are the units of **J**?

**15.4** Using the Klein–Gordon equation, the continuity equation, and the expression for **J** from Equation (15.10), derive the Klein–Gordon probability density:

$$\rho = \frac{i\hbar}{2mc^2} \left( \phi^* \frac{\partial \phi}{\partial t} - \phi \frac{\partial \phi^*}{\partial t} \right)$$

**15.5** Write out explicitly the full $4 \times 4$ matrices corresponding to $\alpha_1$ and $\alpha_2$.

**15.6** Multiply out the matrices in the Dirac equation to express the Dirac equation as four coupled differential equations for the four components of $\psi$: $\psi_1$, $\psi_2$, $\psi_3$, and $\psi_4$.

**15.7** Write down the Dirac spinor corresponding to a spin-1/2 particle at rest with spin in the $+x$ direction and positive energy.

**15.8** (a) The general solution for the Dirac equation can be written in the form

$$\psi = \begin{pmatrix} \phi_1 \\ \phi_2 \\ \chi_1 \\ \chi_2 \end{pmatrix} e^{i(\mathbf{p}\cdot\mathbf{r}-Et)/\hbar}$$

where $\phi_1$, $\phi_2$, $\chi_1$, and $\chi_2$ are numbers independent of $\mathbf{r}$ and $t$. To take advantage of this form for the Dirac equation, use the shorthand

$$\phi = \begin{pmatrix} \phi_1 \\ \phi_2 \end{pmatrix}$$

and

$$\chi = \begin{pmatrix} \chi_1 \\ \chi_2 \end{pmatrix}$$

Using this form for the solution, show that $\phi$ and $\chi$ satisfy the coupled equations

$$(E - mc^2)\phi = c(\mathbf{p}\cdot\boldsymbol{\sigma})\chi$$

and

$$(E + mc^2)\chi = c(\mathbf{p}\cdot\boldsymbol{\sigma})\phi$$

(b) Use the results from part (a) to show that the general four-component solution of the Dirac equation may be written as

$$\psi = \begin{pmatrix} \phi \\ c(\mathbf{p}\cdot\boldsymbol{\sigma})\phi/(E + mc^2) \end{pmatrix} e^{i(\mathbf{p}\cdot\mathbf{r}-Et)/\hbar}$$

# Index

Addition
  of complex numbers, 28
  of linear operators, 108
  of vectors, 112
Adiabatic change, 265
Adjoint operator, 117–118
Alkali metals, ground state, 328
$\alpha$ decay, 87
Anderson, Carl, 360
Angular frequency, 42
Angular momentum, 129–141
Angular momentum operator, 130–138
Anharmonic oscillator, 226–227
Anomalous magnetic moment, 182
Anticommuting, 356
Antimatter, 360
Antisymmetric wave function, 317, 319
Argon, ground state, 328
Atomic structure, 326–334
Aufbau principle, 327

Balmer series, 20
Balmer, Johann, 20
Basis sets, 120–122
Bell, J.S., 211–212
Beryllium, ground state, 334
Binary gates, classical computing, 345
Blackbody radiation, 3–11
Bohr atom, 19–22

Bohr magneton, 181
Bohr radius, 155
Bohr, Neils, 20
Boltzmann distribution, 8
Boltzmann's constant, 6
Born approximation, 292–302
  scattering from a delta-function potential, 301–302
  scattering from a repulsive spherical well, 299–300
Born, Max, 51
Boron, atomic structure, 327, 330
Bose–Einstein statistics, 317
Bound states, 88–99
Bra vectors, 172
Bracket, 172

Carbon, ground state, 330–334
Central potentials, 141
Clebsch–Gordon coefficients, 205
Column vectors, 163
Commutators, 109–112
Complex conjugation, 33–34
Complex numbers, 27–34
Complex vector spaces, 113
Composition, 108
Compton effect, 14–17
Compton wavelength, 14
Computing
  classical computing, 345–346
  quantum computing, 344–348

Copenhagen interpretation, 208–211
"Corpuscular theory" of light, 11
Cross section, scattering, 287–292

Davisson–Germer experiment, 18–19
de Broglie wavelength, 17
de Broglie, Louis, 17
Degeneracy, 129
Delta function
    *See also* Dirac delta function
    *See also* Kronecker delta
Delta-function potential, 301–302
Density of states, 295
Derivative operator, 35–36
    adjoint of, 118
Differential cross section, 289
Dipole approximation, 278
Dipole–dipole interaction, 206–208
Dirac delta function, 172–174
Dirac equation, 355–361
Dirac notation, 170–172
Dirac probability density, 358
Dual space, 171

Eigenfunctions, 36–38
    simultaneous, 110–112
Eigenvalues, 36–38
Einstein, Albert
    photoelectric effect, 13–14
    special relativity, 351
Einstein–Rosen–Podolsky paradox,
    209
Electric field
    atom in, 239–241
    due to electromagnetic wave
        propagation, 277
Electric quadrupole transition, 281
Electromagnetic radiation
    electric field due to wave
        propagation, 277
    selection rules for, 276–281
Electron–positron pair annihilation,
    360
Electrons
    in multielectron atoms, 326–334
    spin angular momentum, 180–185

spin precession, 197–201
spin-orbit coupling, 230–237
spins in a magnetic field, 228–229
spins in an oscillating magnetic
    field, 275–276
Stern–Gerlach experiment,
    183–185, 194–197
Everett, Hugh, 212
Exchange operator, 315–317
Expectation value, 57–58

Fermi–Dirac statistics, 317
Fermions, 317
Feynmann, Richard, 182
Fine structure, 229–237
Fine-structure constant, 233
First excited state, 65
Forbidden transitions, 276, 281
Fourier series, 121
Frequency, classical harmonic
    oscillator, 96

Good quantum numbers, 112
Goudsmit, S.A., 183
Ground state, 63
    Hund's rules, 330
    of alkali metals, 328
    of argon, 328
    of beryllium, 334
    of boron, 327, 330
    of carbon, 330–334
    of helium, 256–261, 323–325,
        328–329
    of hydrogen, 149–158, 229–239, 329
    of lithium, 328
    of neon, 328
    of noble gases, 328
    of potassium, 328
    of sodium, 328

H-bar ($\hbar$), 20
Hamiltonian operator, 60–61
Harmonic oscillator potential, 92–99
    ladder operators, 138–141
Heisenberg uncertainty principle,
    99–100

Helium, atomic structure, 328–329
Hermite polynomials, 96
Hermitian operators, 119
Hidden variables formulation, 211–212
Hund's rules, 330
Hydrogen atom
  allowed transitions, 276–281
  energy levels, 149–158
  fine structure, 229–237
  hyperfine structure, 208, 236–237
  wave functions, 149–158
Hyperfine structure, 208, 236–237

Identity matrix, 165
Imaginary number, 27
Infinite square well, 88–91
Inhomogeneous magnetic field, 183–184
Inner products, 113–117
  in Dirac notation, 171
Interacting spins, 205–208

*j-j* coupling, 329

Ket vectors, 172
Kirchhoff's law, 3
Klein–Gordon equation, 351–355
Kronecker delta, 131, 174

*L-S* coupling, 328
Ladder operators, 134–137
  harmonic oscillator, 138–141
Lamb shift, 237–239
Lamb, Willis, 237
Landé *g* factor, 246
Laplace's equation, 149
Legendre polynomials, 149
Light, 11–17
  Compton effect, 14–17
  Newton's "corpuscular theory", 11
  photoelectric effect, 12–14
  wave–particle duality, 2, 17–18
Linear algebra, 107–122
Linear operators, 35–38
  matrix formulation, 163–170

Linear Stark effect, 241
Lithium, atomic structure, 328
Logic gates
  classical computing, 344–345
  quantum, 345–348
Lyman series, 20

Magnetic dipole transition, 281
Magnetic dipole–dipole interaction,
  between two particles, 206–208
Magnetic field
  atom in, 241–246
  electron spins in, 228–229, 275–276
Magnetic moment, 181–183
Magnetic resonance imaging (MRI), 339–343
Many worlds interpretation, 212–213
Matrix formulation, linear operators, 163–170
Matrix multiplication, 164
Matrix, adjoint of, 169
Matter, wave nature, 17–18
Measurement theory, 208–213
Modern physics, 1
Momentum, 45–47
  expectation value of, 57–58
  *See also* Angular momentum
Momentum transfer, 297
MRI (*see* Magnetic resonance imaging)
Multielectron atoms, 326–334
Multiparticle Schrödinger equation, 313–334
  multielectron atoms, 326–334
  wave function for identical particles, 313–326

Neon, ground state, 328
Newton's "corpuscular theory", 11
Noble gases, ground state, 328
Noninteracting spins, 202–205
Nonrelativistic probability current, 353–354
Normalization, 52
NOT gate, 348
$\sqrt{\text{NOT}}$ gate, 348

Nuclear magnetic resonance, 343

Observables, 56
One-dimensional harmonic oscillator
potential, 92–99, 138–141
One-dimensional Schrödinger
equation, 48
solutions, 48–51, 71–100
One-dimensional square well, 48–51,
88–91
Operators, 35–38
linearity of, 35–36
Orbital angular momentum, 133,
143–149
Orthonormal basis, 120

Parity operator, 37–38
Partial waves, 303–309
Paschen series, 20
Paschen–Back effect, 244
Pauli exclusion principle, 317
Pauli spin matrices, 192, 356–357
Perturbation theory
time-dependent, 265–281
time-independent, 219–246
Perturbation, atomic energy levels,
229–239
Phase velocity, 43
Phipps, T.E., 184
Photoelectric effect, 12–14
Photons, 13
Planck's constant, 9–11, 13–14
Planck, Max, on spectrum of
blackbody radiation, 8–11
Positrons, 360
Potassium, atomic structure, 328
Precession
spin precession, 197–201
Thomas precession, 231
Prime numbers, quantum algorithm
for factoring, 348
Principal quantum number, 155
Probability current, 353–354, 358
Probability density, 51, 353–354, 358
Proton magnetic moment, 236,
340–341

Quadratic Stark effect, 239–241
Quantization, origins of, 61–66
Quantum computing, 344–348
Quantum dot, 91
Quantum field theory, 182, 237
Quantum harmonic oscillator, 92–99,
138–141
Quantum numbers
good quantum numbers, 112
orbital angular momentum
quantum number, 155
principal quantum number, 155
Qubits, 345

Radial Schrödinger equation, 146
Radial wave function, 146
for the hydrogen atom, 156
Radiation
blackbody radiation, 3–11
*See also* Electromagnetic radiation
Rayleigh–Jeans formula, 7
Rectangular coordinates, Schrödinger
equation solution in, 126–129
Recursion relation, 95
Reduced mass, 152
Reflection probability, 81
Relativistic quantum mechanics,
351–361
Dirac equation, 355–361
Klein–Gordon equation, 351–355
Rest energy, 352
Retherford, R.C., 237
Russell–Saunders coupling, 328
Rutherford, Ernest, 19
Rydberg constant, 19–20

$s$-wave phase shift, 308
$s$-wave scattering, 303–309
Scattering, 287–309
Born approximation, 292–302
classical, from a hard sphere,
290–292
Compton scattering, 14–17
cross section, definition, 287–290
from a delta-function potential,
301–302

from an infinitely hard sphere,
308–309
one-dimensional, from
step-function potentials,
77–87
partial waves, 303–309
$s$-wave scattering, 303–309
three-dimensional repulsive
spherical well, 299–300
Schrödinger Equation
defined, 48–49
derivation, 42–48
multiparticle equation, 313–334
one-dimensional, 48–51, 71–100
radial equation, 146
solution in rectangular coordinates,
126–129
solution in spherical coordinates,
141–149
three-dimensional, 125–158
time-independent, 58–61
Schrödinger's cat, 209–211
Selection rules, 276–281
Self-adjoint operator, 119
Semiconductor heterostructures, 71,
91
Separation of variables, 49–50
Shell, 326
Shor, Peter, 348
Simultaneous eigenfunctions, 110–112
Singlet state, 187
Sodium, atomic structure, 328
Spherical Bessel functions, 305
Spherical coordinates, 141–142
Schrödinger equation solution in,
141–149
Spherical harmonics, 148
Spin angular momentum, 179–208
evidence for, 180–185
Spin magnetic moment, 182–183,
340–341
Spin operators, 179–180
Spin precession, 197–201
Spin systems with two particles
interacting spins, 205–208
noninteracting spins, 202–205

Spin, matrix representation, 187–194
Spin-1/2 particles, 187–194
Spin–spin interaction in hydrogen,
236–237
Spin-orbit coupling, 229–237
Spin-orbit perturbation, 232
Spinor, 359
Square well
infinite square well, 88–91
two identical particles in, 320
Stark effect, 239–241
Stefan, J., 4
Stefan–Boltzmann law, 4–5
Step-function potentials, scattering
from, 77–87
Stern–Gerlach experiment, 183–185,
194–197
Strong-field Zeeman effect, 244
Subshell, 326
Symmetric wave function, 317, 319

Taylor, J.B., 184
Thomas precession, 231
Thomas, L.H., 231
Three-dimensional delta function, 174
Three-dimensional Schrödinger
equation, 125–158
Time-dependent perturbation theory,
265–281
derivation, 266–273
selection rules for electromagnetic
radiation, 276–281
Time-independent perturbation
theory, 219–246
atom in a magnetic field, 241–246
atom in an electric field, 239–241
derivation, 220–226
Stark effect, 239–241
Zeeman Effect, 241–246
Total angular momentum, 185–187
Total cross section, 289
Total magnetic moment, 241–243
Transition probability, 269–273
Transmission probability, 81
Triplet state, 187
Tunnel diode, 87–88

Tunneling,, 84–88
Two-particle Schrödinger equation, 314
Two-particle wave function, 314–325
Two-qubit system, 345–347

Ulhenbeck, G., 183
Ultraviolet catastrophe, 8
Unbound states, 72–88
    scattering from step-function potentials, 77–88
    tunneling, 84–88
Uncertainty principle, 99–100

Vacuum polarization, 237–238
Vacuum, in quantum field theory, 237
Variational principle, 251–261
    helium atom and, 256–261
    theory, 252–256

Vector spaces, 112–122
    basis sets, 120–122
    dual space, 171
    inner products, 113–117

Wave function
    defined, 41
    for identical particles, 313–326
    meaning of, 51–58
    time-independent, 59–60
Wave number, 42
Wave vector, 43
Wien's displacement law, 5–6, 10

X-rays, MRI advantages over, 343
XOR logic gate, 345–347

Zeeman effect, 241–246
Zero-point energy, 90, 96

Printed in the United States
by Baker & Taylor Publisher Services